Harvard Studies in Business History XXXIII

Edited by Alfred D. Chandler, Jr.
Isidor Straus Professor of Business History
Graduate School of Business Administration
George F. Baker Foundation
Harvard University

Big Business in China

*Sino-Foreign Rivalry
in the Cigarette Industry,
1890-1930*

SHERMAN COCHRAN

Harvard University Press
Cambridge, Massachusetts
and London, England   1980

Library of Congress Cataloging in Publication Data

Cochran, Sherman, 1940-
    Big business in China.

    (Harvard studies in business history; 33)
    Bibliography: p.
    Includes index.
    1.   Cigarette manufacture and trade—China—History.
2.   China—Commerce.   I.   Title.   II.   Series.
HD9149.C42C63   381'.45'679730951   79-23907
ISBN 0-674-07262-6

To Jan

Editor's Introduction

THIS HISTORY of the growth of and the competition between the British-American Tobacco Company and the Nanyang Brothers Tobacco Company in China broadens the scope of the Harvard Studies in Business History. It is not only the first book in the series on Chinese business history, it is the first on any Asian country. And it is more than just a pioneering study in Asian business history, for it is an outstanding example of comparative business history. In this case the comparison is between the activities of one of the oldest and most successful western multinational enterprises and one of the largest indigenous Chinese firms competing for the same potentially vast market.

His comparative approach and his extraordinarily rich source material thus give Professor Cochran's findings a broader relevance than most traditional business histories. The book provides new information on the operations of large industrial enterprises in what remained essentially a preindustrial commercial economy undergoing the initial political and cultural, as well as economic, shocks of early modernization. In addition Cochran provides new information and insights on the ways of duopolistic competition. In analyzing the nature and role of entrepreneurship in the Chinese cigarette industry, he helps to resolve the debate over the impact of the absence or presence of entrepreneurial and organization skills on the modernization of the Chinese economy. Further he examines from a new perspective the fundamental issues of imperialistic exploitation and economic nationalism. His evaluation of these issues casts light on a central problem of economic development—the place of rivalry between indigenous enterprises and outside multinationals in the process of growth and modernization. Professor Cochran has demonstrated most successfully how a detailed, thorough study of business records can enlarge our understanding of institutional and economic change.

Alfred D. Chandler, Jr.

Acknowledgments

THIS BOOK is a study in Chinese business history, a new area of inquiry that will develop into a full-fledged field only if scholars, archivists, business executives, and others are willing to go beyond their usual concerns and give their attention to it. Fortunately for me, many people, from a variety of fields and disciplines, have been willing to take this step. Even when their interests have seemed remote from mine, they have come to my assistance, and I am especially grateful to them.

Among Chinese historians, there are four to whom my debt is greatest: Arthur and Mary Wright, whose graduate seminars stimulated my interest in the history of Sino-foreign interaction; Jonathan Spence, who guided my work on this topic from the beginning, generously sharing his insights and encouraging me at every stage; and Andrew Hsieh, who introduced me to sources and bibliography and showed me how to search. Other Chinese historians, Beatrice Bartlett, Knight Biggerstaff, Lloyd Eastman, Robert Kapp, Lillian Li, Susan Naquin, and Charles Peterson, read all or parts of this book in manuscript form and saved me from errors of fact and interpretation.

Among scholars in other fields who advised me, several also read all or parts of this book and made useful suggestions for improving it: historians of American foreign relations Michael Hunt and Walter LaFeber; economists Peter McClelland and Thomas Rawski; and historians of American business Alfred Chandler and Mira Wilkins.

Among those I consulted about business records, two people were especially helpful: Parker Po-fei Huang, who answered my questions about nuances in Chinese business records; and Mattie Russell, who welcomed me to the Manuscript Department, Perkins Library, Duke University, and gave me access to unpublished papers there.

Among those I contacted within the cigarette industry, several granted me extensive interviews about their experiences in China: Y. L. Kan (Chien Jih-lin) of the Nanyang Brothers Tobacco Company; C. C. Kwong (Kuang Ch'ao-ch'i) formerly of the Nanyang Brothers Tobacco Company; Dorothy Quincy Thomas, widow of James A. Thomas of the British-American Tobacco Company; and John Ephgrave, Brian Hearnshaw, and Brian Pearson of the British-American Tobacco Company.

I wish to thank all of these people and many others for reaching beyond their immediate concerns, taking an interest in Chinese business history, and helping me.

Above all, my thanks go to my wife, Jan, who has worked tirelessly on this book, and, in her tough-minded and loving way, has convinced me to do my part.

Contents

ILLUSTRATIONS

"It's Toasted" on a BAT billboard   *136*
> Julean Arnold et al., *China: A Commercial and Industrial Hand-
> book*, U.S. Department of Commerce, Bureau of Foreign and
> Domestic Commerce, Trade Promotion Series no. 38. Washington,
> D.C.: Government Printing Office, 1926, p. 198.

Women packing cigarettes and processing tobacco leaves in Nanyang's
cigarette factory   *156*
> Shanghai Economic Research Institute, *Nan-yang hsiung-ti yen-ts'ao
> kung-ssu shih-liao* (Historical materials on the Nanyang Brothers
> Tobacco Company; Shanghai, 1958), following p. 288.

BAT neon sign in Shanghai, 1930   *180*
> *B.A.T. Bulletin*, 20.112 (March 1930), 166.

TABLES

Big Business in China

WEIGHTS, MEASURES, AND EXCHANGE RATES

Equivalents of weights and area used in this book are as follows:

1 catty (*chin*)	1.1 pounds
2 catties	1 kilogram
100 catties	1 picul (*tan*)
20 piculs	1 metric ton
2,000 catties	1 metric ton
1 *mou*	0.1647 acre
6 *mou*	1 acre
15 *mou*	1 hectare

Two units of currency are used in this book: the Chinese *yuan*, designated ¥, which foreigners in China referred to as the Mexican dollar, and U.S. dollars. Between 1894 and 1930, the value of the U.S. dollar varied in relation to *yuan* as follows:

1894	¥1.95	1904	¥2.27	1913	¥2.05	1922	¥1.81
1895	1.88	1905	2.05	1914	2.24	1923	1.88
1896	1.85	1906	1.88	1915	2.42	1924	1.85
1897	2.08	1907	1.90	1916	1.90	1925	1.79
1898	2.14	1908	2.31	1917	1.46	1926	1.97
1899	2.05	1909	2.38	1918	1.19	1927	2.17
1900	2.00	1910	2.27	1919	1.08	1928	2.11
1901	2.08	1911	2.31	1920	1.21	1929	2.34
1902	2.38	1912	2.03	1921	1.97	1930	3.26
1903	2.34						

Source: C. F. Remer, *Foreign Investments in China* (New York, 1933), appendix, p. 172.

1.

Introduction: Sino-Foreign Commercial Rivalries

Is IT POSSIBLE to compete with big and expanding multinational corporations? This question has been raised by contemporary economic commentators who have begun to doubt whether any organization—even the nation-state—is capable of limiting the growth of Western- and Japanese-based multinational giants. According to recent studies, by 1985 a few hundred multinational corporations will control 80 percent of all the productive assets of the non-Communist world, and by the turn of the century they will own goods and services valued at $4 trillion—54 percent of the gross product of the entire world.[1] Such predictions about the future need to be based on an understanding of the past. To provide a fuller basis for understanding these businesses in history, scholars have begun to document the historical records of multinational corporations,[2] but unfortunately, little attention has been paid thus far to the interaction between these corporations and their local rivals in business. This book will, I hope, contribute to our understanding of interaction between Western-based multinational corporations and their local competitors outside the West, for it focuses on a rivalry between two early twentieth century businesses that fit this description. The Western-owned multinational corporation was the British-American Tobacco Company (BAT) with headquarters in London and New York, and its local competitor was the Chinese-owned Nanyang Brothers Tobacco Company based in Hong Kong and Shanghai. The central aim of this study is to analyze the battle for the Chinese cigarette market between these two big businesses.

This rivalry between BAT and Nanyang originated early in the twentieth century. Before 1915, BAT, founded in 1902, became the premier business in the industry, overcame all competition, and maintained a lucrative monopoly of the market. But between 1915 and 1919 Nanyang, a Chinese-owned newcomer, entered China and extended its operations into the country's major regional markets despite competition from BAT. In the early 1920s, with Nanyang well established in China, the competition between the two companies reached its peak. But in the mid-1920s, the rivalry declined as Nanyang weakened and suffered large losses at the end of the decade, bringing an end to an era of serious commercial rivalry in China's cigarette industry by 1930. In the 1930s and 1940s Nanyang continued to trade in China, but neither it nor any other Chinese company competed as successfully with BAT as it had before 1930. In decline for two decades, commercial rivalry in the cigarette industry ended altogether in the early 1950s when BAT and Nanyang both began the transition from private to state ownership after the founding of the People's Republic of China.

The theme of this story, Sino-foreign commercial rivalry, is one that can be traced throughout most of China's history between the Opium War of the early 1840s and the founding of the People's Republic in 1949. In the nineteenth century, while treaties between China and the West banned foreign manufacturing within China, Sino-foreign rivalries were confined almost entirely to mercantile rather than industrial operations, as foreign traders and Chinese middlemen competed to determine who would manage China's domestic markets and foreign trade. In this contest, as research by several scholars has shown, the Chinese tended to defeat their foreign rivals, refusing to relinquish control over domestic markets and gradually acquiring control over foreign trade as well.[3] Rhoads Murphey has characterized this commercial rivalry between foreigners and Chinese in the nineteenth century as one in which the foreigners "were attempting to invade a traditional Chinese system which was fully able to meet and beat them at their own game of commerce, on home grounds."[4]

In 1895 the nature of Sino-foreign commercial rivalries began to change with the signing of the Treaty of Shimonoseki because this treaty for the first time legally authorized foreigners to build factories in China's treaty ports. Signed after China's defeat in the Sino-Japanese War of 1894-95, the treaty granted this privilege to Japanese manufacturers, but Westerners also took advantage of it by invoking the most-favored-nation clause. Thereafter, foreigners gradually increased their investments in China. Whereas foreigners had owned no more than 100-odd small factories in China before 1895 (all of them illegal under the treaties), by 1938 they had invested $2.56 billion in China (including Manchuria)—a larger sum than foreigners at the time had invested in any other underdeveloped territory in the world except two, Argentina and India-Burma-Ceylon (the latter of which was taken as a unit for accounting purposes). The large extent to which this capital was used to finance foreign-owned enterprises is reflected in the fact that the predominant share—78 percent in

1931—was in the form of direct investment, leaving only a small amount in the form of loans which might have been used to finance Chinese-owned enterprises.[5]

As foreign-owned industry in China expanded in the early twentieth century, so did Chinese-owned industry. Prior to 1900 there had been only about a dozen sizable Chinese-owned industrial enterprises in China, all of them either initiated or supervised by officials, and this officially sponsored program for industrialization, as analyzed by Albert Feuerwerker, had failed to make the "institutional breakthrough" needed to start "a genuine 'industrial revolution.' "[6] But in the twentieth century, especially after the establishment of China's Ministry of Commerce in 1903, Chinese industrial enterprises were founded in greater numbers. One scholar, Hou Chi-ming, has gone so far as to argue that Chinese firms collectively grew as rapidly as foreign ones. As foreigners increased their industrial investments during the period from 1895 to 1937, according to Hou, "there was a persistent coexistence between Chinese and foreign enterprises in the modern sector of the economy . . . [and this] co-existence . . . implies that [Chinese enterprises] were able to grow as fast as [foreign ones] over the long run." In actual fact, as Hou admits, there is no quantitative evidence available to indicate precisely "the trend of the Chinese share for manufacturing as a whole."[7] Accordingly, the concept of "persistent coexistence" may perhaps best be tested in case studies of particular rivalries between foreign and Chinese firms.

The largely unresearched field of twentieth century Chinese business history abounds with cases of Sino-foreign commercial rivalry worthy of study. Among big businesses (not to mention small ones), the potential for Sino-foreign commercial rivalry existed in several industries: in the textile market between the Chinese-owned Shen-hsin Company and its Japanese rival the Naigai Wata Corporation; in the market for textile machinery between the Chinese-owned Ta-lung Machinery Works and two American manufacturers, the Saco-Lowell Shops and the Whitin Machine Works; in the market for matches between the Chinese-owned China Match Company and competitors of two nationalities, the Swedish Match Corporation and the Japanese Suzuki Company; in the market for books and printed matter between two Chinese-owned publishing houses, the Commercial Press and the Chung-hua Book Store, and the British-owned Calico Printers Association; in the soap, alkali, and fertilizer markets between the Chinese Yung-li Chemical Company and Wu-chou Soap Company and the British-owned corporations, Lever Brothers, Brunner, Mond and Company, and Imperial Chemical Industries (with which Brunner, Mond and Company merged in 1926); in the market for cement between the Chinese-owned Canton Cement Works and the British-owned Green Island Cement Company of Hong Kong;

and in the market for iron and steel between the Chinese-operated Han-yang Ironworks and the Japanese-operated South Manchurian works at Anshan and Penchi. Some of these firms have been studied individually, but the Sino-foreign rivalries within each industry all await research.[8]

Scholars have ventured generalizations about the nature and significance of such rivalries, and, as often happens when scholars speculate about subjects not yet thoroughly researched, their conclusions differ markedly. The primary point at issue has been the extent of Sino-foreign rivalry. Did foreign companies have insuperable advantages over their Chinese rivals or was there an even competition? Historians such as Cheng Yu-kwei and Jean Chesneaux have argued that the competition was uneven and have supported their case by listing advantages which they say foreign firms possessed over Chinese rivals: access to foreign capital, technological sophistication, entrepreneurial talent, managerial efficiency, and political and diplomatic privileges (the latter as a result of unequal treaties which China had signed with the West in the nineteenth century).[9] In reply, other historians, notably Hou Chi-ming and Robert Dernberger, have concluded that, on the contrary, Sino-foreign competition was even. Foreign firms may have had the advantages noted above, Hou and Dernberger have acknowledged, but these advantages were neutralized, they say, by advantages that Chinese firms had over foreign ones. According to Hou, whereas foreign firms may have had greater "capital availability and intensity, technical know-how, and immunity from official interference," Chinese firms had the advantages of "local knowledge, nationalism, and mobility"; and, according to Dernberger, foreign companies may have had "cheaper capital costs, . . . but this advantage could be offset by cheaper labor costs and greater labor intensity in Chinese enterprises."[10]

This debate has raised important issues which deserve to be tested in historical research. Not until historians have taken full advantage of the available sources (and if possible, sets of internal business records that are currently unavailable) will enough be known to make definitive judgments about whether Sino-foreign rivalries were even or uneven. One aim of this book is to take a first step in this direction by showing the efforts that foreign and Chinese businesses made to gain advantages over each other in one rivalry.

As the debate over Sino-foreign commercial rivalry has developed, it has often been conducted broadly in terms of imperialist exploitation (as a characterization of the foreign competitor's behavior) and economic nationalism (as a characterization of the Chinese competitor's behavior). Usage of these terms has attracted considerable attention, stirred heated controversies, and caused people—specialists and generalists alike—to take sides against each other. This debate has immense significance for it bears not only upon Chinese history but upon international economic relations throughout the world today. Unfortunately, as participants in the debate have become polarized, the terms imperialist exploitation and economic nationalism have been used more and more loosely, sometimes almost to the point of meaninglessness. Mindful of this problem of definition, historians in China have sought concrete illustrations of these concepts in the historical record, and they have identified the two cigarette companies as exemplary cases. The history of BAT, they have suggested, exemplifies imperialist economic exploitation (*ti-kuo chu-i ti ching-chi ch'in-lueh*), and the history of Nanyang exemplifies the development of a national capitalist enterprise (*min-tsu tzu-pen ch'i-yeh*).[11] Their characterizations have led me to reconsider the meaning of these terms with reference to the history of the cigarette industry. Did Sino-foreign rivalry in the cigarette industry pit a firm that embodied imperialist exploitation against one that embodied economic nationalism? Answers to this question depend upon definitions of these terms. Accordingly, it might be helpful to bear in mind some of the definitions that have been advanced for imperialist exploitation and economic nationalism.

Of the various possible definitions of imperialist exploitation and economic nationalism, several major ones may be tested by asking questions about foreign and local businesses within five different categories. The first category concerns investment. Was a foreign company imperialistic in the sense that it drained revenue out of a poor country for use in its home country or did it make investments that contributed to the poor country's welfare? Was a local company nationalistic in the sense that it relied only on capital invested by local people or was it financially dependent on foreign investors? The second category concerns relations between management and labor. Was a foreign company imperialistic in the sense that it paid local workers less than the "going wage" in the industry or were the wages that it paid its workers and the prices that it paid peasants for raw materials higher than those paid by its local competitors? Was a local company nationalistic in the sense that it was identified with anti-foreign strikes, boycotts, and other popular demonstrations or was it as much a target of such demonstrations as its foreign rival? The third category concerns businesses' procurement of raw materials in agrarian economies. Was a foreign company imperialistic in the sense that it introduced an extractive industry which distorted development and left the local economy dependent upon foreign markets or did it adapt to the existing market structure and use raw materials to manufacture products for local consumption? Was a local company nationalistic in the sense that it used local raw materials or did it import these from abroad? The fourth category concerns relations between businesses and governments. Was a foreign company imperialistic in the sense that it consistently benefited from the backing of its own government against weaker local governments or did it fail to win its government's unwavering support? Was a local company nationalistic in the sense that it formed alliances with its own country's government and benefited from its government's retaliation against foreign companies or did it operate without governmental support? The fifth category concerns relations between foreign and local businesses. Was a foreign company imperialistic in the sense that it barred local firms from entering the industry or did its local competition grow as fast as it did? Was a local company nationalistic in the sense that it competed directly with foreign rivals in the industry or did it avoid confrontations with foreigners and facilitate foreign economic penetration?

As the debate over these concepts of imperialist exploitation and economic nationalism has developed, so too has a separate debate over the significance of entrepreneurship in China's industrial history. Until recently, there was little debate over this subject because existing research all pointed to the conclusion that few if any entrepreneurs were to be found in China's industrial history. Adopting Joseph Schumpeter's definition of entrepreneurial innovation in its most general form, scholars looked for evidence of entrepreneurship among economic leaders in the history of China's industrialization but reached the conclusion that it was conspicuous by its absence. Albert Feuerwerker, for example, found that nineteenth century Chinese official-industrialists had a "willful absence of initiative" which resulted in "the relative absence of the entrepreneurial spirit among those men who led the industrial effort of late Ch'ing China." These industrialists' lack of entrepreneurship, Feuerwerker has concluded, was traceable to their ties to the institutions of the Chinese elite—especially the experiences of receiving a classical education, taking state examinations, and serving in the official bureaucracy, all of which stifled the development of entrepreneurship by discouraging Ch'ing industrialists from taking risks.[12]

Approaching the subject from a sociological perspective, Marion Levy has also emphasized the failure of entrepreneurship to develop in China, but he has attributed this failure to the structure and attitudes of the Chinese family. Merchants, Levy has argued, cut off the development of entrepreneurship in their families by investing their capital in the pursuit of gentry status and by insisting that their sons do the same. So eager were Chinese merchants for gentry status, Levy has maintained, that they and other members of Chinese society felt that "a merchant was most fully successful to the extent that he and his ceased to be merchants" and became members of the Chinese gentry instead. "There was, therefore, a flight of talent from the merchant field in China and with it a flight of funds." As a result, when opportunities for industrialization arose, China, by contrast with Japan, lacked the requisite "administrative talent" needed to carry out the process of industrialization. Moreover, Levy has asserted that the strength of family ties further hindered entrepreneurial activity because it forced businessmen to "find sinecures for a host of relatives, friends, or neighbors or face social ostracism." In this way, would-be Chinese entrepreneurs were victimized by nepotism which decreased the efficiency and increased the financial burden under which their businesses operated.[13]

These important scholarly studies have advanced a thesis—the absence of entrepreneurship in China's industrial history—that stood without qualification until recently. But in the past few years, case studies of particular businessmen have shown that China was not devoid of entrepreneurs. Hao Yen-p'ing, for example, has concluded that the typical Chinese who worked as a manager within a foreign firm in the nineteenth century—the Chinese comprador—"was a Schumpeterian entrepreneur *par excellence*." According to Hao, compradors, who numbered as many as 10,000 by 1900, fit this description to the extent that they dared to invest risk-capital in industrial enterprises on a short-term, if not a long-term basis.[14] Another instance of Chinese entrepreneurship has been discovered by Thomas Rawski who has described the origin and early development of the Ta-lung Machinery Works in early twentieth century China as a case of "classic Schumpeterian entrepreneurship."[15] And Dwight Perkins, perhaps with the findings of Hao and Rawski in mind, has formulated a generalization that is the antithesis of the earlier scholarly consensus concerning entrepreneurship in China. "China's premodern [pre-1949] development created values and traits," he has suggested, "that made the Chinese people on the average effective entrepreneurs, workers, and organization men when given the opportunity."[16]

In short, a scholarly debate over the existence and significance of entrepreneurship seems to have been joined. This and other studies of Sino-foreign commercial rivalries may contribute to such a debate by helping to refine our understanding of entrepreneurship and by applying this concept to the record of foreign as well as Chinese businessmen in Chinese history. With this aim in mind, it might be preferable to adopt a more elaborate set of criteria for judging entrepreneurship than specialists in Chinese economic history have generally used. Joseph Schumpeter's concept of entrepreneurial innovation offers the proper beginning point for virtually all studies of entrepreneurship (including this one), but for the sake of precision it seems worthwhile to go beyond his most general formulation in favor of the specific criteria he has established for judging whether a businessman's action was innovative. To be so judged, according to Schumpeter, an action must have brought about one of the following five consequences: the introduction of a new good or a new quality of a good; the opening of a new market for goods; the conquest of a new source of supply of raw materials; the introduction of a new method of production; the organization of an industry along new lines, whether creating or breaking up a monopoly.[17] These criteria are worth bearing in mind, for they may serve as a basis for evaluating the performances of the various businessmen in the cigarette industry—both foreigners and Chinese.

If the study of Sino-foreign commercial rivalries has the potential to illuminate such a variety of issues in Chinese social and economic history, one might wonder why no previous books have been devoted to this subject. The answer surely lies in the scarcity of primary sources on businesses in Chinese history. Other historians have noted the difficulty of gaining access to extant records of companies that have done business in China, and there is no need to belabor discussion of the problem here. Suffice it to say that without access to internal business records, case studies in business history are extremely difficult to do, and without case studies, the field of Chinese business history will remain very limited.[18]

On my topic, the records published in Chinese are unusually rich in material for the study of a Sino-foreign commercial rivalry in China. By far the most intimate and revealing of these records are contained in massive compilations of Chinese documents that were skillfully and carefully edited by historians and published in China during the early stages of the "Four Histories Movement" (*Ssu-shih yun-tung*) in the 1950s.[19] I have found that these documentary collections are fully reliable, and I have drawn upon them for data on the development of mechanized industries, handicraft industries, agriculture, and other subjects. I am particularly indebted to an extraordinary volume on the cigarette industry which contains financial records, private correspondence, memoirs, interviews with managers and workers of the Nanyang Company, and relevant excerpts from newspapers, periodicals, and other sources. The result is remarkably detailed documentation on Nanyang's internal history, its management's investigations of BAT, and, to a lesser extent, Chinese public reactions to BAT, Nanyang, and their rivalry. The richness of this volume's contents makes it unique as a source for the study of Chinese business history.[20] I have profited not only from the data in this and other Chinese compilations but also from the citations in them which have guided me to valuable books, newspapers, and periodicals—items which I probably would not have discovered otherwise. Judging from the notes to an article on BAT by the historian Wang Hsi which was recently published in China (and which I have cited in this book), there is additional material in China on the history of the cigarette industry which has not been published and has not been available to me, but I am indebted to historians in China for publishing as much documentary source material as they have. Without it, this study would not have been possible.

Next to the records published in Chinese compilations, my richest sources have been found in Western archives. Unfortunately, I did not have access to archival records at BAT's world-wide headquarters in London, but when I visited there in September 1971 I was granted interviews with executives who had served the company in China earlier in their careers. Most valuable of all the sources that I have used in the West are unpublished records which are housed in three locations: the personal papers of James A. Thomas, Edward J. Parrish, and Richard Henry Gregory (each of whom represented BAT in East Asia early in the twentieth century), which are in the Duke University Archives, Durham, North Carolina; the archives of the United States Departments of State, Commerce, and Agriculture, which are in the National Archives, Washington, D.C.; and the archives of the British Foreign Office, which are in the Public Records Office, London. These Western archives contain little on Nanyang or other Chinese companies but have yielded considerable material on BAT. Particularly valuable for my purposes has been private correspondence that was exchanged either within BAT; or between officials of the Chinese, British, or American governments and BAT; or among officials of the Chinese, British, and American governments about BAT.

On the basis of these and other sources, it is possible for a historian to enter the world of business in early twentieth century China. This world was by no means self-contained, for, in the case of big cigarette companies, it extended deeply into almost every aspect of Chinese life. In fact, in following these businesses as they became involved in boycotts, labor disputes, student protests, tobacco growing, family feuds, controversies in local society, and a host of other activities, I have felt drawn across disciplinary boundaries—from economic and social history into political, diplomatic, agricultural, and other varieties of history. But these boundaries have been well worth crossing, for the excursions have led me to one of the most revealing discoveries about the two big businesses in this Sino-foreign commercial rivalry—namely, that they had significant effects not only on each other but also in the larger arenas of China's economy, society, politics, and foreign affairs.

2.

Penetrating the
China Market

HISTORIANS of Sino-American relations have generally maintained that American businessmen who attacked the market for industrial goods in China between 1890 and 1915 found it impenetrable. According to the "myth of the China market," a generalization accepted by many historians, Americans expected an almost inexhaustible demand for their goods to develop in China but had to settle for a very small volume of trade—so small, compared to the Americans' high expectations, that it amounted to little more than a "myth." Preoccupied with this myth, historians have paid little attention to the activities of American companies within China and have left the impression that American economic penetration of China was so insignificant between 1890 and 1915 as to be unworthy of attention.[1] And yet, some American-owned companies succeeded in penetrating the China market and making it pay before World War I. If historians are to analyze the full impact of American trade in China—not only on its economy but also its politics and society—these commercially successful companies need to be taken into account, for they, not the failed American companies, had the greater impact.

Like other American businessmen supposedly imbued with the "myth of the China market," American cigarette manufacturers spoke of China as though it would provide an endless demand for their product simply by virtue of its large population, but in their case the market proved to be no myth. James B. Duke (1865-1925), an American tobacco tycoon from North Carolina, for example, had had China's population figures on his mind since the invention of the cigarette machine in 1881—a time when cigarettes were still new and unfamiliar in America, much less China. According to a story told by executives in his company, Duke's first words upon learning of the invention were: " 'Bring me the atlas.' When they brought it he turned over the leaves looking not at the maps but at the bottom, until he came to the legend, 'Pop.: 430,000,000.' 'That,' he said, 'is where we are going to sell cigarettes.' "[2] And "that" was China. Duke never visited China, but Americans he sent there also were excited by the prospect of tapping such a large market. James A. Thomas (1862-1940), a representative of Duke's company overseas as early as 1888 and its managing director in China from 1905 to 1922, originally calculated possible sales by imagining that each of "China's population of 450,000,000 . . . in the future . . . might average a cigarette a day."[3]

These were high estimates of China's potential as a market for cigarettes, but Duke, Thomas, and other Americans involved in the trade with China were not disappointed. Exporting the first cigarettes to China in 1890, their sales at first grew gradually and then rose meteorically in the early twentieth century, increasing from 1.25 billion cigarettes in 1902 to 9.75 billion in 1912 and to 12 billion in 1916—ten times as many as in 1902. By 1915 (and every year thereafter in the 1910s and 1920s with only one exception) more cigarettes were exported annually from the United States to China than to all other nations of the world combined. And as early as 1916, China consumed at least four fifths as many cigarettes as the United States (where 15.75 billion were consumed in 1916).[4]

As the premier firm in this booming market, Duke's company enjoyed sales in 1916 valued at $20.75 million with a net profit of $3.75 million. Such high sales and handsome profits delighted Duke. "We have made big progress in China," he reported to the press at the time. "The possibilities . . . there can hardly be overestimated."[5]

If other American businessmen and traders failed in their pursuit of a "mythical" China market between 1890 and 1915, why did Duke's company succeed in finding a real market and extracting profits that fully satisfied its American management? This question is explored here with respect to three aspects of the American cigarette trade in China: the investments of Duke's multinational corporation, the dependence of the company on Chinese, and the company's dealings with its East Asian opposition.

An American Multinational Corporation in China

Part of the explanation for the commercial success of Duke's company in China lies in its transfer of cigarettes, tobacco leaf, capital, technology, and managerial techniques from the West to China. Of these, little reached China prior to the founding of the British-American Tobacco Company in 1902, but between 1902 and 1915, the branch of this company in China invested on a large scale.[6] BAT, which became a multinational corporation with world-wide operations, originated in an Anglo-American alliance between Duke's American Tobacco Company and the

James B. Duke

James A. Thomas

Imperial Tobacco Company of England after a "tobacco war" between the two companies in 1902. The war had been fought largely over international markets, and in the truce which followed, BAT was formed to serve, in effect, as the two former rivals' international division—making it one of the first organizational units ever created by an American multinational corporation to administer international business.[7] Duke and his American Tobacco Company and its American affiliates, as winners of the war, took as spoils 12 of the 18 positions on BAT's board of directors and two thirds of the $24 million (6 million British pounds sterling) worth of stock at which BAT was initially capitalized. In this arrangement, Duke remarked to a friend, he made "a great deal with British manufacturers covering the world."[8]

Like many American-owned businesses abroad at the time, including other American multinational corporations in China, BAT was registered in England and legally based in London, but it had a distinctly American identity. The American Tobacco Company held controlling interest until the dissolution of Duke's "tobacco trust" in 1911, stockholders in the United States (not necessarily in the American Tobacco Company) owned the majority of BAT shares at least until 1915, and an American (Duke) was chairman of BAT's board until 1923.[9] Americans were made managing directors of BAT's branches in parts of the world where the American Tobacco Company had been well established before 1902, and English directors managed branches where Imperial had held stronger positions. In China, site of the largest of BAT's overseas operations, the first managing directors were both Americans, C. E. Fiske from 1902 to 1905, and James A. Thomas from 1905 to 1922.

Immediately after the founding of BAT, James Duke began to extend its reach deeply into China using an approach that earlier had been the key to his swift rise to the top of the cigarette industry in the United States. According to Alfred Chandler's incisive analysis, Duke had succeeded in the United States because he had recognized the commercial possibilities of continuous-process cigarette machinery, which had been invented by James Bonsack in 1881—especially the possibility of creating mass markets to absorb the greater quantities of cigarettes made available by mass production. In Chandler's words, Duke's "success resulted from his realization that the marketing of the output of the Bonsack machine required a global selling and distributing system . . . Duke became the most powerful entrepreneur in the cigarette industry because he was the first to build an integrated enterprise."[10] After building up this enterprise in the United States during the 1880s and 1890s, Duke sent salaried American executives to extend it to China in the early twentieth century. They had the responsibility for integrating mass cigarette production with mass cigarette distribution there, following the example he had set in the United States.

   Duke's determination to integrate mass production with mass distribution in China was reflected in BAT's decision to manufacture cigarettes there. Duke probably made this decision himself, for he took a keen interest in BAT's China branch. As one of Duke's first managing directors in China later recalled, "Mr. Duke kept in close touch with what was going on. Not only did he have long talks with the men who had been in the Far East, but he regularly read the reports that were sent in . . . In discussing a proposition in far-away China or India, the question would always come up as to whether Mr. Duke would approve of the policy chosen. If the project were submitted to him, even after it had been put into effect, his advice and counsel were always helpful."[11] As the final sentence implies, Duke did not originate all policies for BAT's China branch, but a decision to manufacture on a large scale abroad was too significant to have been made without his approval, and, in this case, he gave his whole-hearted approval, urging that the company's factories in China be built and expanded even faster than his American managing director there had anticipated.[12]

   With Duke's support, the first two managing directors of BAT's branch headquarters in Shanghai carried out an investment policy in China almost identical to the one that Duke had used earlier during his rise to the top of the cigarette industry in the United States. Even as Duke had first bought and consolidated existing cigarette firms and then introduced machinery capable of mass production in America in the 1880s, so BAT went through the same two-phase sequence in expanding its industrial base in China. In 1902 and 1903, BAT completed the first phase, consolidation, by bringing under one management the two Western factories that had produced almost all of the small number of cigarettes previously manufactured in China. One had been operated since 1891 by Duke's agent in China, an import-export house registered in England called Mustard and Company, and the other, the American Cigarette Company, had been founded in 1890 and purchased by the Imperial Tobacco Company of England in 1901. In 1902, under the terms of the agreement between Duke and Imperial, both came under the management of BAT.[13]

   In 1905, with the old factories in hand, James Duke selected a new managing director to supervise the company's policy of large-scale investment and expansion in China. For this major undertaking he chose one of his most trusted subordinates, James Thomas. Duke had several reasons for trusting Thomas. First of all, the two men talked easily, perhaps because they came from similar backgrounds. Like Duke, Thomas was the son of a tobacco belt farmer; in fact, he was born in the same county of North Carolina as Duke.[14] Second, Thomas had a personality which undoubtedly appealed to Duke as it did to others in BAT. On the one hand, he was gracious—"an impressive, soft-spoken North Carolinian," as one American BAT representative described him, "with the natural directness and good manners some topnotch people have."[15] On the other hand, he was tough—"as tough as a hickory nut and as square as a dye" in the words of one of his fellow North Carolinians in China.[16]

Besides possessing personal qualities Duke could appreciate, Thomas had experience as a businessman that qualified him for the position of managing director. At age forty-three in 1905 Thomas had been selling cigarettes since the age of nineteen.[17] Still more important, he had begun selling cigarettes across the Pacific as early as 1888. His explanation for why he took his first assignment overseas, which sent him to Australia and New Zealand, reflected an attitude that surely appealed to Duke: "It was the chance, the life that drew me . . . As a missionary of this new American industry I went out to the East . . . I knew not a soul. I used to walk the streets alone at night, possessed about equally with the longing for a friend and with the idea of how to market cigarettes. In the end I made a life-long friend of my idea. I married it."[18] Although Thomas' omission of any reference to the profit motive may cause cynics to smile, this passage shows a determination to seek out and take advantage of unexploited commercial opportunities that undoubtedly contributed to Thomas' success. Once Thomas began to work for Duke's company in 1899, he rose rapidly from one post to the next. He managed its branch in Singapore from 1900 until 1903 and then its branch in India from 1903 to 1904 before accepting the job as managing director in China in 1905. Perhaps the first sign that Thomas had Duke's complete confidence came in 1902 when Duke chose him to act as the company's courier during the trans-Atlantic negotiations that led to the founding of BAT. Thereafter, as Thomas was promoted to higher positions his salary rose too. As managing director in China, he became the best paid foreign businessman in Asia, reportedly receiving between $60,000 and $100,000 per year.[19]

Thus, before BAT made its first large investments in China, James Duke appointed an American managing director whom he knew and trusted to form the managerial link between his own headquarters in New York and the headquarters of BAT's branch in Shanghai. Serving in this capacity, James Thomas reported to Duke and Duke's closest associate, George G. Allen, took orders from them, and secured their approval for any changes in the China branch's major policies. Subordinate to them, Thomas, in turn, had an administrative staff subordinate to himself. In 1906, within one year after being appointed managing director in China, he had thirty-three American and British executives under him at the headquarters for BAT's China branch in Shanghai.[20] Like Thomas, they were all salaried BAT employees sent by the company to Shanghai, but, unlike Thomas, they worked within specialized departments, concentrating on manufacturing, marketing, or purchasing.

Once the administrative linkages between New York and Shanghai had been established, BAT entered the second phase in its plan for achieving mass production in China by making large investments there between 1905 and 1915. Initially capitalized at $2.5 million, the company's branch in China was valued at $16.6 million by 1915.[21] According to an estimate by one historian, Wang Hsi, BAT's reinvestment rate in China was especially high between 1902 and 1912; unfortunately Wang does not quote a figure, but if high, it was perhaps at least 30 percent, the median figure for a majority of foreign firms in China between 1872 and 1936 according to the estimates of another historian, Hou Chi-ming.[22] If the $16.6 million at which the company valued its China branch in 1915 was all direct investment, then BAT was responsible for no less than 15 percent of the total direct investment ($110.6 million) made by all foreign manufacturers in China at the time.[23] And even if the $16.6 million was not all direct investment (for it might have included the cost of inventories), BAT was unquestionably one of the leading foreign investors in early twentieth century China.

A large share of the capital that BAT invested in this period was used to expand its production capabilities in China. In 1906 and 1907, the company built huge new plants in Shanghai and Hankow: the one in Shanghai—part of a BAT complex at Pootung (across the Whangpoo River from the International Settlement) that included 160 buildings on 200 *mou* of land—had the capacity to make eight million cigarettes per day; and the one at Hankow could produce at an even faster rate, turning out as many as ten million per day. Soon BAT added factories in two Manchurian cities, building one in Mukden in 1909 with a capacity of two million cigarettes per day and buying another from Russian tobacco men in Harbin in 1914 with a capacity of about a quarter of a million per day. These factories employed more workers—13,000 by 1915—than any other industrial enterprise in China except Kawasaki, the Japanese textiles company in Dairen.[24] By 1916 BAT relied as much or more on these factories to produce cigarettes for China as it did on its factories in the United States and England, manufacturing in China between one half and two thirds of the 12 billion cigarettes it marketed annually there.[25]

Along with Western capital and technology that BAT exported to manufacture cigarettes in China, it sent Western salesmen to distribute them there. Again following Duke's example (and presumably Duke's orders), Thomas set out to create an organization of Western sales representatives that would give the company mass distribution in China of the kind Duke had achieved in the United States. Since 1885 Duke had transformed the cigarette industry in America by setting up branch sales offices and hiring salaried managers to supervise each one.[26] Beginning in 1905 Thomas sought to do the same in China by creating an administrative hierarchy staffed by salaried young Westerners. Thomas recruited and trained only bachelors under age twenty-five—perhaps, as one of the recruits observed in retrospect, because Thomas and the other directors "believed that only inexperienced and adventurous young men would be fools enough to risk what they proposed."[27] Men of various nationalities and backgrounds were hired, but, according to Thomas, the majority were from the American South. As he later recalled, "Most of the Far Eastern representatives of the company in the early days were recruited from North Carolina and Virginia. From infancy [they] had cultivated, cured, and manufactured tobacco, so that it was second nature to them. In addition, these farm-bred boys were healthy, well-reared, and had a background of good character and good habits."[28] Representatives were hired at a salary of $1,200 per year plus living expenses for a four-year term—during which they were not permitted to marry—and were then granted a one year leave-of-absence. Only two out of five remained after their first year, and many returned home complaining that China was incomprehensible and that their lives there had been unbearably lonely.[29] Those who stayed were urged to overcome barriers between themselves and Chinese merchants by learning to speak colloquial Chinese and were given an incentive to learn it well, a $500 bonus awarded to anyone who passed BAT language examinations which were held once every six months. Representatives were expected to become familiar with the dialect of the region to which they were assigned, and by 1915 a number of BAT representatives achieved a degree of fluency in the spoken language uncommon among Westerners in China.[30]

This corps of young sales representatives took up assignments in all parts of the country. They were based, according to the recollection of one North Carolinian, in "divisional and territorial offices" which numbered between twenty and twenty-five when he started to work for BAT in 1916.[31] Within these offices the representatives were responsible for the company's warehouses, which were so extensive by 1912, according to the British minister to China, Sir John Jordan, that there was "probably not a city of any size in the eighteen provinces where such warehouses have not been established by the British-American Tobacco Company."[32]

Besides their work in these urban-based offices, some BAT representatives "roughed it" (in their words) outside China's cities. They went on caravans into China's "interior," slept in Chinese country inns, and rode boats up the Yangtze, mules in the mountains of Yunnan, horses outside the Great Wall, and camels across the Gobi desert.[33] They may have romanticized the tales of their exploits, but the fact remains that they contributed to a BAT marketing system that developed the capability to distribute cigarettes on an enormous scale.

C. E. Fiske selling cigarettes in China, 1902

Regular shipments of BAT cigarettes reached all parts of China. Aware of the obstacles to empirewide or nationwide distribution, BAT management adopted a regional approach. Through its territorial sales "divisions" it established centers for distributing cigarettes not only around its factories at Shanghai, Hankow, Mukden, and Harbin but in all of China's regional marketing systems: North, Northwest, Lower Yangtze, Middle Yangtze, Upper Yangtze, Southeast, South, and Southwest.

BAT entered most regions by using railroads or foreign-owned steamers which, unlike foreign carriers in sovereign nations, enjoyed the privilege of access to inland waterways in China.[34] Cigarettes were first transported to BAT's division centers in the metropolitan cities, which lay at the core of each of China's regional marketing systems and were then disseminated from the cities to the suburbs and ultimately to the periphery of each region.[35] In North China, BAT cigarettes were shipped up the

Grand Canal from Shanghai to division centers at Tientsin, Peking, and Paoting. For Northwest China, the company's division centers were Chengting and T'ai-yuan, and for markets in Mongolia, Sinkiang, Ili, and elsewhere beyond the Great Wall, BAT representatives dispatched monthly camel caravans from two border cities, Kalgan and Ta-t'ung. Along the Yangtze, BAT cigarettes traveled by steamer to river ports and then overland to inland cities with major division centers in the Lower Yangtze area at Shanghai, Hangchow, Chinkiang, Nanking, and Anking; in the Middle Yangtze at Chiukiang, Sha-shih, Nanchang, Wuhan, and Changsha; and in the Upper Yangtze at Chungking and Chengtu. Along the Southeast coast, the division center was Foochow. In South China (Lingnan), BAT goods were sent from the Canton delta east to Swatow and west up the West River to Wuchow in Kwangsi province. In Southwest China (Yun-Kwei), BAT reached Kunming from three different angles by 1909. From the east steamers carried its cigarettes as far up the West River as possible at which point they were loaded onto mules and packed through the mountains; from the west, BAT goods were brought overland via Bhamo in Burma; from the south they came as far as Mengtzu on the newly completed French railroad that originated in Vietnam and the rest of the way by mule. And in the Northeast, the one area where BAT faced much Japanese and Russian competition, it used trains, carts, wheelbarrows, and men's backs to reach from Mukden and Harbin throughout the region—even to the Amur River along the Russian border —and captured 70 percent of the Manchurian market.[36]

To support this vast distributing system and attract attention to its product, BAT produced advertising that was disseminated as widely as its cigarettes. In this as in other areas of the business, James Thomas emulated his mentor, James Duke. *Collier's* magazine had reported early in the century that Duke "was an aggressive advertiser, devising new and startling methods which dismayed his competitors; and he was always willing to spend a proportion of his profits which seemed appalling to more conservative manufacturers."[37] Encouraged by Duke to use similar methods in China,[38] Thomas invested heavily in advertising. In 1905, as new BAT factories were constructed in Shanghai, the company built beside them new print shops equipped with imported printing presses that provided some of China's most sophisticated printing facilities.[39] Between 1905 and 1915 these presses turned out an enormous quantity of advertising matter.

The BAT advertising system left no region of China untouched. In 1905 in Manchuria, for example, BAT put up 2,000 large paper placards and 200 large wooden or iron signboards in the city of Ying-k'ou—creating an effect that reminded an American journalist of the sensational billing that the Barnum and Bailey circus arranged in advance of its arrival

BAT advertising team at work

in American cities.[40] In North China a newspaper correspondent in Kai-feng reported in 1907 that "the whole city has been placarded with thou-sands of staring [BAT] advertisements," and another correspondent in Sian writing in 1911 described "huge [BAT] posters on the city gates, city walls, on every vacant piece of wall or board in the street, on the brick stands supporting the masts in front of the yamens, in fact anywhere and everywhere."[41] In South China an observer commented in 1908 that "the walls of Canton City and the delta towns are literally covered with the brightly coloured [BAT] advertisement posters."[42] And west of Canton a Western diplomatic official noted in 1904 that BAT agents had canvassed the area along the West River "in a houseboat gaily decorated with flags and other emblems [from which they distributed] picture placards and samples of their wares, with the result that their cigarettes are now on sale in every town and village along the river."[43] In the Southeast, ac-cording to a report from the British consul at Foochow in 1909, BAT agents drummed up business by "preaching the cult of the cigarette and distributing millions gratis so as to introduce a taste for tobacco in this particular form into regions where it was as yet unknown . . . The streets of Foochow are brilliant with [BAT's] ingenious pictorial posters, which are so designed to readily catch the eye by their gorgeous coloring and attractive lettering, both in English and Chinese."[44] In the Southwest (Yun-Kwei) the British consul general in Kunming observed in 1910 that there was "hardly a bare wall in the town that is not brightened by

BAT cigarette posters on walls of Chinese buildings, including a temple

the [BAT's] flaming posters," and a journalist making a tour across Szechwan, Yunnan, and Burma in the same year found that he was "rarely
out of sight of the flaring posters in Chinese characters advertising the
[BAT] cigarette."[45] Along the Lower, Middle, and Upper Yangtze, BAT
inundated every city from Shanghai to Chungking with advertising,
crowning its campaigns with a large billboard placed prominently in the
river gorge. (Although directors had expressed pride in the sign, they removed it in 1914 in response to complaints that it defaced the gorge.)[46]

By 1915 BAT's advertising had so impressed Julean Arnold, U.S. commercial attaché in China, that he urged all American companies interested in advertising to contact and learn from BAT. While he praised
BAT advertising publicly, he teased James Thomas about it privately.
"You have gotten so accustomed to the advertising game that you seem
unable to pull off anything without it from one end of China to the
other," he wrote to Thomas in 1915. "Yes, when it comes to advertising
you have them all skinned."[47]

As BAT's manufacturing, distributing, and advertising systems grew,
its management showed an interest in growing tobacco in China. The
great bulk of its tobacco continued to come from the American South—
which exported over 11 million pounds to China in 1916, more than to
any other non-Western country—but as early as 1906 Duke sent American agricultural specialists from North Carolina to conduct experiments
with American tobacco seed in China. By 1913 BAT procured the first
Chinese-grown harvest of a type of American tobacco commonly used in
cigarettes, bright tobacco, a mild and fragrant leaf characterized by a
bright golden color and a low nicotine content, which had been developed in Virginia and North Carolina just prior to the American Civil
War.[48]

American BAT agricultural specialists induced Chinese peasants to
plant bright tobacco by giving away free tobacco seed imported from
Virginia and North Carolina, lending curing equipment such as ther-
mometers and iron pipes, promising to pay cash for the entire first harvest regardless of its quality, and even assuring peasants that they would
be reimbursed for any damages or losses that they might suffer from
planting American seed. In Hupeh province, where peasants were still
wary, the Americans also agreed to hand over after the harvest, free of
charge, four curing barns used in the initial BAT experiment. After two
years of building up its operations in this way at two locations, Wei-hsien
in Shantung province and Lao-ho-k'ou in Hupeh, BAT in 1915 procured
over 2 million pounds of bright tobacco grown from American seed,
435,000 pounds of flue-cured, almost all from Wei-hsien, and 1,750,000
of sun-cured, most of it from Lao-ho-k'ou. This tobacco was good
enough to be used in the most expensive brand of cigarettes made by

R. H. Gregory of BAT inspecting tobacco, 1906

BAT in China, and a study by the United States government of efforts to plant bright tobacco in many parts of the world showed that the Chinese variety came closest of all to duplicating the color and texture if not the aroma of the American original. Though the total yield supplied no more than 10 percent of the bright tobacco used by BAT in China during 1915, BAT's management was pleased with the savings on taxes, labor, and transportation costs arising from this agricultural operation, and production of bright tobacco expanded rapidly after 1915.[49]

By introducing bright tobacco, BAT started a purchasing system to integrate backwards and achieve vertical integration of its operations in China. That is, by linking tobacco purchasing with cigarette manufacturing, distributing, and advertising, BAT management intended to form a chain reaching from the tobacco fields of Shantung and other provinces to factories in Shanghai and Hankow, and ultimately to customers all over China. Again following the example of Duke, who had vertically integrated the American tobacco industry with remarkable speed in the 1880s,[50] the American-led management of BAT sought to achieve vertical integration of the Chinese tobacco industry by 1915.

These findings suggest one answer to the question posed at the beginning of this chapter. BAT successfully exploited the Chinese market because it did not merely dump surplus goods on China to relieve a glut in Western markets (although Duke, like many American manufacturers, may initially have turned to foreign markets in the early 1890s because of overproduction in the United States). Instead, BAT invested heavily in

China in order to create an efficient, well organized, vertically integrated operation in the image of James Duke's American Tobacco Company. It created economies of scale, cut costs, and effectively utilized Chinese resources to manufacture goods within China for sale to Chinese consumers.

Such an interpretation is helpful in that it highlights Western features of BAT's operation, but it is at best only a partial explanation for the company's commercial success. Was BAT manufacturing efficient because of superior capital-intensive technology or because of low labor costs? Was BAT marketing effective because of its American organizational techniques or because of its relations with Chinese intermediaries? Was bright tobacco successfully grown because of American agricultural expertise or because of BAT's reliance on Chinese cultivators? Before judging the efficiency and organizational strength of BAT and the extent of its vertical integration, it is necessary to explore these questions by taking into account the role that Chinese played in the company.

The Role of Chinese in BAT

Impressed by BAT's American business practices, Western officials, journalists, and other observers in China tended to maximize the importance of Americans and minimize the importance of Chinese in the company. James Thomas was dubbed "the most brilliant American taipan" and was personally credited with bringing "American cigarettes within the reach of China's ragged millions."[51] The sophistication of BAT technology was lauded, particularly the Hankow factory, which was said to be "second to none of its kind."[52] BAT was praised for initiating the "scientific cultivation of tobacco [which] has been one of the greatest blessings that ever happened to Shantung."[53] And BAT's young American "pioneers" and other Western representatives were applauded for breaking away from "the old, established method of conducting business through import houses and compradors [and for putting in] their own elaborate dealer system throughout the interior, supervised directly by foreigners in branch offices located at strategic points."[54] BAT, it was said, "had no compradore—just a few young Chinamen to take care of the money."[55]

Such accounts leave the impression that Chinese contributed nothing more to BAT than a market of consumers—a land of "400 Million Customers." But this impression is misleading, for BAT depended on Chinese workers, peasants, compradors, and merchants in every phase of its operation.

Chinese industrial workers, for example, gave BAT a large, efficient, and comparatively inexpensive labor force. According to Thomas, James Duke found the prospect of manufacturing in China "particularly interesting" precisely because labor was cheaper there than in the United States.[56] In America Duke had relied heavily upon capital-intensive technology; in fact, a study of the American Tobacco Company indicates that he led the industry from labor-intensive to capital-intensive technology in the United States and concludes that "Duke's introduction of machine production was clearly the most significant innovation he made in the industry."[57] But in China the company hired large numbers of unskilled laborers—13,000 by 1915—to perform simple tasks such as preparation of tobacco leaves and packaging of cigarettes by hand.

The company relied on a chain of Chinese intermediaries to recruit workers who, like their counterparts in other Chinese industries, were women. A Chinese comprador (employed by the company) contacted a Chinese labor contractor (not employed by the company) who retained the women on a daily basis. The women received their wages—as little as ¥.20 to ¥.50 per day—through the Chinese comprador and contractor rather than directly from BAT and had little if any job security.[58] Although the cost of employing the Chinese women (including the shares of their wages taken by compradors and labor contractors) was much lower than labor costs in the United States and Great Britain, the quality of the women's work fully satisfied BAT's management. "We have an abundance of good, cheap and efficient labor which works eighteen hours a day without the assistance of labor unions," Thomas enthusiastically reported in 1915.[59] Pleased with the low cost and high performance of Chinese workers, he urged his friend Willard Straight, formerly a China specialist in the U.S. diplomatic corps who had joined the banking firm of J. P. Morgan and Company in 1915, to encourage other American businessmen to take advantage of the Chinese labor force as BAT had.[60]

In growing tobacco as in manufacturing cigarettes, BAT relied on Chinese workers and used Chinese middlemen as recruiters. Duke approved this investment in a China-based tobacco purchasing system for the same reason that he approved investing in BAT's China-based manufacturing system: to lower production costs on goods made for distribution within the country. According to Thomas, "Mr. Duke maintained that if Chinese could produce [bright tobacco] cheaper than we could over here [in the United States], we ought to buy it from them, thus enabling them to buy more cigarettes"—and thus enabling BAT to expand sales, lower costs, and raise profits.[61]

The key to low production costs was cheap agricultural labor at appropriate locations in the Chinese countryside, and BAT gained access to this labor through Chinese intermediaries. Despite the American tobacco specialists' vaunted expertise, they had to rely on Chinese to guide them to the best sites for tobacco growing. It was Chinese middlemen like Jen Pai-yen, later BAT's purchasing agent at Hsuchang, not Americans, who selected locations for BAT's experiments. American "experts" merely confirmed the choice made by Jen and other Chinese familiar with agricultural conditions. Moreover, it was the Chinese middlemen who bought land for BAT experimental stations and compounds because under the treaties a foreign company did not have the right to own land in the interior. Americans also relied on Chinese middlemen to act as bilingual liaisons with the local community, especially peasants.[62]

To appreciate the role that peasants played in BAT's agricultural system, it is important to recognize that they did not by any means learn their first lessons in tobacco growing from Americans. By the time BAT "experts" arrived, Chinese had been raising tobacco for more than two centuries. It was introduced in the late Ming period, probably from the Philippines, and by the eighteenth and nineteenth centuries this valuable cash crop was grown in virtually all of China's provinces—despite numerous imperial decrees prohibiting cultivation of it to prevent it from supplanting needed food crops. By the late nineteenth century, Chinese local gazetteers had already recorded tobacco growing on a large scale in the very provinces where BAT American agricultural specialists later conducted their experiments. In addition, a study done by a Chinese scholar at the end of the nineteenth century confirms that Chinese peasants were already using sophisticated techniques for growing and drying tobacco prior to the arrival of BAT.[63] In fact, American agricultural advisers could not improve upon many of the Chinese techniques and technology. Chinese fertilizer, for example, proved to be as effective as fertilizer used in the West. The company imported ten tons of chemical fertilizer from the United States and tested it in the fields side by side with Chinese beancake fertilizer (tou-ping) for a season in 1912 before deciding not to import any more. It is true that BAT distributed new seeds and curing equipment, but even these seem to have been used by Chinese peasants with a minimum of guidance from the Americans. In private correspondence, the North Carolinian in charge of BAT's collection station in Shantung expressed amazement at the speed with which Chinese mastered the "new" American techniques. They learned, he observed, "even quicker than the average farmer at home."[64]

These reports suggest that BAT benefited from agricultural labor in the Chinese countryside that was not only cheap but highly skilled at tobacco growing and well organized for the purpose of tobacco production. Members of local society in tobacco growing areas did not need to make many adjustments to supply BAT's needs. They were already accustomed to marketing indigenous tobacco as a commercial crop, and they sold American tobacco to Chinese BAT agents in the same manner that they had sold indigenous tobacco to middlemen who had represented Chinese tobacco merchants in the years before 1913. Peasants planted new imported American seed but otherwise accommodated BAT without changing their agricultural techniques or departing from established patterns of intensive farming in small plots. Making adjustments as slight as these, peasants produced high quality bright tobacco, and middlemen sold it to BAT at a far lower price than the company paid for imported tobacco.[65]

In cigarette marketing even more than in cigarette manufacturing and tobacco purchasing, BAT depended on Chinese intermediaries and workers. Despite all the Americans' boasting about the independence of BAT's daring young American "pioneers," they and other Western BAT representatives, like Western merchants throughout the nineteenth century, were dependent on (salaried) Chinese compradors and (nonsalaried) Chinese agents. To understand the significance of the contributions that Chinese made to BAT's marketing, it is necessary to question the assumption —commonly held by American observers at the time—that American "pioneers" enabled BAT to create its own distributing system for a new product. It is true that BAT representatives were among the first Western merchants to travel inland from the coastal treaty ports and deal directly with major Chinese agents, but only in this very limited sense do they deserve to be called "pioneers."

The true "pioneers" of tobacco products in China long antedated BAT. They were Chinese who had introduced tobacco as early as the seventeenth century and had widely distributed it and opium—often mixing the two—by the end of the nineteenth century. These Chinese paved the way for BAT's later introduction of cigarettes by distributing tobacco through an elaborate marketing system and persuading Chinese to smoke pipes on a gigantic scale.[66] During the first decade of the twentieth century, this market became particularly receptive to alternate forms of smoking because anti-opium campaigns cut many consumers off from some of the most popular smoking mixtures, creating a void in the smoking market at the very time that BAT began distributing cigarettes in China.[67] Evidence of Chinese switching from opium to cigarette smoking in 1910 was noted, for example, by a Western journalist in Southwestern China: "In Yun-nan, especially since the exit of opium, this common cigarette is smoked by high and low, rich and poor. I have been offered them at small feasts, and when calling upon high officials at the capital [Kunming] have been offered a packet of cigarettes instead of a whiff of opium, as would have been done formerly."[68] The number of people who substituted cigarettes for opium is difficult to estimate, but observers at the time were struck, as one put it in 1915, by the "astonishingly rapid" spread of cigarette smoking among men and women "of all classes and ages, from ten years up."[69] As evidence that it had penetrated the lower classes and become commonplace among the poor as early as 1911, another noted that bearers and carters and other haulers of heavy loads gauged the distance between two points not by miles but by the number of cigarettes smoked en route.[70] The earlier history of pipe smoking does not fully explain the speed with which cigarette smoking spread in China, but it is a partial explanation for this phenomenon, and it serves as a reminder that BAT did not introduce a new social habit so much as it offered a variation on the already popular habit of pipe smoking.

The term "pioneer" was a misnomer for Western BAT representatives not only compared to Chinese distributors of previous tobacco products but also compared to Chinese within BAT's own sales organization, for Western BAT representatives did not generally introduce cigarettes into China's regional markets. Instead, they left the job of penetrating new markets to Chinese compradors and agents who took responsibility for distribution at local, regional, and eventually national levels. By 1907, before BAT had begun to employ many Western representatives in China, its Chinese compradors and agents had already established numerous agencies for the company in all of China's major population centers—twelve Yangtze River ports, twelve towns in the South and Southwest, eight in the Southeast, thirty-five in the North, and six in Manchuria.[71]

The kind of man who served as a liaison between Westerners in BAT and Chinese markets is evident in the career of Wu T'ing-sheng. First hired by James Thomas at the age of twenty, Wu was destined to become BAT's leading comprador. He was from Chekiang—the home province of many Chinese compradors and financiers at the time—and he had grown up in close proximity to English-speaking Westerners because his father had been a Chinese Christian minister and because his father had arranged for him to be educated at the Anglo-Chinese College in Shanghai.[72] Thomas recollected that he was attracted to Wu because Wu was "ambitious . . . and was of good address and pleasing personality. He understood also how to approach a Chinese merchant and gentleman."[73] Other Westerners were impressed by the range of his Chinese connections and the aggressiveness of his manner. Wu acted as a trouble-shooter and mediator, representing BAT in disputes and negotiating agreements with people from various strata of Chinese society—not only "merchants and gentlemen" but officials, militarists, businessmen, and peasants. He served BAT for more than twenty years and remained in the tobacco industry until murdered by a competitor in 1935.[74]

Wu and other Chinese salaried employees of BAT (generally called "interpreters" rather than compradors in BAT's terminology), not Westerners, performed the crucial task of recruiting Chinese merchants to serve as (nonsalaried) BAT agents and distributors. To get Thomas started, for example, Wu attracted potential Chinese distributing agents by acting as cofounder of the Shanghai Tobacco Trade Guild in 1898. At first it had only ten members, but between 1898 and 1909, the period in which BAT's distributing network took shape, its membership rose to three hundred, all of whom, according to a Western observer, were "leading merchants doing a large business with all the principal towns in China."[75] Even if BAT "induced the formation" of this guild, as a British official suggested in 1906, it was still a traditional form of Chinese organization, adapted to serve BAT's purpose.[76]

Wu seems to have worked mainly within his home region, the Lower Yangtze River Valley, but BAT hired compradors in other regions as well. Li Wen-chung, for example—a graduate of a Methodist missionary school in Peking and an experienced bilingual interpreter—was a leading BAT comprador in North China. Within that region he persuaded Chinese merchants to become BAT dealers in nineteen new locations during a single year, 1905. He recruited Chinese merchants to accept BAT dealerships in Paotou and Suiyuan and was said in his career to have "traveled into ice and snow even as far north as the Russian frontier and as far west as the vicinity of Tibet." Still other BAT compradors led mule trains over rugged mountains in Yunnan and plied rivers in Fukien. When Western BAT representatives finally began to go on such expeditions, Chinese compradors always accompanied and guided them.[77] In performing all these tasks, BAT's Chinese compradors were more like "pioneers" than were any of its Western representatives.

Even more essential to BAT's distributing network than Chinese compradors were Chinese merchants whom compradors recruited to act as agents for the company. These were "pioneers" in the sense that they made the critical decision whether to permit cigarettes to enter China's markets at the regional and local levels. But they were by no means "pioneers" in the sense of outsiders breaking into new or unfamiliar markets. Instead, they were established merchants who operated through well-worn and often complex channels of distribution. Ma Yü-ch'ing, for example, who agreed to act as a BAT agent in Tientsin in 1903, was general manager of the Yü-sheng-ho house, a firm that had existed for more than fifty years and that had branches in Tientsin, Kalgan, and other towns in Chihli and Shantung provinces. He succeeded as a regional distributor not because he "pioneered" or deviated from common Chinese commercial patterns but because he operated through an established business house and traded on its reputation. In similar fashion, other leading merchants represented the company at the core of each region—H. H. Kung (later Minister of Finance under Chiang Kai-shek) in the Northwest at T'ai-yuan; Yu Shao-tseng in the Peking-Tientsin area and Ts'ui Tsun-san around Paoting in North China; Liu Chin-sheng in southern Manchuria and Hsu Lo-t'ing in the northern part; Ch'in Sung-k'uan in the Southeast at Foochow; and numerous merchants in ports along the Lower, Middle, and Upper Yangtze. All these Chinese BAT agents came from merchant families that had been in business for generations in their respective locales, and they were therefore well connected within regional and local merchant associations and guilds that were based on vocations and native place ties and that had served commercial purposes in China since at least the sixteenth century. They received no salary from BAT, and, after becoming BAT agents, they continued to deal in a variety of products besides cigarettes—including other foreign products such as Standard Oil's paraffin candles and coal.[78]

In moments of candor, American BAT representatives admitted that Chinese compradors and Chinese agents were the key to the company's sales organization. One Chinese-speaking American recollected after serving as a BAT representative in North and Northwest China between 1911 and 1914, "We [Western BAT representatives] were called salesmen. But actually we did no selling. The large majority of foreigners in the company spoke no Chinese; [Chinese] interpreters and dealers took care of that end. The foreigners were really inspectors, overseers, advisers. More than anything else, it seemed to me, our job reduced itself to advertising."[79] Another American who was a BAT representative in Northeast China during the 1910s made the point even more strongly. "I was simply window dressing for our Company," he conceded. In public his Chinese comprador left the impression that the foreigner managed the market, for, "being a good Chinese, he always kept the Foreigner puffed up and believing himself [to be] a man of great affairs." But the American recognized that it was the Chinese who actually supervised all the "business affairs" in his division of BAT's selling organization.[80]

With this Chinese sales network in mind, it is possible to evaluate critically James Thomas' and other Westerners' claims for the originality of BAT's distributing system. For example, Thomas spoke of having spent fifteen months in Hong Kong in 1903 "getting my goods established in the maritime provinces of South China," but what he did was hand the goods over to Chinese compradors and regional agents who, in turn, arranged for Chinese merchants from Macao, Canton, Swatow, and Kunming to act as local agents. The Chinese, not Thomas, actually managed the market.[81]

In addition to working through Chinese regional agents, BAT also distributed cigarettes through two Chinese nationwide distributing agents, the Yung-t'ai-ho (Wing Tai Vo) Company and the Mao-i Company. Like most of BAT's other leading agents, Cheng Po-chao (Cheang Park Chew), the president of the Yung-t'ai-ho Company, was a well-established merchant before becoming affiliated with BAT; his company had been in existence for thirty years and its management was mostly in the

Cheng Po-chao

hands of his family before he began to act as a BAT agent in Hangchow in 1912. Unlike other BAT agents, Cheng was given exclusive rights to handle one or two brands throughout China. According to his agreement with BAT, the foreign company designated the prices of his cigarettes, took all returns from his sales, and then paid him a commission—literally "handling charges" (*shou-hsu fei*)—according to the volume of his business.[82]

The strategy employed by Yung-t'ai-ho and Mao-i in Manchuria in 1914 illustrates the selling techniques these distributors used. Upon entering the market in 1914, the Mao-i Company added 1.5 million to the 35 million cigarettes distributed by BAT every month in Mukden, and Yung-t'ai-ho immediately added another 3 million. They marketed goods by spreading their network of Chinese representatives throughout Manchuria and by supplying local Chinese agents. The local agents, in return, submitted inventory reports on their stock and paid up their accounts every three months. According to a spy for one of their competitors, these local agents also led advertising campaigns through "wine shops, tea houses, schools, yamen, military administrative centers . . . down every street and alley, sticking up posters everywhere."[83]

BAT heralded its delegation of authority for nationwide distribution to the Chinese distributing companies, Yung-t'ai-ho and Mao-i, as a grand "experiment,"[84] but the Japanese historian, Ōi Senzo, has offered a more plausible interpretation. Rather than a new experiment, Ōi has argued,

BAT's arrangement with Cheng Po-chao represented a successive phase in the evolution of the nineteenth century comprador system. Cheng differed from nineteenth century compradors, according to Ōi, only in that he was a "capitalist comprador"—a capitalist in the sense that he retained the majority of stock in his business and a comprador in the sense that he represented the interests of a foreign firm in Chinese markets. On the other hand, Ōi's characterization has its limitations, for it makes no distinction between salaried employees of a foreign firm (which I have designated as compradors) and nonsalaried participants in a foreign firm's distributing system (which I have designated as agents); according to this distinction, Cheng was an agent, not a comprador, of BAT. Moreover, Ōi has failed to note the related point that Cheng, like other leading Chinese BAT agents, differed from compradors in that he was an established, independent merchant long before he agreed to represent BAT or any other foreign firm, and therefore, that he need not have been as dependent on foreigners for initial capital as compradors were. This degree of financial independence gave him more power and leverage vis-à-vis BAT than compradors had with their foreign employers.[85] For this reason, Cheng was perhaps more of a capitalist than a comprador.

However he is labeled, Cheng—and for that matter all of the leading BAT agents—bore a resemblance to nineteenth century compradors insofar as they provided a bridge between a foreign firm and China's complex commercial system. Over this bridge BAT hoped its cigarettes would flow into China unimpeded by obstacles that had previously stymied Western businessmen—obstacles such as the variations in regional marketing structures, differences in spoken dialects, and idiosyncrasies in local currencies and standards of weights and measures.[86] As one BAT pamphlet expressed BAT's premise, for Chinese people "to disturb [Cheng Po-chao's and other Chinese BAT agents'] markets, to place difficulties in the way of the cigarettes they sell, is to destroy the business of Chinese companies and merchants."[87] But the business of Chinese distributors was not destroyed. Most of the Chinese BAT agents continued to represent BAT for two decades or longer and distributed huge numbers of cigarettes.[88]

Thus, BAT's marketing system had as its foundation a solid and substantial Chinese sales organization, and once this marketing system delivered the goods to the core of a regional market, Chinese agents and merchants, not Westerners, invariably managed local distribution. Even in Shanghai, the site of BAT's headquarters in China, the company relied on an established Chinese distributing system rather than attempt to "modernize" it or institute close Western supervision over it. A Chinese company that was planning to enter the Shanghai market discovered how this system excluded newcomers. In 1915 one of its representatives tried the shops on every one of the commercial streets of Shanghai, but no merchant would agree to accept his goods, even when he offered them on consignment. Through this frustrating experience, he learned that BAT had formed restrictive agreements with Shanghai's twenty largest mercantile houses (*hang*) to distribute BAT tobacco products and no other brands. The twenty leading *hang*, in turn, controlled 170 smaller *hang* in the city. Together these *hang* formed an intricate network through which BAT utterly dominated the cigarette market in Shanghai. According to a study by one of BAT's potential competitors in January 1915, BAT sold over 100 million cigarettes per month—more than fifty times as many as its combined Shanghai competitors—giving it monthly sales valued at ¥400,000.[89]

In other cities, the Westerners' role in marketing was even less significant. Generally the foreigners dealt with no one lower in the marketing hierarchy than a "division dealer," a Chinese regional distributor like Ma, Kung, Yu, Ts'ui, and others mentioned earlier, each of whom was responsible for a territory usually as large as, but not necessarily coterminous with, a whole province. The division dealers distributed the cigarettes among subdealers. Upon receiving the merchandise, division dealers put up cash security generally in silver dollars amounting to at least half the value. Credit was granted for the remainder on consignment, the balance being secured by written guaranty bond or "shop guaranty" for thirty, sixty, or ninety days. Division dealers were held financially responsible for subdealers and received a rebate ranging from 5 to 10 percent. On retail sales, the subdealers made a profit ranging from 10 to 20 percent.[90] The tendency for BAT agents to rely on established Chinese merchants to market goods in this manner may be documented for cities in several regions: Peking, Tientsin, Chinwangtao, and Changli in North China; Mukden in Manchuria; Chungking on the Upper Yangtze; Shanghai and Hangchow in the Lower Yangtze; Foochow in the Southeast; Canton in the South; Kunming in the Southwest.[91]

How far outside the treaty port cities did BAT's distributing system reach? The question is an important one because of its possible relevance to the larger issue of whether imported Western industrial goods competed with native goods and crushed native handicraft industries. Albert Feuerwerker, for example, has argued that foreign industrial goods— including BAT cigarettes—did not reach rural markets and therefore did not significantly affect Chinese handicraft industries. "The simplistic indictment of 'foreign capitalism' by some contemporary Chinese historians for having progressively 'crushed' and 'exploited' domestic handicraft industry from the mid-nineteenth century onward is belied by the actual state of the Chinese economy as late as the 1930's . . . Anyone who would claim that the Hunan or Szechwan peasant in the 1930's . . .

smoked BAT cigarettes . . . has a big case to prove."[92] The available statistical evidence on the tobacco industry tends to support Feuerwerker's thesis concerning the staying power of China's traditional handicraft industries, for it shows that the gross value of hand-made tobacco products (mostly shredded tobacco smoked in pipes) rose from ¥38.5 million in 1914 to ¥115.6 million in 1916; declined to about ¥70 million per year in 1917-18; fell to less than ¥30 million per year in 1919-20; but then recovered and surpassed previous levels, reaching between ¥128.9 and ¥171.8 million by 1933.[93] Nonetheless, while acknowledging, in light of this evidence, that the distribution of BAT cigarettes seems not to have crushed the handicraft industry for tobacco products, I believe that a case can be made that Chinese peasants smoked BAT cigarettes before the 1930s. The case is based primarily on evidence from a BAT house organ published in Chinese for Chinese compradors, dealers, and agents that described the work of Chinese BAT agents in rural markets.[94]

According to this BAT publication, Hsieh I-ch'u, based in Chinwangtao in northeastern Chihli, proved himself to be an exemplary BAT agent by doing outstanding work for the company in rural markets. He kept informed of activities in the countryside and set up a mobile cigarette concession when villagers gathered for plays, feasts, festivals, and fairs. Accompanying him were attendants who displayed brightly colored posters, played a gramophone, distributed lottery tickets, and gave away prizes. He and his team, it is said, went "everywhere" in the region.[95] Jonathan Spence has pointed out that itinerant vendors were able to widen the market for opium by making cheap sales during such festivities, at which time people from the countryside "in high spirits with loose cash in their pockets (for probably the only time in the year) might well contract a habit that would last a lifetime."[96] Hsieh I-ch'u tried to attract customers from rural areas in the same manner, and, according to reports from Maritime Customs officials, he was successful. In 1909 BAT's business in his territory around Chinwangtao grew, they reported, because in "the surrounding rural area . . . the demand for cigarettes was exceptionally great."[97]

Other Chinese BAT agents used similar techniques to penetrate rural markets elsewhere in China. In Shantung and Anhwei provinces, Wu K'o-chai circulated cigarettes in small market towns along the Tientsin-Pukow railroad. Around Kalgan, just south of the Great Wall, Yang Teh-fu, another Chinese agent, extended BAT's reach into rural areas by hiring carter-hawkers to pull mule carts full of cigarettes into outlying towns and villages. In southern Manchuria, Liu Chin-sheng hired children to do hawking in the streets of Mukden and sent vendors into rural marketing areas around the city. And other Chinese BAT agents in South as well as North China conducted sales campaigns in rural markets.[98]

The results of the Chinese BAT agents' work outside the treaty ports deeply impressed Chinese businessmen. In 1914 a spy for a Chinese cigarette firm reported that BAT cigarettes were selling in small market towns (ch'eng-chen) all over Manchuria.[99] In 1916 and 1917 a Chinese businessman took trips up the Yangtze and overland in North China to study BAT's operation and was awed by what he found. He confided to his Chinese business associates that BAT had canvassed not only treaty ports and market towns but even villages (hsiang) that were "extremely poor." Moreover, in inland areas where "the land is vast and the people are few" shopkeepers had pledged to carry only BAT cigarettes and customers had already become loyal to BAT brands.[100] This report gains credence from an American BAT representative's recollection that he and Chinese BAT agents regularly visited small market towns (chen) in the early 1910s where "the average population was about 1,000 of which at least two thirds were children and babies."[101]

This evidence is not statistically grounded, but it indicates that BAT cigarettes were reaching rural markets in many parts of early twentieth century China. If complete statistics were available, they would probably show that the farther markets were from treaty ports, the fewer BAT cigarettes appeared in them. Nonetheless, whatever the magnitude of this trade,[102] the point is that Chinese agents were able to reach outside the treaty ports and widen the market for their product in rural as well as urban areas.

To enhance the appeal of BAT cigarettes, the company also relied on Chinese to design and disseminate its advertising. Chinese artists and calligraphers were hired to draft advertisements, and they showed a talent for adapting BAT's message to the Chinese cultural setting. Westerners in BAT were inclined to advertise through newspapers, sign boards, wall paintings, posters, and cigarette cards (the latter introduced in the 1880s in America by James Duke) as they had in the West. BAT advertised extensively in these, but it also introduced a variety of other advertising media—scrolls, handbills, calendars, wall hangings, window displays, attractive and strong cigarette packing cases (whose wood and nails were re-used by the Chinese), cotton canvas covers for the tops of carts, and small rugs to serve as footrests in rickshas—all bearing BAT's trademark. Of the ¥1.8 million which BAT allocated annually for publicity and sales promotion by the late 1910s, only 10 percent was spent on newspapers and publications. The remaining 90 percent was invested in outdoor advertising and material distributed in campaigns.[103]

In the advertising they designed for BAT, Chinese artists presented portraits of legendary and semilegendary figures familiar to almost any Chinese. Subjects included, for example, Yang Kuei-fei, an artful concubine supposedly responsible for the decline of the T'ang dynasty; Yueh

Fei, a patriotic general of the Southern Sung; the White Snake (*Pai-she*), an immortal serpent transformed into a beautiful woman; and whole series of characters from Peking operas and popular novels such as *Water Margin* (*Shui-hu chuan*), *Monkey* (*Hsi-yu chi*), and the "Twenty-four Stories of Filial Piety" (*Erh-shih-ssu hsiao*). Occasionally BAT artists portrayed figures from their own day, but rather than show contemporary celebrities endorsing BAT products (as advertising has tended to do in the West) they pictured nameless representatives of broad social groups: women (represented by "beauties" from Shanghai's *filles de joie*, and, beginning in 1914, by the "new women" in roles as social activists); children at play (based on the "Hundred Children Pictures" [*Pai-tzu t'u*]); and working people engaged in the "360 Trades" (*San-pai liu-shih hang*, an expression used to mean all trades and professions). Sensitive to China's "Little Traditions" as well as its "Great Tradition," BAT artists screened all the company's advertising proposals and adjusted any details in Western advertisements which, because of cultural differences, might have been considered offensive in China (for example, a green hat was taboo, for it signified cuckoldry in the Chinese vernacular). And they kept advertising up-to-date by dropping symbols that became unfashionable; the dragon, for example, was not used after the monarchy, which it had symbolized, was overthrown by the revolution of 1911.[104]

As time passed, less and less language was used in BAT advertisements. Pictographic advertising was believed to be more effective because it helped BAT reach potential customers who were illiterate and because it made posters less subject to embarrassing and damaging puns. The latter was a problem because the Chinese language is rich in opportunities for punning, and the opportunities are even greater as one moves from one dialect to another. As Carl Crow, a veteran journalist in China, remarked in the 1920s, "an advertising phrase which is quite suitable and effective in Shanghai is turned into a vulgar or ludicrous pun in Canton."[105] To prevent punning, some posters had no words at all—except BAT's trademark, a bat (a symbol of happiness in China)[106] and the roman letters "B.A.T." with their Chinese equivalent, *Ying-Mei yen-ts'ao kung-ssu*.

To design appealing advertisements, BAT quickly learned to rely on Chinese who had a knowledge of Chinese traditions, an awareness of local customs, and a sensitivity to the popular imagination—cultural sensibilities which Americans and other Westerners on the staff simply did not possess. At first the company had entrusted Westerners with the responsibility for advertising, but they chose pictures (for example, from German fairy tales) and introduced slogans (for example, "beauty has a short life") that made no sense in the Chinese context and became objects of derision and sources of amusement among Chinese customers.[107] Em-

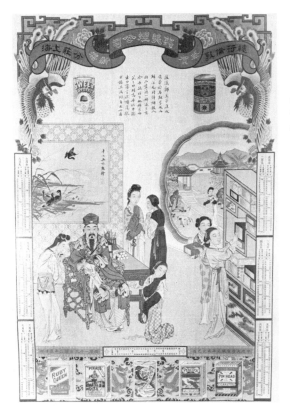

After using the European scene in China, BAT introduced the Chinese scene by 1909

barrassed, BAT management began to rely on Chinese artists and protect itself against future mistakes by arranging for its Chinese distributing agents to check the advertising staff's work. To prepare its annual calendar, for example, the company paid retaining fees to several leading Chinese artists for submitting sketches and circulated these sketches in rough draft among all Chinese BAT agents throughout the country who evaluated them, checked to see that the characters in them carried no potentially damaging local double meanings, and cast votes for their favorites. The winning entry was reproduced in full-color calendars in time for Chinese New Year. In the words of an American BAT representative who later returned from China to become a successful copywriter with the American advertising agency in New York, J. Walter Thompson and Company, BAT's calendar became its "big advertising smash every year" and was distributed "in every nook and corner of the nation."[108]

Chinese in BAT's marketing system not only distributed advertising matter but also adjusted cigarette packaging and initiated promotional campaigns appropriate to their own locales. To make the product accessible to the poor, for example, hawkers broke down packages of the cheapest BAT brands, which were priced at $0.02 for a pack of twenty cigarettes, and sold as little as one cigarette at a time. To attract crowds, Chinese BAT distributors organized parades in which BAT banners waved, held lotteries in which first prize was a gold watch, and paid itinerant story tellers to weave into their tales appealing references to brands of BAT cigarettes. BAT's management supplied prizes and rewarded anyone showing initiative in promotional work.[109] In this way, the company encouraged Chinese distributors and dealers to supervise their sub-dealers and hawkers more closely and penetrate into local markets more deeply.

Thus in all aspects of its business—manufacturing, distributing, purchasing, and advertising—BAT's success depended on Chinese. Without the work they did and the links to Chinese society they provided, BAT would not have been able to manufacture its product so cheaply, distribute it so widely, and profit from the trade so handsomely.

These findings require modification of earlier conclusions concerning the efficiency, organizational strength, and vertical integration of BAT. First, the role of workers and peasants in BAT shows that the company's "efficiency" derived not merely from its transferral of a superior capital-intensive technology to China but also from its reliance on cheap labor. Second, the prominence of Chinese compradors and agents in BAT's marketing system suggests that the company's "organizational strength" derived more from the established Chinese network into which Duke and Thomas injected their product than from changes in marketing patterns introduced by American "pioneers." BAT pioneered in the sense that it (along with the Standard Oil Company) was one of the first businesses ever to distribute a trademarked product on a national level in China (that is, in all of the country's regional markets), but it did so by relying heavily on Chinese compradors and agents who, in turn, operated through their connections within the existing marketing structure. And third, if attributing BAT's success to "vertical integration," it must be understood that only the Americans and other Westerners in its organization were tightly centralized and departmentalized; the Chinese who actually controlled labor, purchasing, and distributing were generalists and—except for their incorporation of new advertising—they continued to rely on business practices that were not radically different from the ones used in the past.

Americans and other Westerners in the company depended so heavily on Chinese compradors that the foreigners seem to have been only dimly aware of the commercial processes beneath the highest levels of the operation. Looking back on his career, James Thomas admitted that he had not been able to impose his own authority and that he had accepted Chinese commercial practices as they were because he had felt powerless to change them. "When I landed in China," he recalled in the early 1920s after retiring from BAT,

> I felt that someone had taken me by the heels and thrown me into the Pacific Ocean. I had to swim out. The thing was overwhelming. Here was a country of four hundred million people extraordinarily true to their civilization . . . Their ways, unlike though they were to mine, were worthy of my respect.
>
> My conviction, therefore, was that I ought to trade with these people as nearly as possible according to their ideas . . . I knew that not in my lifetime could I educate a handful to my ways; I must adapt myself to theirs.
>
> That is the word in either cigarettes or civilization—adaptability.[110]

Overwhelmed by Chinese "civilization" (by which he seems to have meant China's traditional commercial system), Thomas sought to accommodate Chinese marketing conditions by delegating a large measure of authority to Chinese compradors. The compradors were salaried employees who recruited Chinese nonsalaried agents, who, in turn, worked through other Chinese networks that enabled BAT to reach deeply into Chinese economic life. These Chinese intermediaries—compradors, agents, and others—not Westerners, were between BAT's Western management, capital, and technology on the one hand, and Chinese labor, agriculture, and marketing structure on the other. The chains of Chinese intermediaries that linked the company with Chinese commercial institutions were held together by distinctive Chinese social and commercial relationships that were neither directly controlled by BAT's Western management nor formally integrated into the company's organizational structure.

Thus summarized, the complex process by which BAT penetrated markets in China may appear to have been completed more simply or more smoothly than it was. Even if, as I have suggested, BAT succeeded at gaining access to Chinese markets in these ways, it nonetheless aroused hostility among many Chinese. To evaluate the full consequences of BAT's commercial penetration, it is necessary to consider the sources of this hostility, analyze the resistance that the company faced, and explain why—despite resistance to it as a foreign intruder—its operations expanded and its sales and profits rose in these years.

East Asian Opposition to BAT

In assessing opposition to BAT in China, it is helpful to consider earlier opposition to the company in Japan, for the contrast between the two is illuminating. In Japan, by contrast with China, the strongest opposition to James Duke's company in the late nineteenth and early twentieth centuries came directly from the national government. The Japanese government mounted opposition to Duke's company as part of an effort to revise the unequal treaties, and it achieved success along these lines much earlier than governments did in China. Whereas Chinese governments did not recover tariff autonomy until 1929 and did not abolish extraterritoriality until 1943, the Japanese government began negotiating agreements with the Western powers as early as 1894 and eventually eradicated extraterritoriality by 1899 and retained tariff autonomy by 1911. In the late 1890s and early 1900s the Japanese government followed up on these diplomatic initiatives by passing laws to limit commercial penetration by foreign companies, including Duke's.

The Japanese government first acted against Duke in 1897, promulgating a higher tariff on cigarettes that was scheduled to go into effect in 1899. The significance of the proposed tariff did not escape Duke who responded by more than doubling exports of cigarettes to Japan in 1898, an increase that brought his company's total sales in Japan for the years 1894-98 to 985 million cigarettes. After the higher tariff went into effect in 1899, he than adopted the same strategy that other American exporters had used to overcome high tariffs at the time in continental Europe, jumping the new barrier by investing in local factories.[111] Duke made his investment in Murai Brothers Company, largest of the many private Japanese cigarette firms founded in the 1890s. He initially paid $2 million in cash for stock in Murai Brothers and subsequently secured controlling interest by buying 60 percent of the $5 million worth of stock that Murai

Brothers issued. This investment, the largest by any foreign company in late Meiji Japan, gave Murai a total capitalization of ten million *yen*—a huge sum compared with the financial backing of its Japanese rivals.[112] By the beginning of the twentieth century, Murai Brothers, with Murai Kichibei as president and American Tobacco's Edward J. Parrish as vice-president, produced on a large scale, advertised widely and extravagantly, and enjoyed high and rising profits in Japan—772,641 *yen* in 1900; 1,408,623 in 1901; and 1,687,691 in 1902—which during the latter year alone added almost $500,000 to the treasury of Duke's company in the United States. Early in 1903, just after Murai Brothers had become a subsidiary of the newly founded BAT, Parrish reported to Duke that within a few months Murai's most popular brand would be selling at the rate of 100 million cigarettes per month, and Duke calculated that Murai's net income would reach 2 million *yen* for the year, but contrary to these expectations, the life of this Japanese-American venture was cut short.[113]

Murai Brothers was undone by Japanese opponents who demanded governmental action against it on patriotic grounds. Japanese cigarette firms carried out nationalistic advertising campaigns in newspapers attacking Murai Kichibei for his capitulation to foreigners and accusing him of permitting his foreign backers to drain revenue out of Japan. Although these native competitors remained tiny by comparison with Murai and posed little threat to it, the message in their advertising appealed to political leaders who had never been enthusiastic about direct foreign investment in Japanese industry and who had taken pride in the nation's independence, especially since the victory over China in the war of 1894-95. Japanese political leaders accused Duke of being "the capitalist [who] is intending to monopolize the whole world," and, on these grounds, they ended American involvement in the Japanese cigarette industry in 1904 by expropriating every existing private cigarette firm, including Murai Brothers. In thus nationalizing the industry, the government compensated Duke for his investment but forced him to dissolve his interests in Japan.[114] Thereafter the Japanese Imperial Tobacco Monopoly, capitalized at 10 million *yen* (5.8 million paid), imported tobacco directly from the United States and controlled the cigarette industry not only in Japan but also in Korea (which was annexed by Japan in 1905). Forced to withdraw from Japan and Korea, Duke shifted his attention and his capital investments to China.[115]

While strong action by the Japanese government decisively ended American penetration of the cigarette market in Japan and Korea, the absence of direct governmental opposition to Duke's company in China encouraged him to invest there. By contrast with his experience in Japan, for example, Duke was virtually unhindered by tariffs in China. Prior to

1902 tobacco and tobacco products entered China duty free. Under the Treaty of Tientsin of 1858, the tariff on almost all goods had been 5 percent ad valorem with an additional duty of 2.5 percent on goods reshipped to a second port (the maximum rates under the treaties),[116] but the cigarette had been exempted from these low duties on the grounds that it was for the personal use of foreigners in China. In the list of duty-free items the treaty did not actually specify cigarettes, a product unimportant in the West and nonexistent in China when the treaty was signed. It did, however, specify "tobacco (foreign), cigars (foreign)" in the English text and "foreign prepared tobacco and tobacco leaf" in the Chinese text. These items were construed to include cigarettes until 1902.[117]

In 1902 the Chinese government, still lacking tariff autonomy, attempted to begin taxing BAT at the maximum rate under the treaties of 5 percent ad valorem upon arrival in a Chinese port and 2.5 percent ad valorem for reshipment, but the attempt was only partially successful. Despite the fact that the new tax was clearly intended to apply to all of the Western company's goods and any other machine-made cigarettes sold in China, BAT argued that all its cigarettes manufactured in China were subject not to this tax but only to a tax on "native prepared tobacco." The latter tax was designed to cover pipe tobacco that was shredded or pulverized by hand, and it was payable at the much lower rate of 0.675 taels per picul on goods when first marketed in a city and 0.3375 taels per picul for transshipment, or a total of 1.0125 per picul, the equivalent of no more than 1 or 2 percent ad valorem. Though BAT's cigarettes were all machine-made, its management insisted that this tax be applied to all the goods it manufactured in China because these deserved to be considered "native" products. In 1904 Prince Ch'ing (I-k'uang), the highest ranking official in foreign affairs in Peking, challenged BAT's contention and proposed that cigarettes be taxed at the higher rate. But BAT, with the support of the American and British ministers in China, convinced him to accept the company's position on the condition that foreign cigarettes not be exempted from transit duties (which were paid at the rate of half the basic tax, that is, 0.3375 taels per picul), likin, and other local taxes outside ports. In January 1905 (six months after BAT's ouster from Japan), BAT obtained formal assurance from the Ch'ing central government that the company would be considered a "native" manufacturer and that cigarettes it made in China were taxable at the lower rate.[118]

With this favorable tax arrangement in hand, BAT made its first substantial direct investments in China, inaugurating construction of its largest factories at Shanghai and Hankow in 1906 and 1907, but as its building program continued with the addition of a plant in Mukden in 1909, the company encountered stronger opposition from officials at the provincial level in Manchuria than it had faced from the Ch'ing central government in Peking. BAT had expected cigarettes made in Mukden to be taxed at the same low rates as in Shanghai and Hankow. The company's proposal to this effect was approved by the Chinese Foreign Ministry (*Wai-wu pu*) in Peking on August 1, 1909, but, to BAT's surprise, it was rejected by Chinese provincial officials in Mukden. Hsu Shih-ch'ang, the governor-general of Manchuria who was experienced at dealing with Americans, proposed that BAT pay taxes at Mukden at 5 percent ad valorem. One of Hsu's subordinates in Mukden pointed out that a rate of 1 or 2 percent ad valorem was "ridiculously small" and the rate of 5 percent ad valorem was "reasonable and generous" in light of the taxation imposed on cigarettes in other countries.[119] Hsu's successor, Hsi-liang, who resolutely resisted foreign encroachment here as he had in other provinces, continued the campaign and finally convinced the Ch'ing Foreign Ministry to reverse its position. One month after approving BAT's proposal for lower taxes, the Foreign Ministry endorsed Hsi-liang's demands for a tax of 5 percent ad valorem. BAT loudly complained, claimed the company had been led to build the factory under false pretenses, and curtailed production there. It paid the tax between 1909 and 1911 but finally threatened to close the factory if taxes were not reduced.[120]

In 1911, rather than lose tobacco revenue in the region altogether, officials in Mukden and the Foreign Ministry in Peking accepted a compromise. On its side, BAT agreed to pay 1 tael per picul at the factory in Mukden (as compared with 0.675 tael per picul in Shanghai and Hankow), 0.45 tael per picul in Manchuria on cigarettes exported to China proper, and 0.225 tael per picul at the point of entry in China proper; the cigarettes were declared exempt from likin and other local taxes in Manchuria but not in China proper. And on its side, the Foreign Ministry agreed not to use this as a precedent for changing the 1904 agreement concerning taxation of BAT factories in Shanghai and Hankow. Although this arrangement was less favorable for BAT than the one in Shanghai and Hankow, the rate of taxation was still very low by comparison with tobacco tariffs and taxes in other countries, and the exemption from local taxes gave BAT advantages over its Chinese competitors in Manchuria.[121]

After thus keeping the Ch'ing government at bay, BAT sought still greater political and fiscal concessions after the fall of the dynasty in 1911. Within months after a Republican government replaced the Ch'ing, the company "voluntarily" (in James Thomas' words) sent "two experts to Peking to study the question of taxation on tobacco and cigarettes."[122] By 1914 these "experts" and Chinese officials worked out a scheme whereby the young Republican government would obtain a new source of revenue and BAT cigarettes would be exempted from likin or any other taxes beyond the import tariff. According to BAT's version of the scheme, the Chinese government would create a bureau in which two Chinese government officials, two BAT representatives, and one "independent party" would preside jointly over tobacco taxation. The function of this board would be to organize tobacco taxation and, "without hurting the tobacco industry," gradually increase revenue. The board would be responsible for regulating the industry and issuing excise stamps indicating that cigarettes so marked should be taxed no further. The Chinese government in Peking was to benefit from the increased revenue from loans which BAT would help it negotiate with other Westerners and from a loan which BAT agreed to make directly to the government. The arrangement would have given BAT a long-term monopoly over the cigarette industry in China because it granted the company power to determine which cigarette firms were to be licensed and which cigarettes taxed. In addition, BAT proposed that it be given a fixed fee (the amount to be negotiated) plus 10 percent of the revenue accruing to the bureau in return for the company's "experience and services in this matter." BAT executives anticipated that they and the Chinese government would preside jointly over tobacco taxation for twenty-five years.[123]

BAT's scheme aroused opposition both from government officials and businessmen. Western governments protested that such a bureau would violate the Treaty of Tientsin, which forbade the creation of any new monopolies in China after 1858, the Japanese government objected to BAT control over a market where the Japanese Imperial Tobacco Monopoly was involved, and the Association for the Protection of Chinese Tobacco Interests published letters and circulars complaining that such an arrangement would ruin Chinese tobacco businesses.[124]

BAT replied that the proposed agreement would result not in a monopoly but in a stamp tax that would provide an alternative to likin and other inland taxes. This was the line of argument taken by Wu T'ing-sheng, a BAT comprador who put himself in a position to further BAT's interests by securing an appointment from the Peking government as Special Commissioner for the Investigation of the Tobacco Tariff in the Ministry of Finance in 1913. But even working within the Peking government and as a member of it, Wu was not able to make a case for BAT's monopoly scheme that was acceptable to the government's representatives with whom he held talks, Chou Tzu-chi, Minister of Finance, and Liang Shih-i, director of the tax administration and personal secretary to President Yuan Shih-k'ai. A year after Wu's appointment as Special Commissioner, he still had not persuaded Chou and Liang to approve the monopoly, and in June 1914 his talks with them stopped. In 1915 BAT and the government resumed their discussions briefly, but the company's monopoly scheme was never put into practice.[125]

Failing to secure a tax monopoly, BAT continued to resist the government's attempts to impose new taxes. In 1915, when the Peking government instituted a new tax on tobacco products made in China's "native industry" (*t'u-yeh*), some bureaus in the provinces levied it on BAT goods. Despite BAT's earlier eagerness to secure tax privileges as a "native" company, its management in this instance protested that it deserved to be exempted from the tax because it was Western and was protected from the tax by China's treaties with the West. The company seems to have had no legal grounds for invoking the treaties, but it never did pay the tax.[126]

In this series of negotiations over taxes prior to 1915, Chinese governments clearly failed to offer Duke's company any direct opposition comparable to that which Japan had given it. Within this period, the late Ch'ing and early Republican Chinese governments did not nationalize the cigarette industry and oust Duke's company as the Japanese government did in 1904, nor did they regain tariff autonomy as the Japanese government did in 1911. Lacking tariff autonomy, each Chinese government had a legal limit on the tariffs that it could impose on BAT, and, failing to overcome BAT's legal arguments, financial strength, and diplomatic support, Chinese officials yielded to the company's demands for taxes that were even lower than the legal limit. Insofar as the Ch'ing government checked BAT's commercial penetration at all, it did so not directly (as the Japanese government did) but only indirectly by inspiring or tolerating popular movements that threatened BAT. Of these movements, by far the strongest before 1915 was the anti-American boycott of 1905 and 1906.

The Ch'ing government indirectly encouraged this boycott by refusing in 1904 to renew an American treaty excluding Chinese laborers from the United States. The Ch'ing leadership made this decision as part of a reform program designed to promote nationalistic opposition against foreign aggression and save the dynasty. As part of the same program, between 1901 and 1904 the Ch'ing had approved the opening of new schools and military academies to give Chinese youth the expertise to cope with foreigners and had authorized the creation of chambers of commerce to enable merchants to resist foreign commercial penetration. Patriotic merchants and students who joined chambers of commerce and attended the schools were determined to bring an end to a period of national humiliation which they traced from the Opium War of the early 1840s to the Boxer Uprising of 1900.[127] After the negotiations over the treaty dragged on first in Washington and then in Peking for several months in 1904 and 1905, these merchants and students lost patience with the slow-moving processes of diplomacy and decided to take direct action against the Americans. Organizing within their new government-approved chambers of commerce and schools, they urged the Chinese people to show their support for a more equitable treaty by boycotting American goods.

Chinese merchants in the Shanghai Chamber of Commerce inaugurated the boycott on July 20, 1905, and BAT immediately became its prime target. At the meeting called to announce the boycott, two Chinese spokesmen for BAT, the comprador Wu T'ing-sheng and the tobacco merchant Hao Chung-sheng, rose to defend the company, but, according to reports circulated by boycott leaders, Wu and Hao "had not said more than a few words when they were removed by unanimous consent of their angry hearers."[128] In the months that followed, participants in the boycott identified BAT as an American firm and demonstrated against it more widely and spectacularly than against any other company. Boycott leaders placed the names of its brands at the top of blacklists and singled out its product in one of the movement's most popular slogans, "Don't use American goods, don't smoke American cigarettes."[129] Merchants in the Lower Yangtze River Valley and in South China responded to this plea, especially in Shanghai and Canton. Ironically, leaders of the boycott in these cities beat BAT at a game in which the Western company believed itself supreme—advertising.

In Shanghai, red posters appeared with messages written in big, black characters. The signs all urged patriotic Chinese not to buy American goods, and many specified BAT cigarettes as contraband. In one poster, a dog—an animal that Chinese disesteemed—was pictured smoking a cigarette labeled Pinhead (one of BAT's most popular brands). "Those who smoke American cigarettes," read the caption, "are of my species." The poster was signed by "the Fraternity of Chinese Merchants." Another purported to be an edict issued by the "God of Thunder." It congratulated "virtuous men and women" for refusing to buy the Standard Oil Company's kerosene, the American Trading Company's soaps, and BAT cigarettes (naming the three most popular BAT brands). It also warned that any person violating the boycott should "beware of my thunderbolt." Another poster named seven BAT brands of cigarettes and one BAT brand of plug tobacco to be boycotted "to maintain the dignity of our 400,000,000." In yet another, Buddhist priests declared several BAT brands contraband.[130]

In South China boycott leaders attacked BAT in a variety of ways. They persuaded Chinese newspaper publishers in Canton not to carry BAT advertisements. They circulated, according to BAT, "divers pamphlets . . . throughout the entire provinces" of Kwangtung and Kwangsi. They issued handbills, one of which, in the words of a BAT complainant,

pictured "one of our Canton dealers as a disgusting reptile or animal, thereby intimidating him against continuing the sale of our goods."[131] They made persuasive posters such as one that showed a hearty common laborer accosting a blasé young scholar. Snatching a Western cigarette from the lips of the "idle young man" and replacing it with a Chinese cigarette, the laborer demanded, "Sir, you must have more pride!" The poster's caption exhorted the people of Canton to be sincere like the worker and not an enemy of the people like the young scholar.[132]

Cantonese also sang anti-American songs. Composed in the form of Cantonese love songs, their distinctive idiom, rhyme, and form surely enhanced their appeal.[133] One satirical song in this genre might have been the most influential anti-BAT boycott device of all. Called "Elegy to Cigarette," its lyrics playfully mourned the decline of BAT's popularity in 1905.

> You are really down and out,
> American cigarette.
> Look at you down and out.
> I think back to the way you used to be
> In those days when you were flying high.
> Who would have rejected you?
> Everyone loved you
> Saying you were better than silver dollars
> Because your taste overwhelms people
> And is even better than opium.
> Inhaling it makes people's mouths water.
> We've had a relationship
> In which up to now there has been no problem.
> I thought our love affair would remain
> Unchanged until earth and sky collapsed.
> Who would have expected that the Way of Heaven would not be
>   as always,
> That human things might change.
> Then this movement against the treaty got underway
> And spread everywhere
> Because America mistreated our Overseas Chinese
> Degrading us like lowly oxen and workhorses.
> Therefore everyone has united to boycott America,
> And that means opposing Americans.
> What is the most ideal way?
> People say it is best not to sell American goods
> And to this end we must all unite into a collective body.
>
> Ah cigarette,
> You have the word American in your trademark for everyone to
>   see

So I must give you up along with my bicycle.[134]
Our love affair (*chiao-ch'ing*)
Today must end.

Ai,
Cigarette please don't harbor resentment.
Perhaps a time might come when we meet again,
But it must be after Americans abrogate the treaty.
Then as before I shall be able to fondle (*ch'an-mien*) you.[135]

How many smokers sang or heard this song is impossible to say. Shortly before the boycott, a student of Chinese music said of Cantonese love songs, "Today they are known to high and low, rich and poor: they are sung alike by 'toys of paint and powder' [prostitutes] on board the gilt and scarlet flower boats, by blind minstrel-girls in the houses of wealthy men, and by the dirty beggar in the suburban slum."[136] Spread widely over this popular network, "Elegy to Cigarette" might well have reached a huge Cantonese audience. (If Arthur Waley is justified in saying that Po Chu-i's ninth century ballads which were sung in the streets and aroused popular interest in political questions became "the T'ang equivalent of a letter to the *Times*," then it is possible to say with equal validity that this song was a late Ch'ing Cantonese equivalent of a letter to the *Daily News*.)[137] Behind this propaganda were organizations formed by students and merchants. While student leaders denounced BAT in public speeches, merchant leaders persuaded shopkeepers to throw their stock of BAT cigarettes onto bonfires at boycott rallies and to make pledges not to place new orders.[138]

The campaigns brought results, and BAT immediately felt the effects. Within a week after the boycott began, BAT's management in Shanghai complained that the company's posters had been damaged, that Chinese clients had been intimidated, and that business was suffering. An eyewitness report by Julean Arnold, interpreter for the American consulate, lends credence to the last of these complaints. After translating posters and observing consumers' behavior in the streets of Shanghai, Arnold concluded in early August that the boycott against American goods was 90 percent effective. Maritime Customs officials in the Lower and Middle Yangtze regions reported that BAT sales declined outside Shanghai as well. In Chia-hsing and Wu-hu, BAT cigarettes completely disappeared from the market, and in Hankow, BAT and Standard Oil were the main targets of the boycott.[139]

In South China BAT's business also suffered. According to newspaper reports, smokers from every social class in cities throughout the region gave up BAT brands. And BAT representatives fell victim to the spirit of the boycott in rural as well as urban areas. One was attacked near Wuchow across the western border of Kwangtung province, and another was stoned and beaten near Swatow in eastern Kwangtung. According to

a representative in the South China region, within the first month the
boycott caused a 90 percent decrease in the company's sales, an immediate loss of over $200,000 to the company, and a "stupendous" change in
the company's future prospects in the region. He probably exaggerated
the company's losses, since other sources estimated that BAT sales fell off
50 percent rather than 90 percent in South China, but the company
clearly suffered substantial losses.[140]

BAT representatives and other foreign businessmen made every effort
to end the boycott, including appeals to American diplomats. Partly in
response to these complaints, the American minister, W. W. Rockhill,
with the backing of President Theodore Roosevelt, ordered his consuls in
China to exert maximum pressure on local officials. Nonetheless, many
Americans—businessmen, journalists, politicians—grew impatient when
the boycott continued throughout the summer of 1905. At the end of
August, Senator George C. Perkins of California requested that President Roosevelt take advantage of Secretary of War William Howard
Taft's presence in the Far East. He urged the President to instruct Taft to
take time out from his "good will" tour of China for a visit to Canton, the
"backbone of the boycott." Roosevelt denied this request and rebuked
Perkins for meddling in the affair.[141]

At this point, BAT approached Roosevelt. Within a few days of the
President's riposte to Perkins, a BAT representative made virtually the
same proposal, requesting that Taft be sent to Canton to "call officially
on the viceroy . . . with a view to settling [the] boycott."[142] On the same
day, the Standard Oil Company also urged Roosevelt to send Taft to
Canton. Judging from Roosevelt's previous pronouncements, one would
have assumed that these messages had little hope of success, for he had
frequently·expressed low opinions of the parent companies of BAT and
Standard Oil, specifically citing them as examples of evil trusts in America.[143] But this time, the combination of demands from politicians, diplomatic officials, and businessmen, culminating in the telegrams from BAT
and Standard Oil, persuaded him to act as they wished. The very day,
September 2, that Roosevelt received the requests from BAT and Standard Oil, he cabled Taft to meet with officials in the Canton area.

Taft's visit to South China on September 3 did not calm Canton. All
through the city, he was greeted by anti-American posters, some of
which insulted Alice Roosevelt, the President's daughter, who was accompanying Taft. After Taft failed to stop the boycott, BAT representatives and other foreign businessmen contravened directions from the
American consul-general in Canton, went directly to the leaders of the
boycott, and pledged to support Chinese demands for changes in United
States immigration laws. But still the boycott continued in South
China.[144]

In the meantime, an imperial edict, issued on August 31, ended the boycott against most American goods outside South China but not against cigarettes. By mid-September, customers in Shanghai and the Lower Yangtze region, for example, ceased to boycott American goods but continued their campaign against BAT products into the spring of 1906. In the Middle Yangtze region at Hankow and Wuchang, as late as December 1905 and January 1906 weekly meetings were held at which the chief denunciations were directed at BAT. Placards charged that BAT masqueraded as a British firm when it was in fact American, that it gave away free cigarettes out of "bestial motives," and that BAT cigarettes contained poison. And in Canton the boycott against BAT cigarettes persisted longest of all. Cantonese students continued to draw large crowds near the Canton waterfront and exhorted people to continue or to revive the boycott of American cigarettes and other goods. When local police tried to silence them, they claimed to be agents for Western cigarette companies concerned only with making sales, not political activists promoting the boycott. This ruse was apparently effective until March 1, 1906, when the Canton municipal police headquarters issued an order to arrest anyone posing as an agent for a Western cigarette company and calling for a boycott of American goods. Revived repeatedly, the boycott did not finally run its course in Canton until the end of 1906.[145]

Why were BAT cigarettes boycotted for so long and with such intensity? Part of the answer lies in BAT's lack of experience with Chinese anti-foreign movements. Later BAT's management learned to shunt Chinese anti-foreign sentiment or even turn it against the company's rivals. Part of the answer lies in the nature of BAT's product. As clearly labeled consumer goods, cigarettes (like American kerosene, flour, soap, and textiles) were more tempting targets than industrial goods (such as American coal, cement, paper, or machinery) because a large segment of the Chinese population could learn to recognize and reject such a commodity. And part of the answer lies in the availability of local substitutes. In the early stages of the boycott, there were only four small Chinese cigarette companies, but by the end more than twenty had come into existence.[146]

The boycott stimulated Chinese to enter the cigarette industry (and mining, textiles, shipping, railroading, banking, and other Western-style businesses) because it temporarily freed them from foreign competition. All the Chinese cigarette companies were small—indeed tiny by comparison with BAT. One had a capital base of ¥300,000, four had between ¥100,000 and ¥140,000, and the others had less than ¥100,000. (Capitalized at $2.5 million [approximately ¥5.25 million], the financial resources of BAT's Shanghai headquarters were then over seventeen times greater than the largest of these.) Almost all were managed exclusively by merchants (shang-pan). The only exceptions among the largest companies were the Peiyang Tobacco Company and the San Hsing Tobacco Company, which were jointly managed by merchants and officials (*kuan-shang ho-pan*).[147]

The Chinese entrepreneurs who opened these businesses did not originate, organize, or lead the boycott, but once the boycott was underway, they actively tried to capitalize on the nationalistic sentiment it had generated. Several of them identified their products with the movement to "recover economic rights" from foreigners, some emphasized that they used Chinese rather than imported American tobacco leaf, and one of the most successful, the San Hsing Company, advertised in newspapers and posters in several Lower Yangtze cities using the slogan, "All enthusiastically supporting the nation should smoke Chinese cigarettes." In their advertising naturally none mentioned that several of them relied on foreign advisers and imported foreign supplies, particularly Japanese technicians and technology and American-grown tobacco.[148]

While BAT was beleaguered in 1905 and 1906, some of these small companies grew rapidly, but as soon as the boycott ended BAT drove almost all twenty existing Chinese cigarette firms out of business or out of the market. According to Sheng Hsuan-huai (1844-1916), a leading Chinese official and industrialist and a stockholder in the San Hsing Tobacco Company, the Western company resorted to unethical (though not illegal) tactics to force these companies to close. Lodging a protest in 1909 with Tsai-tse, president of the Ch'ing Ministry of Finance, Sheng complained that BAT had, for example, drastically reduced prices, starting price wars that its smaller Chinese competitors could not hope to win; it had intimidated Chinese officials into banning many Chinese cigarettes for allegedly imitating BAT brands; and it had formed exclusive dealing arrangements which, in Sheng's words, bound "every single retailer in the interior as well as the coastal ports" not to carry any cigarettes except those of BAT.[149] Sheng's charges gain credence in light of the past record of BAT's owners, for Duke and his American associates had relied on similar tactics to pound their rivals into submission in the West, forcing other companies in the American cigarette market to merge with Duke's company in the 1890s, coercing the Imperial Tobacco Company of England into forming BAT in 1902, and subverting a rival American cigarette company's plans to enter the China market in 1903.[150] With these precedents to follow, Duke's BAT subsidiary in China probably behaved as ruthlessly toward young Chinese cigarette firms in the aftermath of the boycott as Sheng claimed. Victimized by BAT's aggressive methods, the Chinese cigarette industry could not withstand BAT's return to the market. Though the boycott had protected the Chinese firms briefly (almost as a higher tariff might have), it did not last long enough to give them time to experiment, overcome financial problems, and develop economies of mass production and technological efficiency that might have enabled them to continue to compete with their better financed and more experienced foreign rival.

After BAT disposed of the new Chinese firms between 1906 and 1908, it periodically faced additional boycotts and anti-smoking campaigns but encountered no serious competition from rival companies before 1915.[151] Following the boycott, as earlier noted, BAT built its largest factories at Shanghai and Hankow, completing them in 1906 and 1907. These factories gave BAT economies of scale that also discouraged new firms from entering the market because the Western company's cost advantages over potential Chinese rivals grew still larger.

By 1915 BAT's monopoly in China seemed unstoppable. As early as 1904 the American executive at Duke's subsidiary in Japan had predicted that the company would overcome any obstacle it might face in China. As he had prepared to return to the United States after his company's ouster from Japan in 1904, he had written to his counterpart in Shanghai, "You have a great field and untold possibilities are in your reach. The situation in China, with prospective business, makes it an inviting field, requiring large views and gigantic movements . . . I shall expect to hear that your Sales in China alone have passed the *One Thousand Million Cigarettes per Month mark* . . . As I see it . . . your 'ball of snow' [is rolling] and it will not be long before it is so large as to crush everything in its pathway."[152] In 1915, little more than a decade later, his prediction came true: BAT cigarettes were selling at the rate of a billion per month. Its manufacturing system, which employed 13,000 workers at factories in Shanghai, Hankow, Mukden, and Harbin, produced between half and two thirds of these cigarettes within China. Its marketing system distributed cigarettes and advertising in every region of the country. Its purchasing system in China procured less than 10 percent of the bright tobacco it used there in 1915 (importing the remainder from the United States), but with reception stations recently built in Shantung and Honan provinces, it was prepared to take advantage of the much larger harvests of Chinese-grown bright tobacco that were soon to come. With these systems all in place and functioning effectively, BAT seemed indeed to have crushed "everything in its pathway."

To BAT's potential Chinese competitors, the company's operations in China in 1915 surely must have seemed formidable. If they were not intimidated by its Western capital, technology, managerial expertise, and legal privileges under the treaties, then they might still fear that they could not match its Chinese marketing, advertising, and purchasing systems. The latter problem was particularly vexing for, as an Overseas Chinese businessman pointed out in 1915 to a Chinese cigarette manufacturer who was vying with BAT for the cigarette market in Southeast

Asia, "If most of the wolfish country's [BAT's] manufactured goods are made from native [Chinese] products and most of the wolfish country's goods are distributed and sold by our fellow Chinese, how can we love your product and despise that country's [BAT's] product?"[153] For any Chinese hoping to enter China's cigarette market, this was an arresting question. By using Chinese workers and resources as well as foreign advantages, BAT had established a foundation for its monopoly that seemed unshakable.

And yet in 1915, at the very time when Duke's and Thomas' optimism was at its height and their modern, powerful multinational corporation's monopolistic position in China seemed perfectly secure, BAT's directors suddenly found themselves facing competition from an unexpected source. It came from a Chinese firm, Nanyang Brothers Tobacco Company, which posed the first serious threat to BAT's domination of the Chinese cigarette market.

3.

The Rise of
Commercial Rivalry

In 1915 BAT encountered its first serious competition in China when Nanyang Brothers Tobacco Company entered the cigarette market at Canton. Although BAT had built no factories south of the Yangtze, it had developed an extensive marketing system in the vicinity of Canton. In 1906 BAT had six Western representatives in Hong Kong and one in Canton; and by 1908 Chinese agents also represented it in major towns on the Canton delta—Sheklung, Fatshan, Kongmoon—at which time it was reported that BAT cigarettes were sold in "immense quantities all over the provinces of South China."[1] To protect this investment, a BAT executive in the company's headquarters in the West responded to Nanyang's entrance into the market in 1915 by ordering his men in Shanghai to buy the Chinese company's one factory immediately and eliminate the competition before it became more alarming.[2] Prior to 1915 BAT had effectively barred Chinese and Western competitors from the market or driven them to the wall and crushed or absorbed them. Accordingly, BAT's directors in London asked, why couldn't the company's subsidiary in China dispose of Nanyang as it had disposed of these other potential competitors?

Nanyang, however, proved to be different from BAT's previous rivals. Despite opposition from BAT, it secured a portion of the market in Canton in 1915, and from this base in South China, expanded into the Lower Yangtze region and North China in 1916 and 1917. The explanation for why Nanyang penetrated these markets and became competitive with BAT in China—a feat which no other company, Chinese or Western, was able to accomplish in this period—lies in the history of Nanyang outside China before its invasion of the mainland in 1915 and its survival in commercial warfare with BAT in China between 1915 and 1918.

East Asian Origins of a Chinese Industrial Enterprise

The origins of the Nanyang Brothers Tobacco Company were radically different from the origins of BAT. Whereas BAT imported capital, technology, and personnel from the West and tried to introduce Western manufacturing, purchasing, marketing, and advertising systems to China, the founders of Nanyang were Chinese who had never been in the West, had never been employed by BAT or any other Western company, and had no direct access to Western capital, technology, or business practices. They were members of a Cantonese family named Chien (pronounced Kan in Cantonese) who found the original elements of their business in East Asia rather than the West.

Born into a family near Canton in the 1870s that was at least as impoverished as the one into which James and Benjamin Duke were born in Orange County, North Carolina, two decades earlier, the Chien brothers who eventually founded Nanyang began their careers inauspiciously.[3] At the time of their birth, their father was a carpenter in Fatshan (*Fo-shan*) near Canton, and he died while his three sons and two daughters were still children. After his death, his wife took in sewing to sustain the family through hard times, and, before long, her sons began to work in local shops. Responsiblity for playing the father's role fell upon the eldest son, Chien Chao-nan (1870-1923), who was in his early teens at the time of his father's death and was regarded by his brothers and sisters as the daring and decisive (*kang-ch'iang kuo-kan*) leader of the family.[4] At sixteen, after only a few years of formal education and a little experience as a merchant, he, like many other Cantonese in the nineteenth century, decided to leave China in search of economic opportunities abroad through which to fulfill his familial and other financial responsibilities at home in Canton. A restless young man on the make, he sailed northward to Japan in the late 1880s where he began to acquire experience, capital, and technology that were vital to the eventual founding of Nanyang.[5]

The capital Chien Chao-nan raised in Japan did not come easily or quickly. Investing his savings in a shop in Kobe, he first dealt in high-risk and high-profit items such as drugs and marine products; and, once established, he began to export Japanese porcelain, pottery, and earthenware through a network of fellow Overseas Cantonese to Southeast Asia via Hong Kong in the late 1880s and early 1890s. With the outbreak of the Sino-Japanese War in 1894, he moved to Hong Kong but continued in the same import-export business, handling porcelain from Japan that was sold to Cantonese distributors in Hong Kong and Southeast Asia. When this business failed around the turn of the century, he moved to Bangkok, another link in the chain of Overseas Chinese contacts through which his import-export firm had operated, and went to work for his

Chien Chao-nan

uncle, Chien Ming-shih, who had led one wing of the Chien family from
Canton to Thailand in the last years of the nineteenth century and had
subsequently become successful in the grocery business. In Bangkok
Chien Chao-nan and his brother, Chien Yü-chieh, who had followed him
after he became established in Kobe, Hong Kong, and Bangkok, worked
as salesmen for their uncle's firm, the I-sheng Brothers Company.[6]

Not until Chien Chao-nan traveled to Japan for the second time in
1902 did he begin to accumulate the capital that made possible the founding of Nanyang Brothers Tobacco Company. At this time he established
the Sheng-t'ai Navigation Company in Kobe, which proved to be sufficiently successful to help him begin to build up a capital base. Carrying
Japanese textiles along the same shipping routes to Hong Kong and Southeast Asia that his porcelain trade had earlier followed, his shipping line's
profits were ¥3,000 in 1902 and increased sharply within the next few
years.[7] By 1905 when Chien Chao-nan founded the Nanyang Brothers
Tobacco Company in Hong Kong, his success in Japan enabled him to
use ¥24,000 of his profits to invest in the new firm—almost one fourth
of its total initial capital.[8]

On this same trip to Japan in 1902 Chien Chao-nan also began to learn
about the technology of cigarette making. The available evidence does

not indicate why he was attracted to the cigarette industry, but, living in Japan at this time, he was undoubtedly impressed not only by BAT's penetration of the market but also by the rise of its Japanese competitors and the success of its Japanese collaborator, Murai Brothers Tobacco Company.[9] He made the decision to start his own cigarette factory after visits to the Kawai Tobacco Company in Kobe where he was introduced to tobacco technology. On the basis of this experience, he decided to build a factory of his own in Hong Kong, and he bought cigarette-making machinery and recruited technicians for it in Japan. He later wryly remarked to his friends, "I stole my tobacco technology from Japan."[10]

These were the valuable assets that Chien Chao-nan acquired in Japan —capital, technology, and technicians—but Japan did not provide all the prerequisites he needed in the cigarette industry. To achieve success, Chien Chao-nan had to raise more capital than his business interests in Japan would yield, and he had to manufacture and market his goods outside Japan, for (as shown in Chapter 2) the Japanese Imperial Tobacco Monopoly, established in 1904, barred all foreign cigarette companies from manufacturing in that country. For markets and additional capital, he turned once again to South China and Southeast Asia.

Chien Chao-nan initially raised a total of ¥100,000 to start the new company among several Cantonese in South China and Southeast Asia. Besides his own investment of ¥24,000, another ¥76,000 came from his brothers in Hong Kong, fellow Cantonese (t'ung-hsiang) in the Chiens' home area around Canton, and Overseas Chinese friends of the family living along the Southeast Asian archipelago. Chien Chao-nan and his brothers were able to appeal to the latter through the "Northern and Southern Hang" (Nan-pei hang), an exclusively Cantonese merchant association with headquarters in Hong Kong and with a network of commercial connections that extended throughout Southeast Asia. With this orientation to Southeast Asia, one might expect that they named their company Nanyang (literally "South Seas") with reference to that part of Asia, but in fact they meant for the name to refer to South China (complementing the Peiyang Tobacco Company of Tientsin which had been founded in 1904).[11]

Drawing upon their capital base of ¥100,000, the Chiens built a factory in Hong Kong and began to manufacture cigarettes in 1905. It was small and unsophisticated, equipped only with a furnace, a hot-air curing room, an electric generator, two blade-sharpening machines, and four cigarette-rolling machines. Japanese technicians—seven men and four women—taught sixteen Chinese men to operate the machinery and 100 Chinese women to handle and package the finished product. The plant was capable of turning out about 300,000 cigarettes in one ten-hour day, "half by machine, half by hand."[12]

Thus manufactured, most of Nanyang's good were distributed in those Southeast Asian countries where Nanyang was least threatened by high tariffs or BAT competition. In 1912, for example, the Chiens sold 18 percent of their cigarettes locally in Hong Kong, 7 percent in Thailand, and 75 percent in Singapore; after reshipment from Singapore, the latter reached additional markets in Malaya, Indonesia, Borneo, and the Celebes. In these areas, Nanyang was not barred from the market by a strong governmental tobacco monopoly like the one in Japan or by prohibitive import duties on tobacco like the one imposed by the French in Indochina in 1906.[13] Nor did it face competition there from a BAT organization comparable to the one in China. Although the Western company distributed in Southeast Asia on a large scale—almost 740 million cigarettes valued at more than a million dollars were exported there from the United States in 1915—it did not build factories, transplant bright tobacco, or develop a comprehensive, integrated operation in any Southeast Asian country before 1915 as it did in China.[14] Comparatively free from governmental tariffs or BAT domination, these Southeast Asian countries furnished Nanyang with its marketing base for a decade after its founding in 1905.

Though Nanyang's distribution was confined to this area outside China between 1905 and 1915, the success or failure of its advertising and promotional efforts seemed to depend on events within China. When diplomatic incidents or political events in China triggered Chinese boycotts or other patriotic demonstrations, Nanyang's sales in Hong Kong and Southeast Asia rose, and when these mass protests subsided, Nanyang's sales fell. The peaks in this cyclical pattern came for Nanyang during the anti-American boycott of 1905 and 1906, the *Tatsu Maru* incident of 1908, and the Chinese revolution of 1911 and 1912.

During the anti-American boycott of 1905 and 1906 Nanyang grew fast enough to warrant doubling the number of its workers, but in 1907, less than a year after the boycott ended, the company was forced to close down. The Chiens did not reopen until their hopes for a larger share of the market were revived by another popular outburst following the *Tatsu Maru* incident of 1908.[15]

In the *Tatsu Maru* incident, Cantonese demonstrators started a boycott against Japanese goods to protest the Ch'ing government's willingness to release a Japanese ship, the *Tatsu Maru II*, that had been caught smuggling.[16] This boycott did not spread as widely as the boycott of 1905, but it had an impact on commerce in the South, including the tobacco trade. As the boycott began to take hold in April 1908, a correspondent for the Japanese newspaper *Asahi* complained that "Japanese tobacco, which obtained a hold at the time of the boycott of American goods, is being ousted by the American article."[17] By July observers in

South China reported that the import of Japanese cigarettes into Canton had "almost totally ceased" and BAT "reaped a corresponding harvest."[18] Meanwhile, in Hong Kong Nanyang resumed production and benefited from the boycott too, but, unfortunately for the Chiens, the boycott of 1908 was not effective for as long as the one in 1905 and 1906. It lasted only a few months, ending in the summer of 1908, and in early November, Japanese goods were already being quietly sold in South China. By January 1909 the Japanese cigarette trade in South China and Hong Kong, though never very large, was fully restored. While benefiting from the *Tatsu Maru* boycott, Nanyang suffered from its brevity. Apparently the Chiens had expected the anti-foreign sentiment generated by this incident to last longer and had invested more in reopening Nanyang than was prudent. During the aftermath of the incident, the Chiens lost over ¥100,000—more than the total capital originally invested—and were forced temporarily to close down once again.[19]

After shutting down the factory in 1908, Nanyang was on the brink of bankruptcy. With debts accumulating faster than the Chiens could pay them, it probably would have disappeared, as many other small Chinese tobacco companies did at the time, if it had not been rescued by the Chien brothers' uncle, Chien Ming-shih. Through his business in Thailand and Vietnam, he raised ¥90,000 and lent it to his nephews. This ¥90,000 together with ¥40,000 which Chien Chao-nan drew from his shipping business in Japan gave Nanyang ¥130,000 in working capital and enabled the company to resume production once again in February 1909. Saved from bankruptcy but saddled with heavy debts, Nanyang's business was still shaky, according to Chien Yü-chieh, and it lost "several tens of thousands" of *yuan* in 1910.[20]

Then in 1911 and 1912, political events in China again stimulated demand for Nanyang's cigarettes. This time the turn upward in its sales coincided exactly with the Chinese revolution of 1911 and 1912. The Chiens themselves supported the revolution, making substantial cash contributions to Sun Yat-sen's republican movement, and other Overseas Chinese seemed equally enthusiastic about it. According to Chien Yü-chieh, many Overseas Chinese in Southeast Asia expressed their enthusiasm as customers by buying more Chinese-made goods. In Java alone, he noted, after the Chinese revolution, sales of one Nanyang brand shot up to 50 million in a single month.[21]

Riding this wave of enthusiasm for the newly founded Chinese Republic, the Chiens rose out of the trough of 1909 and 1910 up to a new crest generated by the revolution. In 1912 Nanyang showed surplus earnings for the first time in its history, and the Chiens immediately reorganized its financial and managerial structure to bring virtually all of its stock into the hands of the Chien family and to take advantage of the opportunity for expansion. Thereafter, sales grew from 237.9 million cigarettes in 1912 to 347.4 million in 1913 and 526.7 million in 1914, and profits jumped from ¥52 thousand in 1912 to ¥117 thousand in 1913 and ¥175 thousand in 1914. To meet the growing demand, the Chiens expanded their manufacturing system to include 24 machines, 100 staff members and 800 women workers by 1915.[22] As these figures indicate, by 1912 the Chiens had achieved a high level of production and sales, and they did not backslide from it as they had following the anti-American boycott of 1905 and 1906 and the anti-Japanese boycott of 1908. The demand for their goods grew steadily between 1912 and 1915, and, by effectively employing their capital, technology, and marketing skill, they succeeded in exploiting Southeast Asian markets.

Thus, before entering the China market in 1915, the resourceful Chiens developed effective manufacturing and marketing systems without the assistance of Westerners. They used capital raised in Japan, South China, and Southeast Asia to build a factory in Hong Kong staffed by Japanese technicians and Chinese laborers who produced cigarettes on Japanese machines for distribution through Overseas Chinese agents in Southeast Asian markets. Ironically, the only ingredient these Chinese businessmen took directly from the "industrialized" West in these years was raw materials. With bright tobacco unavailable nearby, the Chiens imported it from the American South beginning in 1904 and 1905. They probably mixed some Chinese-grown tobacco with it although, according to their records, all their tobacco came from the United States before 1915.[23]

With this record of success as owners of a non-Western company in a Western-style industry, the Chiens decided in 1915 that the time had come to make their entrance into the China market at Canton. By then, they could rely on a stable manufacturing and marketing base in Hong Kong and Southeast Asia from which to extend their operations into China, but, more than that, they had reason to believe that their goods would be particularly well received in China at this time because of recent events. It was encouraging, for example, to see the growing opposition to foreign commercial penetration in South China before 1915. This development was evident in many ways: in several modern enterprises which eschewed foreign capital and competed with foreign rivals; in the policies of the government of Kwangtung province which helped establish Chinese-owned factories and ease their tax burdens; and in actions taken by Cantonese merchant associations and other societies formed to promote Chinese-made goods in direct competition with foreign imports.[24] If this opposition to foreign penetration turned out to be a powerful force in South China and if it stimulated demand for Nanyang's products in South China as it had earlier in Southeast Asia, then the Chiens could expect to reap high profits from their expansion into China.

Another development that appeared to make this an opportune time to enter the China market was the outbreak of World War I, for it seemed likely to work to the Chinese businessmen's advantage insofar as it caused Western shipping and industrial production to concentrate on the war effort in Europe and thus forced BAT to cut back its operations in China. The Chiens were well aware of BAT's hold on the China market,[25] and they undoubtedly hoped that any disruption of trade and communications between China and Europe during World War I would weaken their Western competitor.

Prompted by a combination of these considerations, the Chiens plunged into the China market in 1915. More than twenty-five years after leaving home in search of business opportunities in colonies and countries around the rimland of China and more than ten years after inaugurating their cigarette company, they returned to South China and set up their first Chinese office in Canton. Upon arriving they found that anti-foreign feeling was as intense in the market of South China as they could have hoped, but, despite World War I, BAT was there in full force too.

Commercial Warfare in South China, 1915

At the time of Nanyang's entrance into the market, BAT remained strong despite World War I by taking advantage of its resources as a multinational corporation. For example, when the war curtailed shipping from Europe, BAT continued to receive its full quota of imported cigarettes in China simply by shifting responsibility for production from its factories in England (which was embroiled in the war) to its factories in the United States (which was not yet involved in it). Prior to the war, English factories had concentrated on producing cigarettes for China and had supplied more than any other exporting country; in 1913 almost two thirds of all cigarettes exported from England had been shipped to China, and these English-made cigarettes had outnumbered others imported into China three to one. After the war broke out, however, the burden of producing exports for China passed from English to American factories; by 1915 more than half of all cigarettes exported from the United States were sent to China, and, as these American-made cigarettes rose into billions per year, they came to outnumber cigarettes imported from other countries to China three to one in the late 1910s and nine to one in the early 1920s.[26]

At the same time that BAT shifted responsibility for manufacturing China-bound cigarettes to the United States, it anticipated the possibility that the United States might enter the war and that American shipping might be cut off as European shipping had been. To prevent America's future entrance into the war from affecting its trans-Pacific trade, BAT secured control over its own ships by becoming agents for the Garland Steamship Company, and it used this company's three ships to transport cigarettes and tobacco from Tacoma to Shanghai and Hankow throughout the late 1910s, even after the United States declared war against Germany in 1917.[27]

By these measures, BAT solved problems of production and transportation and maintained or even raised the level of its trade between the West and China during World War I. In August 1915, after the hostilities had been underway for a year, James Thomas commented that World War I had "not to any extent disturbed our business in China."[28] He subsequently made similar remarks, and he later concluded, in retrospect, that wars throughout his career—among which he counted the Spanish-American War of 1898, the Boxer Rising of 1900, the Russo-Japanese War of 1904 and 1905, and the Chinese revolution of 1911 as well as World War I—"each gave new impetus to trade between China and the United States."[29] Thus, while other Western companies may have curtailed their operations or withdrawn from China with the advent of World War I and may have opened the way for their Chinese competitors to enjoy a "golden age" of industrial growth, this was not the case in the cigarette industry.[30]

Unimpaired by World War I, BAT responded to Nanyang's arrival at Canton in 1915 by aggressively opposing it through the use of sabotage, advertising, and the law. In the market place, BAT started price wars in Canton as Duke's company had against rivals elsewhere in the world. Well aware that Nanyang made expensive as well as inexpensive cigarettes, BAT distributors called for undercutting Nanyang's brands at all price levels either by lowering prices on equivalent BAT brands or by introducing BAT equivalents in cases where none existed.[31]

BAT agents also resorted to more vicious tactics. At one point, they bought up virtually all the Nanyang cigarettes on the market, held them until they became moldy, and then dumped them in large quantities and at low prices on unsuspecting customers. This stratagem created serious problems for Nanyang. It prevented Nanyang's cigarettes from reaching the market during the Chinese company's first promotional campaigns; it cost Nanyang the expense of reimbursing outraged customers; and, of course, it left the impression that Nanyang's goods were inferior and thereby damaged the company's reputation.[32]

BAT's advertising campaigns were more subtle. One attempted, ironically, to turn anti-foreign sentiment in China and Southeast Asia against Nanyang. The opening of Nanyang's Canton branch in 1915 coincided with an anti-Japanese boycott organized by Chinese nationalists to protest against the Twenty-one Demands, a set of demands imposed by

Japan on China which threatened China's national sovereignty. The aim of the BAT's campaign was to capitalize on this coincidence by associating Nanyang's name with Japan. Behind Nanyang's patriotic Chinese facade, said BAT propaganda, lurked tainted Japanese financial and technological backing. In a letter to the editor published in the highly nationalistic Canton newspaper, *Citizens' Daily News* (*Kuo-min jih-pao*),[33] for example, an unidentified correspondent charged that Nanyang was dependent on Japanese raw materials, machinery, and technicians. The author of this anonymous letter referred to himself as a patriot and implored his countrymen to regard Nanyang as unworthy to participate in the "national goods movement" (*kuo-huo yun-tung*) to promote the sale of indigenous products, a campaign that was just beginning in China.[34] In Canton a BAT agent took a similar complaint to the British consulate, claiming to have "certain knowledge that a preponderance of [Nanyang's] capital is Japanese."[35] The moving force behind this campaign in Canton seems to have been Chiang K'ung-yin, a leading member of the Cantonese gentry who aligned himself with BAT against Nanyang and started a feud between his family and the Chiens that lasted more than a decade.[36] In Singapore another BAT manager rebuked the Chinese consul for informing local Chinese that Nanyang was a Chinese company, free of Japanese affiliations. To document his claim that Nanyang had Japanese connections, he produced a copy of a letter in which Nanyang and a Japanese firm appeared to be corresponding about business—evidence which neither the consul nor Nanyang's representatives in Singapore could controvert.[37]

At a more popular level, BAT sent hawkers out to buy Nanyang's cigarettes and then resell them, loudly proclaiming that they were made by Japanese. Men hired by BAT—who were, according to the Chiens, members of local street gangs (*ti-p'i*)—posed as customers in opium dens, asked for cigarettes, and when offered Nanyang brands, refused to accept them on the grounds that they were made by the detestable Japanese. Upon hearing the cigarettes condemned in this way, retailers reportedly vowed to stop stocking Nanyang products. In all of Nanyang's major distributing points—Canton, Singapore, Bangkok—the company's distributors reported that rumors abounded in 1915 concerning Nanyang's reliance on Japanese capital and personnel.[38]

BAT promoters were shrewd to select this theme, for it had some basis in fact. As indicated earlier, Nanyang had indeed been dependent on Japanese technicians and technology in its early years. Moreover, Chien Chao-nan had made legal as well as economic commitments in Japan that were still in effect. On April 2, 1902, he had become naturalized as a Japanese citizen using the name Matsumoto Shonanshi, and in the same year he had established the Sheng-t'ai Navigation Company based in

Kobe. He later argued that he had acquired Japanese citizenship purely for business reasons; he needed it to register the shipping company in Japan and thereby gain maritime legal protection for it, which he did not believe the Chinese government could give.[39] In addition, the Chiens relied on Japan for equipment and materials. In 1915 they imported trademarks and packaging material, tin foil, cigarette paper, bamboo mouthpieces, blending spices, vats, and picture cards (for advertising) from Japan, which altogether cost Nanyang ¥534,029, approximately 21 percent of the year's budget.[40] On the other hand, there is no evidence that Nanyang was financed by Japanese investors (unless Chien Chao-nan is counted as Japanese because of his citizenship). No Nanyang stock was registered under a Japanese name—94 percent was in the hands of the Chien family—and no available evidence indicates that the Chiens were ever in debt to Japanese creditors.[41]

However dependent the Chiens were on Japan, their connections made them (like Sun Yat-sen and other Chinese with Japanese affiliations)[42] vulnerable to criticism from anti-Japanese Chinese nationalists provoked by the Twenty-one Demands. One of Nanyang's customers in Singapore, Ch'en Tse-shan, for example, was shocked to discover that Chien Chao-nan had ties with Japan. In April 1915, when Nanyang moved into the Canton market and BAT launched its anti-Nanyang campaign, he responded to a story in the *Citizens' Daily News* on Chien Chao-nan's Japanese citizenship by writing to Chien Chao-nan personally. "Alas!" he exclaimed, "How can an honorable Chinese become a citizen of another country and give up his father to adopt another father? Up to now people have respected you, but now they . . . are disgusted with you. We are severing all ties with you and will try to ignore you, for it is painful for us to have to look at you." Chien Chao-nan had replied to the accusations in the *Citizens' Daily News* with his own letter to that newspaper. In it he had defended Nanyang, saying that all its stockholders were Chinese, that all its directors were committed Chinese patriots, and that the company used Chinese tobacco and employed Chinese laborers. But to his outraged Singapore customer, this rebuttal seemed lame. Chien Chao-nan had not denied that he was a naturalized Japanese citizen running a Japanese-registered shipping firm. To advance other arguments without clearing himself of these charges was pointless, said Ch'en, "like trying to scratch an itch through a boot." As long as Chien Chao-nan retained his Japanese ties, Ch'en said that there was no basis for considering Nanyang more Chinese than BAT. Unless Chien Chao-nan agreed to publish a newspaper notice announcing the cancellation of his Japanese citizenship, Ch'en threatened that he and his Singapore compatriots would cease to do business with Nanyang and would ostracize the Chiens from their patriotic fellowship.[43] Though this criticism touched upon a sensitive issue that was to plague Chien Chao-nan later, he did not feel compelled to relinquish his Japanese citizenship or ownership of his shipping interests in Japan in 1915.

In BAT's third major campaign in 1915, it brought legal charges against Nanyang for counterfeiting BAT trademarks. Seven years earlier BAT had made similar complaints to the Hong Kong government, and, under legal pressure, the Chiens had burned large quantities of the alleged imitation. In 1915 BAT, now even more insistent about putting an end to the practice, distributed handbills which warned that anyone selling Nanyang's counterfeit goods would be "apprehended by officials and prosecuted" (*na-kuan chiu-pan*). Its lawyers demanded that if the Chiens expected to avoid prosecution, they must immediately withdraw the brand from the market in China, recall it from everywhere in Southeast Asia, burn it in front of BAT's lawyers, and publish a full confession of guilt in all Chinese and Western newspapers.[44]

While tacitly admitting that he had imitated BAT trademarks, Chien Chao-nan was appalled by the severity of these conditions and by the pompousness and condescension with which "big brother" (*lao ko-tzu*), BAT, had imposed them. Rather than meet the conditions he decided to opt for a legal settlement—a bold decision, for Chinese companies had rarely dared to confront BAT in court.[45] After consultation with lawyers, the Chien brothers felt that they knew why others had been reluctant to contest BAT in court: Western lawyers operating under Western law before a Western judge in the British colony of Hong Kong appeared to work only for BAT against Chinese and not conversely. "All of the lawyers that we have hired have been eaten by the tiger and have come, one by one, to the tiger's assistance," the Chiens in Hong Kong informed their branches in Southeast Asia. In other words, BAT seemed to have intimidated Nanyang's lawyers and turned them against Nanyang. Such collusion was not surprising, the Chiens reflected, in light of the fact that their lawyers and the directors of BAT were of the "same race." The Chiens lost the case.[46]

Unable to fight BAT in courts of law and reluctant to enlist the services of Western lawyers, the directors of Nanyang felt deprived of institutional support for their business. In August 1915, after enduring four months of BAT sabotage, smear campaigns, and legal harassment, Chien Chao-nan concluded that anyone who would battle with BAT was forced to ask the 'most basic questions: "How can the small compete with the big? How can one alone stand up against the multitude? How can the weak defeat the strong? Since ancient times the principle [that the strong prevail over the weak] has never changed."[47]

Chien Chao-nan devised no effective response to two of the three types of BAT's campaigns; against BAT's attempts to sabotage Nanyang's product, he had no recourse, and against BAT in court, the best he could manage was an announcement, after losing the case, that the legal proceedings were part of an evil BAT plot to destroy him.[48] But against the third type of BAT campaign, in the realm of advertising, the Chiens by no means felt defenseless. Like many other newcomers trying to overcome barriers to entry in consumer industries elsewhere in the world, they learned to use advertising to differentiate their product from that of their established rival.[49] With this aim in mind, they struck back with their own advertising campaigns, and they proved to be exceedingly resourceful promoters.

One of the Chiens' most elaborate and expensive investments in publicity was made in the form of philanthropy. To project an image of themselves as civic-minded men concerned for the welfare of their community, they made philanthropic contributions to needy Cantonese in South China. Like the gentry in traditional Chinese society (including Confucian-trained twentieth century Chinese leaders such as the scholar-industrialist Chang Chien),[50] they contributed to flood relief—and consciously derived as much favorable publicity from this philanthropy as they could. On the ten steamboats which they sent to the aid of flood victims in Kwangtung province, for example, they flew huge flags on which were emblazoned the words "Nanyang Brothers Tobacco Company Relief." For several days before embarking on their mission, the boats were moored conspicuously in front of a Honan theater on Canton's waterfront, a popular promenade where crowds of people gathered. Chien Chao-nan did not make public the exact cost of this act of benevolence— his friends estimated that over ¥100,000 was spent and his critics thought that it was less—but he allowed privately that such philanthropy paid fat dividends. Flood relief alone, he was convinced, made Nanyang so popular in 1915 in South China that the company became invulnerable to BAT propaganda against it. He felt gratified to be lauded by academicians and others capable of shaping public opinion. His only regret was that press coverage of Nanyang's flood relief could not have been more extensive. In 1915 the Chiens commissioned investigators to ascertain whether this and other philanthropy (which included donations to schools, orphanages, and hospitals) brought enough publicity to be worth the investment. The report assured them that their "altruism" was profitable.[51]

Nanyang claimed another advertising victory over BAT in a battle for the skies. Through family contacts, the Chiens secured exclusive rights to manage the concession at a flying exhibition—a big attraction in Kwangtung province in July 1915. BAT also made a bid for the concession, but the Chiens had better contacts. One of the Chiens' cousins knew the aviator, a Chinese-American and United States citizen named Tom

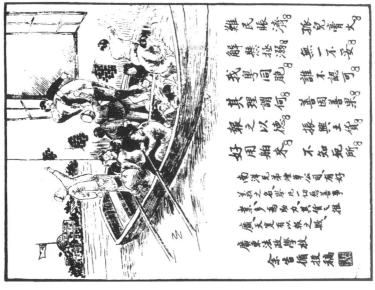

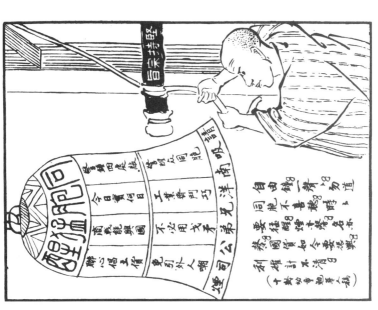

Nanyang advertising calling for commercial war against foreigners and showing the company's philanthropy

Gunn, who made the decision to give concession rights to Nanyang. Envious, BAT dispatched men to fly big kites in the shapes of eagles at the airfield. The police, recognizing that Nanyang had exclusive rights, barred BAT from the area. BAT's Western managers in Kwangtung—their plans frustrated—set up their own concession and tried to salvage some advertising benefits. The police were afraid to oust the foreigners, but, fortunately for Nanyang, Gunn's manager, an American named Mike Newman, persuaded the Western BAT representatives to remove their concession. In the end, Nanyang took full credit for the exhibition. Its advertising staff wrote all the press releases and its banners waved at the airfield and in all the movies taken. Reflecting on the events of the day in a letter to his brother, Chien Chao-nan acknowledged that he was "most pleased" with the outcome.[52]

Nanyang also innovated in advertising by playing on local superstitions. BAT's Chinese agent complained that Nanyang agents bribed coffin bearers and other members of funeral parties to accept BAT cigarettes and hired "riff-raff" (wu-lai) to follow the funeral party and shout that the coffin bearers smoked BAT brands.[53] Since many Chinese believed that carrying coffins brought one into contact with terrifying ghosts and since only people of the lowest stratum in Chinese society would consent to serve as professional mourners, Nanyang thus besmirched BAT's name by associating it with devils and a despised class of people.

These victories were satisfying for the Chiens, but the central theme that emerged in their advertising was their appeal to Chinese national pride. As noted earlier, the Chiens had profited from this appeal in Southeast Asia even before entering the market in China. In the spring of 1915 their prospects for benefiting from it in China were enhanced by a new campaign, the "national goods movement" (kuo-huo yun-tung), which was just then starting in Shanghai as part of the Chinese protest against Japan's Twenty-one Demands. This campaign originated in March 1915 when the Chinese leaders of twenty major guilds met in the International Settlement, formed the Society to Encourage Usage of National Goods (Ch'üan yung kuo-huo hui), and resolved to boycott Japanese goods. Though officially banned by the Peking government, the campaign immediately won endorsements from Chinese political leaders outside the government (such as Wang Cheng-t'ing, an American-educated young man who later served as China's Minister of Foreign Affairs), and it quickly spread throughout the country, resulting in the establishment of offices in seventy locations by May 1915.[54]

In June 1915, only two months after Nanyang's Canton branch opened, the company's representative noticed that consumers there were showing a preference for native goods (t'u-huo), and he expressed the hope that Nanyang's cigarettes might qualify as such. In July the Chiens' nephew,

Wang Shih-jen, who was testing new markets for Nanyang in the North, was amazed to discover some of the potential profits a boycott might yield there. In Tientsin, he gloated, "The shops in the Japanese concession are as quiet as if nobody were there and almost every day wagons filled with goods trundle along the river banks with hawkers shouting 'Chinese must use Chinese products' and other expressions of indignation [over the Twenty-one Demands], (and our cigarettes are in crates on the wagons—how grand!) . . . I dare say that our goods could be sent north at the rate of 2,000 cases [100 million cigarettes] per month and could easily be disposed of."[55]

In hopes of turning this sentiment to their advantage, the Chiens consciously tried to identify Nanyang with nationalism. They first joined and then mustered the support of Cantonese merchant organizations such as the Hong Kong Overseas Merchants Association, the Canton Overseas Merchant Association, the Canton General Chamber of Commerce, and the Canton Press Club. They associated Nanyang's name with nationalistic commercial slogans, popularized during the anti-Japanese boycott, and they adopted as their own motto, "Chinese should smoke Chinese cigarettes!" To disprove BAT's allegation that Nanyang was a lackey of the Japanese, the Chiens staged "investigations" intended to show that Nanyang was a genuinely Chinese company and circulated literature documenting the findings of the "investigators." Nanyang was the true nationalist, they argued, and BAT was the foreign interloper. They even turned BAT's own advertising theme against it, alleging that it had Japanese as well as Western affiliations.[56]

Westerners and Chinese alike testified that Nanyang's advertising produced results. In July 1915 the British consul in Canton reported that Nanyang's newspaper advertisements were effective at "urging patriots to smoke their cigarettes only 'and support native industries.' "[57] In October, six months after the Canton branch was opened, a Nanyang investigator detected the first recorded expression of popular sentiment that linked Nanyang's name with anti-foreign feeling and with the burgeoning "national goods movement" which stimulated the demand for products made by Chinese manufacturers in several industries—candles, soap, watches, towels, cotton cloth, cotton underwear, boots and shoes, mirrors, sugar, umbrellas, and other commodities as well as cigarettes.[58] According to the Nanyang employee's report, BAT's success at forcing Nanyang to withdraw a brand from the market "angered and aroused both merchants and smokers, and they have resolved [to buy only] national goods. This is a great opportunity for [Nanyang] to expand [its market]."[59]

After Nanyang became identified with the national goods movement, sales soared in Canton. By October 1915 the new branch could not supply distributors, stall vendors, and hawkers fast enough. In response to the meteoric rise in demand, the Chiens completely reorganized their factory and built a second factory in Hong Kong. They increased the amount of American tobacco leaf they imported by 37 percent (from 300 to 400 hogsheads per month) and almost doubled the number of their machines by adding 21 American-made ones, all of the latest model. Thus augmented, the company's factories produced 930,540,000 cigarettes in 1915, almost four times as many as in 1912.[60]

The great bulk of Nanyang's increased production was used to meet the growing demand in South China. Nanyang's Southeast Asian distributors placed higher orders for cigarettes too, but Nanyang was rapidly shifting its attention away from Southeast Asia and toward the China market. In 1912 over 80 percent of Nanyang's goods had been distributed in Thailand, Singapore, Malaya, and Indonesia, 18 percent in Hong Kong, and virtually none in China. By 1915 only 55 percent were sold in the archipelago, 27 percent in Hong Kong, and 16 percent in China. Thereafter the trend continued and the amount of cigarettes Nanyang marketed in Southeast Asia remained relatively unchanged while the amount in China steadily increased.[61]

Commercial Warfare in the Lower Yangtze Valley and North China, 1916-1917

With a foothold in South China, Nanyang moved northward, launching a kind of commercial northern expedition from Canton that reached some of the same cities in the Yangtze River Valley and North China as Chiang Kai-shek's military Northern Expedition did a decade later. Before setting out from Canton, the Chiens sent agents to explore markets in the Lower Yangtze and North China regions and made contacts with groups in these regions such as The Society for the Protection of Trade of Shanghai.[62] By 1916 they realized and admitted privately that Nanyang's cigarettes were inferior to BAT's—less tightly packed, hotter, and more bitter[63]—but they were convinced, in light of their reports from the North and their successes in the South, that they should extend their territorial reach. So in 1916 they established new distributing centers in the Lower Yangtze region at Shanghai, the Middle Yangtze at Hankow, and North China at Tientsin.

BAT greeted this extension of Nanyang's sales organization with the same hostility as it had the founding of Nanyang's first Chinese branch at Canton. In North China Chinese BAT agents deliberately sabotaged Nanyang's initial promotional campaigns by pulling down Nanyang's outdoor posters and even persuaded prostitutes to let them carry off Nanyang's free cigarettes, premiums, and calendars from brothels. In

Shanghai BAT tried to cut short the life of Nanyang's branch by recalling cigarettes from other parts of the Lower Yangtze region and flooding the Shanghai market, driving prices of BAT brands far below Nanyang's. Nanyang's representative in Shanghai, Wang Shih-jen, was at first afraid that this tactic would destroy the new branch before it could get started because Nanyang's advertising had not had time to take hold. In the end, Wang—like everyone in the Chien family already a veteran of wars with BAT—reacted coolly and ingeniously. He shipped Nanyang's cigarettes to other ports near Shanghai from which BAT managers had withdrawn much of theirs, supplying recently depleted markets where demand was high. A potential disaster was thus turned into a victory for Nanyang in Shanghai.[64]

Though Nanyang's Shanghai and North China branches survived these initial tests, stiff day-to-day competition from BAT frustrated their growth. In Tientsin, as in Canton, BAT introduced new brands to compete with Nanyang's which were priced lower than Nanyang's, and when Nanyang adjusted its prices to match them, BAT cut prices still further, always keeping its prices below Nanyang's. In Nanking and other ports on the Yangtze and, beginning in early 1916, throughout South China as well, BAT also consistently underpriced Nanyang.[65] In Shanghai Nanyang had difficulty breaking into the market because BAT had not only cut prices but also had formed restrictive agreements with a number of the leading *hang* which bound them to distribute BAT goods exclusively. Similarly, distributors in suburban markets around Shanghai agreed not to carry cigarettes except those of BAT in exchange for BAT's willingness to pay shipping costs from Shanghai to the suburbs. (This practice of signing exclusive dealing arrangements to monopolize markets was not illegal in China at the time.)[66]

As in Canton, Nanyang managers fought back with extensive advertising campaigns, using premiums and calendars to popularize their most successful brands and lining some of their cigarette containers with dollar bills to attract attention, but their advertising and products, they discovered, were received differently in Shanghai than they had been in Canton.[67] Nanyang needed to adjust its promotional campaigns, according to its Shanghai representative, Wang Shih-jen, to appeal to the peculiarities of the "Shanghai mind." Like every executive in the Nanyang organization, Wang was a Cantonese, and he explained the "Shanghai mind" to his kinsmen in South China as follows: "Shanghai men are most complicated. The love in their hearts for national goods is thin and weak. Those among the upper classes who frequent houses of ill-repute smoke expensive cigarettes to appear elegant . . . Therefore if we can produce a brand of cigarettes better than Three Towers [the BAT's most expensive brand] and price it even more expensively, Shanghai men will take it more seriously."[68] Since nationalistic advertising seemed less effective in Shanghai than in Canton, Wang called for a new brand whose appeal would lie not in its patriotism but in its high expense and concomitant prestige. Probably on the basis of Wang Shih-jen's reports, Nanyang brought out Liberty Bell at ¥0.50 to compete with BAT's high-priced Three Towers, which sold for ¥0.80.[69] Thus in Shanghai as in Canton, the two companies competed directly for the trade in expensive as well as inexpensive brands.

Undaunted by the disappointing response to Nanyang's nationalistic advertising in Shanghai, Chien Chao-nan continued the search for markets up the Yangtze, but the farther Nanyang penetrated, the more difficult it became to extend its sales organization, advertise effectively, and show a profit. The Chiens' sales organization was less effective in regions outside South China because they tended to distribute through their fellow Cantonese wherever they marketed goods—a strategy less effective outside South China than within it.[70] Seeking to gain a foothold in the interior, Chien Chao-nan departed from this Cantonese-oriented approach in the autumn of 1916 and spent an extravagant sum of money to lure away a Chinese BAT agent who was responsible for areas outside treaty ports in the Lower Yangtze region, a successful merchant who owned a big, well-endowed business in Chinkiang. But even this bold bid for the market in the interior failed. A year later the vaunted new agent had lost about ¥9,000 on cigarette sales, further frustrating the Chiens' hopes for deep commercial penetration in the Lower Yangtze Valley.[71]

In 1917 Chien Chao-nan himself traveled throughout this region to find out why Nanyang was making no progress at penetrating it. Upon close inspection, he discovered that BAT had, with characteristic thoroughness, canvassed Soochow, Chinkiang, Nanking, and other cities and towns, and even inland villages where "the land is vast and the people few." He reported that wherever he went the streets and by-ways were covered with BAT advertisements, shopkeepers were committed to sell BAT cigarettes exclusively, and customers were loyal to BAT brands —a pattern becoming all too familiar to the Chien brothers.[72]

Unfortunately for Nanyang, Chien Chao-nan also found that the tactics which Nanyang had used to loosen BAT's grip on the market in South China were even less effective inland than in Shanghai. Part of the problem was organizational. Nanyang's largely Cantonese sales organization was not able to make contacts or distribute cigarettes and advertising very effectively in regions outside South China and the Lower Yangtze. BAT had solved this part of the problem by recruiting natives of every region to act as distributors, but the Chiens did not generally follow this pattern—perhaps because of their inability to establish rapport with Chinese from outside their home province. Another part of

Nanyang's problem outside South China and the Lower Yangtze was economic. Even large cities that were not along the China coast seemed to Chien Chao-nan impoverished, and he doubted that their purchasing power was great enough to generate a significant demand for cigarettes. "Nanking," he observed with contempt, "is even less sophisticated than Fatshan," a provincial town near Canton.[73] And still another part of Nanyang's problem north of the Yangtze River was fiscal. BAT enjoyed tax breaks that gave it advantages over its Chinese rivals. Before entering the China market, the Chiens had heard from BAT's competitors in the North about the foreigners' tax privileges there. In 1915, when Nanyang's owners began to ship goods north into Shansi, Honan, and Hopei provinces, their experience confirmed the reports. High taxes were levied on their goods every time they reshipped to a new province or even to a new city—the cost of entering Peking alone was ¥7 per case (of 50,000 cigarettes). The more deeply Nanyang became involved in the northern market, the more numerous and onerous taxes became. In a survey of the North, Wang Shih-jen found, for example, that in Ta-t'ung in Shansi province a tax of ¥15 per case was levied on all cigarettes except those of BAT, which were exempt. He also reported that BAT had "contracted" (pao-pan) with the central and provincial governments to pay them a total of ¥2 per case in taxes on its cigarettes. Nanyang, by comparison, had to pay between ¥3 and ¥10 per case.[74]

While all these problems impeded Nanyang's penetration of new markets, most discouraging to the Chiens was the absence of patriotic support for their product. Chien Chao-nan was shocked to find that "none of the people in the interior know what national goods (kuo-huo) are or why they are significant."[75] Such attitudes seriously weakened Nanyang's appeal, for customers here were not so responsive to Nanyang's nationalistic advertising as Chinese were in Southeast Asia and South China. In fact, the Chiens felt that the success or failure of a Nanyang branch in China correlated directly with the degree of patriotism in its marketing area: their Canton branch was flourishing because Cantonese were fiery patriots; their Shanghai branch was barely surviving because Shanghai people were tepid patriots; their other branches in the Lower Yangtze and North China regions were sustaining losses because people in these regions were not patriots.[76]

The combination of these problems made Nanyang's extension into the Yangtze River regions and North China in 1916 and 1917 considerably less successful than its original entrance into the China market at Canton. In Hankow Chien Chao-nan found that BAT was outselling Nanyang ten to one, and in Tientsin fourteen to one.[77] "Throughout the North," he reported dispiritedly, "after the foreign devil (K'ung Shan) [that is, BAT] has bombarded people daily with its endless [promotional] barrages, we lose out when its price cutting easily lures customers away."[78] And by the spring of 1917 Nanyang began to lose its hold on several urban markets where it had initially made some headway. Its sales fell sharply in the Lower Yangtze cities of Hangchow and Wuhu, and orders for its goods were canceled altogether in the North China cities of Tientsin, Tsinan, and Tsingtao. "It seems," Chien Chao-nan told Chien Yü-chieh in June 1917, "that the market is too small for two."[79]

Though discouraged, the Chiens did not withdraw from the Yangtze River Valley and North China. On the contrary, they fought back in these regions not only through advertising but also by expanding all components of their business—manufacturing, marketing, and tobacco purchasing—between 1916 and 1918. They found, however, that expansion in each of these areas was limited by their rivalry with BAT.

In manufacturing, as soon as Chien Chao-nan began preparations to open Nanyang's first factory within China, he immediately ran into BAT opposition. He took the first step by making a rental deposit in February 1917 on a warehouse located in the foreign concession area in Shanghai. The very next day a BAT comprador tried to buy the building. BAT's offer started a vehement argument involving Chien Chao-nan, the original owner of the building, and BAT representatives over Nanyang's right to retain the building. The matter was finally settled after one of BAT's own compradors, Ch'en Ping-ch'ien—who was, significantly, a Cantonese—defended Nanyang's position and urged BAT's management not to force Chien to surrender his rights to the building. In the end, Chien Chao-nan refused to break the agreement, rented the building, set up a factory, installed 119 American cigarette-making machines, and later bought the site.[80] Thus, Chien Chao-nan succeeded in opening a new Nanyang factory in Shanghai in 1917 but only after overcoming BAT efforts to subvert his plan.

In marketing, Nanyang had difficulty recruiting agents to represent it outside South China, and it had still greater difficulty retaining them because many of them left Nanyang to accept counteroffers from BAT. In the middle of 1917, for example, just as Nanyang was trying to extend its distributing system beyond South China, it lost to BAT three of its ten agents in Shanghai, several of its distributors in North China, and some skilled workers from its Shanghai factory; and, at the highest administrative level, it almost lost its chief translator, a Greek named Constantine whom James Thomas tried to buy off. Not surprisingly, Nanyang's management was alarmed by the raids on its personnel. One director cited these as illustrations of underhanded actions taken by BAT in its "great war" against Nanyang.[81]

To insure Nanyang's survival, Chien Chao-nan took the position that its distributing system would be less vulnerable to attack if it were expanded. Branch offices needed to be built everywhere, he said, so that China would be filled with them "as the heavens are filled with stars." If so, then BAT might snuff out some Nanyang branches but never all of them. Despite BAT's opposition, his policy was carried out to the extent that Nanyang opened sixteen branches in China by 1919—all in South China, the Lower Yangtze, and North China.[82]

In tobacco purchasing as in cigarette manufacturing and marketing, Nanyang's expansion was limited by BAT's strength and influence in China. The Chiens were aware that BAT had procured bright tobacco grown from American seed at considerable savings to the company in the mid-1910s, and, after entering the China market, they were eager to follow their Western rival's example because tobacco leaf was the single most costly item in their budget, absorbing about 60 percent of Nanyang's annual expenditures throughout the 1910s.[83] Accordingly, in 1918 they began to imitate BAT's agricultural system in North China. They built tobacco reception stations at the same three locations where BAT's experiments with American seed had been most successful, Wei-hsien in Shantung province, Hsuchang in Honan, and Fengyang in Anhwei, and they distributed American seed, lent fertilizer, published and circulated a manual on the techniques of growing bright tobacco, and installed redrying machinery.[84] But this venture yielded disappointing results because, according to Chien Chao-nan, BAT lured away from the Chinese company the agricultural specialists who were needed to make a proper evaluation of the tobacco leaf grown in China. Unable to retain these experts, Nanyang was left with purchasing agents who were easily deceived. At Liufu in Anhwei province, for example, the Chiens paid over ¥30 per bundle for leaves which they thought had been grown from American seed only to discover later that the crop had been grown from Chinese seed and was worth no more than ¥10 per bundle. In Chien Chao-nan's words, "We were swindled because we did not know the leaves." Moreover, without competent specialists to check the leaves at the factory, Nanyang sometimes used tobacco that ruined the taste of cigarettes, making them hot and bitter. As a result, tobacco purchased in China contributed to the failure of several Nanyang brands. Piqued by these failures, Chien Chao-nan insisted in early 1919 that thereafter no Nanyang cigarettes should go on the market unless at least 80 percent of the tobacco in them had been grown in the United States.[85]

Such problems would not have occurred, according to Chien Chao-nan, if Nanyang had kept its share of China's "men of talent" (jen-ts'ai), but no sooner had leaf experts demonstrated their abilities at Nanyang than they abandoned it to accept offers of higher salaries from BAT. By his count, BAT had bought off all except one of Nanyang's most competent leaf specialists and, in the process, had foiled his and his brothers' efforts to create an agricultural system in China that would otherwise have enabled them to cut costs and reap other benefits for Nanyang in the late 1910s.[86] So in tobacco purchasing as in manufacturing and marketing, Nanyang's efforts at expansion were slowed by its inability to prevent its high-level personnel from defecting to BAT.

And yet, despite efforts by BAT to bar Nanyang's goods from the China market and limit the expansion of its operations, between 1915 and 1918 the Chinese company penetrated deeply into South China and somewhat less deeply into the Lower Yangtze Region and North China. Its greater success in the South may be explained in part by the fact that BAT might have been better equipped to resist Nanyang outside the South because of the Western company's agricultural system in North China, its factories along or north of the Yangtze, and its headquarters in Shanghai. But this is at best a partial explanation, for BAT was also well entrenched in Hong Kong, Canton, and throughout South China. Perhaps the difference between Nanyang's record in these regions may be better understood in terms of Nanyang's capabilities and limitations as a Cantonese company.

Nanyang's Advantages and Disadvantages as a Cantonese Company

Nanyang's initial penetration of the China market between 1915 and 1918 illustrates some of its advantages and disadvantages in trying to compete with BAT. As long as the Chiens confined Nanyang's operations to Southeast Asia and South China, they were very effective at turning local knowledge and popular enthusiasm for Chinese-made goods to their advantage. The Chiens' local knowledge gave them contacts with the Northern and Southern Hang in Southeast Asia and with merchant associations in Canton through which they gained access to markets in these areas. And the popular enthusiasm for Chinese goods worked to their advantage, particularly during boycotts, because they skillfully used "buy Chinese" advertising to differentiate Nanyang's product from that of its Western competitor in Southeast Asia and South China.

As they moved north, however, they benefited less from these advantages. As Cantonese they, ironically, possessed less local knowledge about Chinese markets in regions outside South China than their Western rival because BAT distributed through local Chinese agents who were "insiders" within each region whereas the Chiens tended to distribute through their fellow Cantonese who were extra-provincial "outsiders" in regions other than the South. Even more discouraging for the Chiens, Nanyang's nationalistic advertising was less effective along the Yangtze and in North China than in the South, for customers in these regions had not participated so fully in boycotts or the national goods movement and, accordingly, proved to be less responsive to Nanyang's nationalistic appeal. Therefore, whereas local knowledge and nationalistic sentiment may have given Nanyang advantages over BAT in Southeast Asia and South China, these served Nanyang's cause less well in other regions of China in the late 1910s.

As Nanyang expanded outside South China and its advantages became fewer, the Chiens appeared to be in danger of succumbing to competitive pressure from BAT as had the owners of other Chinese firms following the boycott of 1905 and 1906. The Chiens were, assuredly, better prepared to withstand this pressure because they had been in the cigarette business longer, had mobilized more capital, had acquired greater technological expertise, had built larger factories, and had established a more stable marketing base than any of BAT's other Chinese rivals. Nonetheless, without the advantage of local knowledge or a nationalistic appeal outside South China, Nanyang's hopes for expanding its manufacturing, marketing, and purchasing systems were severely limited both by BAT's dominance of the industry in China and by BAT's use of aggressive competitive tactics to undercut Nanyang's commercial and industrial expansion.

Faced with this predicament, the Chiens might well have been inclined to bring their commercial wars with BAT to an end by any available means. Given the limitations of their local knowledge, the regional variations in Chinese attitudes toward national goods, and the strength of BAT, they had reason to cooperate with BAT or at least retreat from commercial warfare against it. But surprisingly, only one of the Nanyang owners, Chien Chao-nan, was persuaded to question whether Nanyang should continue to battle with BAT. And when he suggested to his brothers the possibility of relieving Nanyang from competitive pressure by engaging in financial collusion with BAT, he touched off a debate among the Chiens that threatened to destroy their business and tear their family apart.

4.

Motives for Merger

Prior to the rise of the BAT-Nanyang rivalry in China; other cigarette companies had competed with James Duke, but rarely had they remained competitive with him for very long. Instead, Duke's major competitors in the cigarette industry had almost invariably merged each of their companies with his ever-growing tobacco "trust." In the United States during the 1890s, England at the turn of the century, and China since 1905, Duke's rivals had succumbed, one after the other, to this "urge to merge." Duke's previous rivals had chosen to sell out to him because of experiences with his trust that were very similar to the ones that the Chiens had with BAT upon entering the China market in 1915. Like the Chiens, Duke's other rivals had come to recognize, on the one hand, that Duke's business was superior to their own in the sense that it had a bigger, better financed, and more fully integrated organization capable of manufacturing a better-made cigarette, distributing more widely, and advertising more elaborately. And on the other hand, if his rivals had not been sufficiently impressed by the effectiveness of Duke's management to want to join him, then they had been subjected to aggressive tactics like the ones used by BAT against Nanyang—misleading advertising, sabotage of goods, exclusive dealing arrangements with jobbers, legal prosecution, tax rebates, and persistent price cutting. Virtually none of Duke's major rivals had been able to resist this combination: either they had been persuaded to merge by the prospect of benefiting from Duke's managerial efficiency and sharing in his company's profits or they had been pounded into submission by his competitive methods.[1]

If Duke's other major rivals had freed themselves from competitive pressures by merging with his company, why shouldn't Nanyang merge with BAT? In fact, the Chinese company could have done so more easily than its contemporaries in the United States because China had no anti-trust laws like those introduced by the "trust-busters" in America. And yet, though perfectly legal and potentially profitable, the decision whether to merge with BAT was not so simple for the Chiens. In 1917 all members of the family agreed that Nanyang needed money to finance its expansion, but they sharply disagreed about what changes should be made in Nanyang's organizational structure to give them access to capital. Should they merge with BAT? Or should they join with the Peking government in a semi-public corporation? Or should they form a joint stock company? They debated these alternatives among themselves in 1917 and 1918, and their private debates reveal that in their minds the decision whether to end the BAT-Nanyang rivalry was based not only on economic interests but on political and social considerations as well.

Nanyang's Search for Capital

In their search for capital in China, the possibility of merging with BAT was always in the Chiens' minds because BAT had broached the subject with them even before they had entered the China market. In 1914, shortly after Nanyang had begun to show a profit, BAT had approached the Chiens in Hong Kong, but no settlement had been reached.[2] After Nanyang entered the China market, BAT kept open the possibility of a merger, but there were no negotiations between 1914 and 1916. BAT did not bother to reopen negotiations in these years, as Arthur Bassett, BAT's canny American legal adviser who had practiced law in East Asia almost continuously since 1902, reported to James Thomas in 1916, because the Chiens' "ideas are pitched too high to talk reasonable terms."[3]

Before 1916, despite BAT's coercive tactics, none of the Chiens had felt compelled to sell out on BAT's terms because they had been able to maintain a capital base that had supported Nanyang's growth. As Nanyang expanded northward in 1916, they raised more capital than ever before partly through internal financing (raising the company's paid-up capital to ¥978,000 in 1916—ten times more than in 1912), partly by drawing revenue from their business interests in Kobe, Singapore, and elsewhere outside China, and partly by taking out loans from two of the oldest and most important British banks in China, the Hong Kong and Shanghai Banking Corporation and the Chartered Bank of India, Australia, and China. Since 1912 virtually all Nanyang's stock had been in the hands of the Chien family.[4]

By autumn of 1916, however, Nanyang's expansion began to outrun its capitalization, creating financial problems that demanded more radical solutions. The Chiens' dilemma was evident in a clash at the time between Chien Chao-nan and Chien Yü-chieh. Always eager for growth and technological innovation, Chien Chao-nan ordered twelve new

American cigarette machines, But Chien Yü-chieh, fearing that Nanyang could not afford these new additions, intercepted and canceled the order. Furious with his more prudent brother, Chien Chao-nan demanded, "What shall we do! What shall we do! Whoever works in the manufacturing business must have far-sighted vision and cannot be so niggardly about insignificant amounts of money."[5] In this instance, Chien Chaonan prevailed and reordered the machines, but he along with all the other Chiens recognized that Nanyang could only continue to expand if properly financed. In hopes of finding a solution to this problem in the public sector, the Chiens turned in the fall of 1916 to the Peking government.

Nanyang's Negotiations with the Government

Prior to 1916 the Chiens had not established close relations with the Chinese government. Unlike Chinese economic leaders such as late nineteenth century self-strengtheners and early twentieth century gentry-merchants, the Chiens had not been bound to the political order through education or status. Lacking formal education, they had not received degrees within the imperial examination system, become members of the gentry, or held official posts in the Ch'ing or Republican governments. Whereas self-strengtheners like Sheng Hsuan-huai and gentry-industrialists like Chang Chien, for example, had risen through the imperial examination system to become respected scholars and had then used their official influence and high social status to raise capital for "mandarin enterprises," the Chiens had remained outside the mainstream of the educated elite and out of touch with officials in government.[6]

Nonetheless, the Chiens enjoyed the respect that had been officially accorded to all businessmen since the Ch'ing government's establishment of a Ministry of Commerce in 1903, and they were optimistic about receiving official approval for their proposal of a tobacco corporation to be jointly administered by the government and themselves. When Chien Chao-nan and his cousin Chien K'ung-chao, the financial leaders of the two branches of the Chien family, set out for negotiations in Peking in August of 1916, all the Chiens supported the idea of converting Nanyang from a private enterprise to a semi-public corporation. Chien Chao-nan felt that the arrangement would enable the company to pursue his vision of a nationwide Chinese cigarette empire by providing official restraints against foreigners and giving Nanyang better access to credit. Chien Yüchieh viewed it as an opportunity for the Chiens to enrich and stabilize the young and struggling national government and thus to enhance their reputations as patriotic leaders of the national goods movement. And Chien Ying-fu, fearful that the brothers had overextended their capital goods outlay, supported this and every other strategy designed to enable the company to pay its bills and show an immediate profit.[7] Thus, while the brothers disagreed about why they should work with the government, both patriotic Yü-chieh and money-minded Ying-fu were able to send ambitious Chao-nan to Peking in hopes that he would succeed.

Government officials were equally eager to form a tobacco concession. A year earlier, in 1915, President Yuan Shih-k'ai had authorized the creation of a new government agency, the Wine and Tobacco Monopoly; and on August 28, 1916, three months after Yuan's abortive attempt to restore himself as emperor had ended in his death, the new president, Li Yuan-hung, had issued an order directing that land, buildings, and machinery for a government tobacco factory be purchased in Shanghai. These were clear signs that the Peking government wished to become deeply involved in the cigarette industry by the end of the summer of 1916.[8]

In early September 1916 officials met with the Chiens in Peking and proposed an arrangement between the government and Nanyang that was reminiscent of earlier types of collaboration between the state and merchants in Chinese history. As several historians have noted, the imperial government had developed a variety of techniques for coopting merchants, collecting revenue, and coordinating trade in certain commodities through state monopolies (particularly the salt and copper monopolies) in the Ming and Ch'ing periods; and with the advent of modern industrial enterprises in the nineteenth century, late Ch'ing provincial officials had tried to cast merchants in a somewhat analogous role, using them as managers and investors but limiting their control over the enterprises through arrangements such as "official supervision and merchant management" (*kuan-tu shang-pan*) and "officials' and merchants' joint management" (*kuan-shang ho-pan*). In approaching the Chiens, officials in Peking drew upon this tradition of official involvement in the economy and urged that Nanyang cooperate with the government in accordance with the last of the formulae mentioned above, "joint management" (*ho-pan*)—a slogan that had been used by officials since the 1880s to mean that the government would provide protection for merchant managers and would not interfere in the management of cooperative ventures between merchants and officials. More specifically, Niu Ch'uan-shan, who had been commissioner of the Wine and Tobacco Monopoly since its founding in 1915, proposed that the government and Nanyang form a partnership in business (*liang-ho kung-ssu*) with Nanyang responsible for manufacturing and trade and the government responsible for overseeing the operation (*chien-tu ti-wei*); under this proposal, Nanyang and the government would each invest ¥2.5 million. The alternative, Niu told Chien Chao-nan, was a public joint stock company, but Niu preferred not to sell stock on the open market because in joint stock companies "eight or nine times out of ten" the board of directors clashes with the shareholders, making the arrangement unworkable.[9]

After the first talks, Chien Chao-nan became skeptical of this and the government's other proposals which seemed to give it predominant control over Nanyang, and by the end of a month, he was disenchanted with the whole idea of cooperation between Nanyang and the government. He was concerned about the government's efforts to coopt Nanyang, but he was even more disturbed by the government officials' techniques for persuading the Chiens to accept government control—especially the duplicity of Commissioner Niu. When, for example, Niu first welcomed Nanyang's suggestion to consult a comprador who had supposedly resigned from BAT, Wu T'ing-sheng, and then neglected to invite Wu, Chien Chao-nan became convinced that Niu and other officials were "playing a confidence game."[10] Niu later explained that he distrusted Wu because only two years earlier the comprador had been involved in a scheme put forward by BAT, and Niu suspected that Wu might still be secretly employed by BAT. Chien Chao-nan was not reassured by this explanation; he failed to understand why Niu had not openly opposed inviting Wu at the outset.[11]

When Wu T'ing-sheng eventually did join the negotiations, he seemed to Chien Chao-nan to be no less wily than Niu. Before the negotiations, Wu had encouraged the Chiens and had seemed anxious that they benefit from arrangements with the government. Yet once involved, he drafted contracts which seemed to the Chiens more in the government's interests than Nanyang's. Chien Chao-nan despaired of determining whether Wu was working for BAT, the government, himself, or anyone else. After watching Wu perform in Peking, Chien Chao-nan wrote back to Chien Yü-chieh in Hong Kong, "Wu T'ing-sheng is a gentleman (*chün-tzu*) on the outside and a scoundrel on the inside."[12]

In early October 1916 when the government presented the second draft of its proposal, Chien Chao-nan reached his wit's end. He found it even less attractive than the first draft. Niu had the supposedly untrustworthy Wu help prepare this version, a ploy which provoked the exasperated Chien Chao-nan to remark, "I don't know whether Niu is trying to deceive me or Wu!" At the end of the fourth day of disscussions of this proposed contract, one of Niu's assistants appeared at Chien Chao-nan's residence and blithely announced that the Chiens should ignore everything the government's representatives had said that day. The officials, he said, had deliberately fabricated arguments merely to mislead Wu. Once Wu absented himself, they proposed that negotiations would proceed as though this day had never happened. "Officials are devilish," Chien Chao-nan concluded. "After [working with them], a whole picul of grains of Paradise (*sha-jen*) would not calm my stomach. I am astride a tiger and unable to get off."[13]

This visceral reaction against officials and bureaucratic practices prevented Chien Chao-nan from taking advantage of government connections for his private gain. His rise to prominence as a private businessman outside the realm of Chinese officialdom had left him with little experience at dealing with bureaucrats and little tolerance for their talents at dissimulation. After confiding to his brothers that he was disgusted with politics, he withdrew from the talks in October 1916, excusing himself from the bargaining table as unobtrusively as possible to avoid offending or incurring the wrath of Commissioner Niu, who, after all, continued to be responsible for exacting tobacco taxes from Nanyang.

Thus, Chien Chao-nan preferred to keep Nanyang independent of official supervision or close official ties. Lacking the elite education and bureaucratic connections of official-industrialists Sheng Hsuan-huai and Chang Chien, he did not easily establish rapport with officials or feel comfortable negotiating on their terms. Like many other merchants and businessmen in early twentieth century China, he was skeptical of officials' appeals for partnerships between government and business and, fearful of losing control over his business, he refused to participate any further in the negotiations at Peking.[14]

Despite Chien Chao-nan's withdrawal from the talks, other members of the Chien family, led by Chien K'ung-chao, continued to negotiate for almost a year. Chien Yü-chieh showed Nanyang's factories and other facilities to teams of government inspectors, and Chien Ying-fu introduced a wild scheme whereby Chang Ping-lin, a famous scholar, revolutionary, and friend of President Li Yuan-hung, was expected to circumvent Commissioner Niu and appeal directly to the president to arrange the joint enterprise on terms favorable to Nanyang.[15] Some of these efforts seemed more promising than others, but none resulted in an arrangement acceptable to both sides. Chien Chao-nan's withdrawal had emasculated the negotiations in the fall of 1916, and political events brought them to an end nine months later.

During the summer of 1917, less than a year after the talks had begun, even members of the Chien family who most staunchly advocated a joint venture between Nanyang and the government came to realize that the government had lost too much of its authority and had become too unstable to justify continuing the negotiations. Since the death of Yuan Shih-k'ai a year earlier, first competing cliques had jockeyed for power in Peking, then virtually all the provincial warlords in North China had declared themselves independent of the Peking government, and finally President Li Yuan-hung had been forced to dissolve the National Assembly.[16] In the wake of these events, Chien K'ung-chao reluctantly conceded that the Nanyang owners would have to postpone any plans for a joint venture with the government and would have to defer their hopes for the wealth, glory, and historical legitimacy that he thought government sponsorship might bestow. "Only after the general situation is stabilized," he wrote to Chien Yü-chieh in May 1917, "should we ally with the government to increase our influence, and then we shall be wealthy and famous and our name will be known to future generations and·will be regarded as untarnished."[17] Five weeks later the reputation of the Pekking government sank still lower when a northern warlord announced an end to the Republic and a restoration of the imperial system with the eleven-year-old Manchu heir to the Ch'ing throne, P'u-i, as emperor. Within three days after this debacle for the government of the Republic, Chien Yü-chieh informed Commissioner Niu in Peking that the Chiens were temporarily withdrawing from the negotiations because they "had not expected a change in the government at this time."[18] He, Chien K'ung-chao, and Chien Ying-fu continued to cling to the hope that Nanyang and the Peking government might eventually work together, but it was not to be. P'u-i was soon ousted, and during the era of warlordism which followed, discussions of a joint Nanyang-government enterprise were never resumed.[19]

Chien Chao-nan's Negotiations with BAT

In the meantime, after withdrawing in disgust from negotiations with the government, Chien Chao-nan became persuaded that Nanyang should solve its financial problems by merging with BAT. Chien Chao-nan reached this conclusion only after BAT representatives took the initiative and approached him with proposals for a BAT-Nanyang merger in February 1917. Prior to that time, Chien Chao-nan had not been on friendly terms with BAT. On the contrary, he had spearheaded Nanyang's drive into Chinese markets that had previously been monopolized by BAT, opening new branches, introducing new advertising, and in February of 1917, securing a site in Shanghai for Nanyang's first factory on Chinese soil. After the last of these achievements—his rental of a factory site in Shanghai despite BAT's attempt to buy the building from under him—he proudly remarked, "This scheme and all of the foreign devil's previous ones against us have turned out to be bubbles."[20] But after this BAT bubble burst, the representatives of the Western company who descended upon Chien Chao-nan began to convince him that for Nanyang to continue to do battle with BAT was folly. First BAT comprador Ou Pin badgered him. Next Wu T'ing-sheng (now indisputably working for BAT) and Ch'en Ping-ch'ien (the Cantonese BAT comprador who had defended Nanyang's rights to its factory building) prodded him. Finally, James Thomas and his chief assistant and eventual successor, Thomas F. Cobbs, approached him.[21]

They argued, in essence, that Nanyang should merge with BAT because the merger would enable the Chinese company to maximize profits and increase its rate of growth. Not only would the Chiens be free from price wars and other forms of competitive pressure from BAT but they would also benefit from access to BAT's manufacturing, purchasing, and marketing systems in China. The latter argument was especially appealing to Chien Chao-nan for he had learned between 1915 and 1917 that these BAT systems were far superior to his own. He had observed that in manufacturing, towering BAT factories penetrated the skylines of Chinese cities and seemed, in his words, as numerous as "forests of masts [of ships in a port]."[22] In purchasing, BAT's introduction of American bright tobacco in China seemed to him to have been enormously successful—so successful that Nanyang had begun to bid against BAT for the leaf grown in Shantung and Anhwei provinces in 1916.[23] And in marketing, BAT's achievements struck him as most impressive of all. He had traveled extensively along the Yangtze and throughout North China to study BAT's distributing network, and the more he saw of it, the more he admired it. A tour in 1916 of a BAT distributing center in the border city of Kalgan, northwest of Peking and just inside the Great Wall, for example, left him astonished at the scope and effectiveness of BAT's marketing in China. He was surprised that a company could do business in this area at all because it seemed to him (as an urban Chinese from the South) to be utterly primitive. There were no signs of civilization here, he wrote to Chien Yü-chieh, "No Western inns and even the Chinese inns are vile. Electric lights and jinrickshas have not been introduced. There is no way to get around except on horseback, in mule carts, and on foot. The wind fills the air with horse dung." There were no signs of civilization, that is, except BAT representatives. Even in this "bitter, barren" place, he reported, there were two Westerners, both assigned to this station and both able to speak the local dialect beautifully (which Chien Chao-nan, a Cantonese, could not do). "It is evident throughout," Chien Chao-nan concluded, summarizing his admiration and his envy, "that BAT has the talent to do everything."[24]

These travels in China between 1915 and 1917 permitted Chien Chao-nan to learn first-hand about BAT not only as a victim of its commercial onslaughts but also as an observer of its manufacturing, purchasing, and marketing systems. On the basis of this experience, he was aware that BAT representatives did not exaggerate in claiming superiority over Nanyang. In light of these considerations, he was prepared to talk with James

Thomas about the possibilities of a merger, but on one condition: whatever industrial or managerial superiority BAT may have had, Chien Chao-nan insisted that Thomas not treat the Chiens as his inferiors.

Chien Chao-nan was sensitive about James Thomas speaking down to him and his brothers because of the Chiens' painful experience three years earlier in 1914 when BAT had first broached the subject of merger with Nanyang. At that time the Chiens had feared that they, as Chinese operating a comparatively small business and holding inferior political status in the British colony of Hong Kong, would not be taken seriously. Accordingly, they had gone to great lengths to assure that they would be adequately represented—hiring a Japanese to speak on their behalf—and had made every effort to conduct negotiations with BAT on a reciprocal basis. Whereas they had prepared elaborately for the negotiations, BAT had seemed to take a cavalier attitude. It had not bothered to send a company director like James Thomas or a leading lawyer like Arthur Bassett or, for that matter, any of its foreign executives to Hong Kong for the negotiations; instead BAT had left the task to the comprador Wu T'ingsheng. Disappointed that no foreigner had accompanied Wu, the Chiens had felt further affronted by Wu's supercilious handling of the talks. When they had complied with his request that they propose conditions for the sale, he had flatly rejected the proposal and brusquely terminated the negotiations leaving the Chiens feeling betrayed and humiliated. With this incident in mind, Chien Chao-nan opened the negotiations in 1917 by pointedly reminding BAT representatives that in 1914 they had treated his brothers and him "like children."[25]

By contrast, in 1917 James Thomas showed no trace of the patronizing attitude that had wounded the Chiens' pride in 1914. He took charge of the bargaining himself rather than leaving it to a subordinate, and set a conciliatory tone by immediately apologizing for BAT's treatment of the Chiens in the earlier negotiations. He then asked whether Chien Chao-nan wished to submit the first draft of conditions desired in the agreement as the Chiens had done in 1914, and when Chien Chao-nan refused, Thomas readily agreed to draw up the first proposal.[26]

Thomas proposed an agreement which was, he said, comparable to others BAT had used in forming mergers throughout its history. Thomas was able to speak authoritatively about these other mergers because of the prominent role he had played in many of them. In 1902 he had acted as Duke's courier carrying documents back and forth across the Atlantic during the Anglo-American negotiations that led to the founding of BAT itself. And since then, he had negotiated deals that had resulted in BAT's absorption of more than ten Chinese firms.[27] With these experiences behind him, Thomas was sensitive to Chien Chao-nan's apprehension concerning Nanyang's future as a BAT subsidiary and presented a proposal designed to allay his fears.

A merger with BAT, Thomas emphasized, was devised to enrich and strengthen both parties rather than exalt BAT at the expense of the new subsidiary. Once BAT had invested large sums effecting the merger, Thomas pointed out, it would be foolish to let the Chinese company lose its identity or die. On the contrary, the point of the merger was to end commercial warfare before Nanyang was destroyed so that it could join with BAT and other Chinese BAT subsidiaries in a monopoly of the market that would last indefinitely. As a participant in the monopoly, Nanyang would then be able to help bar the entrance of American, English, and Japanese cigarette firms into the market, which, Thomas warned, would soon encroach on China unless warded off by the combined efforts of BAT and Nanyang.[28]

Following this line of reasoning, Thomas urged Chien Chao-nan to accept BAT's terms as other Chinese firms had. According to these terms, Nanyang's organization would remain essentially intact, but BAT would hold over half the stock, would have the option of using Nanyang brands, and would occupy positions on the board of directors (along with the Chiens). Thomas assured Chien Chao-nan that all the Chinese who had merged their businesses with BAT on this basis had since flourished. Such a financial arrangement was ideally suited to the Chiens' purposes, Thomas said, for it provided a source of needed capital without depriving the Chinese of administrative control over their distributing system.[29]

Chien Chao-nan found this argument very persuasive. It seemed to provide the Chiens with the best of both commercial worlds, giving them a share in monopoly profits and access to the incomparable resources of BAT on the one hand without forcing them to relinquish control over Nanyang's administrative structure on the other. He was reluctant to surrender the benefits of the goodwill that Nanyang had built up, but this problem could be solved, he thought, if BAT would agree to permit the Chiens to retain the name "Nanyang" and would give them exclusive rights to use Nanyang's past brands. Once BAT had agreed to meet these conditions and to sweeten the deal by paying the Chiens a large sum in cash, then the merger, he felt, would solve all the financial problems that were troubling his family and would guarantee rapid growth and maximum profits for his company. "If aligned with this great company [BAT]," he wrote to Chien Yü-chieh at the end of February 1917, "then no one can challenge us."[30]

To Chien Chao-nan's dismay, none of his brothers or cousins agreed. In fact, they were upset with him for even considering collaboration with the "foreign devil." All of Nanyang's key executives—Chien Yü-chieh in Hong Kong, Chien Ying-fu in Singapore, Chien K'ung-chao in Bangkok —vehemently opposed the idea. Chien Yü-chieh led the family's attack on his elder brother's proposal and pointed out two fundamental dangers of such a merger which he identified as the diminution of profits (*li*) and the loss of control (*ch'üan*). If Nanyang merged with BAT, its profits would shrink, Chien Yü-chieh maintained, because the Chiens would lose all the goodwill they had built up as leaders of the national goods movement. The Chinese people would accuse the Chiens of hypocrisy or even treason, would mock Nanyang's nationalistic advertising, and would turn against its product.[31] Other members of the family shared Chien Yü-chieh's fear that a merger with BAT would deprive Nanyang of its constituency. One cousin passionately wrote in a letter to Chien Yü-chieh in March 1917: "The Company has used the two characters 'national goods' (*kuo-huo*) to establish its reputation and virtue, and if we amalgamate with foreign capitalists, popular feeling (*jen-hsin*) against us will spread and the market will fall 10,000 feet . . . Our country's industry is just sprouting and successes in it have been few, but our company's goods really have won the respect of our people and glory for our nation. If we are lured astray [by BAT] then surely we shall be reviled and spit on by our own society."[32] As proof that Nanyang's association with national goods was valuable, another cousin, Chien K'ung-chao, pointed to BAT's desire to buy it out. Why was BAT interested in Nanyang? It was not because of the smaller Chinese company's personnel or capital, he asserted, for "the foreign devil's capital is vast, and even ten companies like ours combined could not compete with it." Instead, BAT wanted to buy Nanyang because it feared the "buy-Chinese" sentiment the Chinese company had generated.[33] Therefore, Chien K'ung-chao concluded, if Nanyang merged with BAT, it would be relinquishing the key to its commercial success—the power of nationalistic advertising—which would cut Nanyang off from the national goods movement, diminish its popularity, and thereby reduce its profits.

A merger would cut not only into their profits, the Chiens feared, but also into their administrative control over Nanyang. Like families owning businesses elsewhere in the world, they were loathe to share control with nonfamily members, and, like other Cantonese, they had especially strong kinship ties.[34] At the same time, their opposition to merging with BAT went far beyond the "natural" reaction of a family—even a Cantonese family—against a partnership with outsiders, for it was based on the conviction that BAT directors, as foreigners, were likely to cheat the Chiens in the long run. Chien Yü-chieh contended that even if the Chiens were permitted to hold six of the ten positions on the proposed board of directors, it would not assure them of permanent control because BAT directors, as Westerners, could not be trusted to keep their word. The agreement would not last, Chien Yü-chieh warned, because the lines were drawn in the BAT-Nanyang rivalry not merely between companies but between nationalities and races. Writing from Hong Kong where

Chinese businessmen chafed under discriminatory laws imposed by a British colonial government, Chien Yü-chieh protested to Chien Chao-nan in Shanghai that Westerners, "not being members of our race, do not think like us."[35] Whatever liberal terms the Westerners might offer initially, Chien Yü-chieh felt sure they would soon invoke Western laws and use other foreign devices to enfeeble Nanyang and ultimately would ease the Chiens out of the operation altogether. If, as Chien Chao-nan argued, merger was the only way for Nanyang to achieve rapid growth on a nationwide scale, then Chien Yü-chieh expressed a preference for limiting Nanyang's growth rather than relinquishing any control over it. "It is better to be a chicken's beak," he told Chien Chao-nan, "than an ox's buttocks."[36]

Beyond the issues of profits and control, Chien Yü-chieh urged Chien Chao-nan to drop the idea of a merger because merging with BAT would violate patriotic principles. Even if such a transaction were legal (which Chien Yü-chieh did not deny) and profitable in the long run (which Chien Yü-chieh doubted), he felt that it would be immoral because it would violate the trust the Chinese people had placed in Nanyang as a leader of the national goods movement. The Chiens had gone far by riding the wave of Chinese opposition to foreign commerce in Southeast Asia and China, he reminded Chien Chao-nan. Shouldn't they now go beyond purely selfish considerations to concern themselves with the "larger interests" (*ta-chü*) of the Chinese people? Chien Yü-chieh felt that all of Nanyang's stockholders would answer this question in the affirmative. They would oppose collusion with BAT because they would consider the money received from such a deal to be unclean. Thus the Chiens should reject proposals for a merger with BAT not only to protect their profits and their control over Nanyang but also to align themselves with the Chinese people against evils perpetrated by foreign intruders in general and by BAT in particular. They should not yield to BAT, Chien Yü-chieh declared, alluding to the Three Kingdoms period, any more than Wu had yielded to Ts'ao Ts'ao.[37]

Chien Chao-nan responded as passionately to the other Chiens' criticism as they had to his proposal, and he rejected each of their arguments against a merger with BAT: their economic argument that Nanyang's profits would diminish if it were cut off from the national goods movement; their racial and cultural argument that BAT's management could not be trusted because it was foreign; and their idealistic argument that the Chiens had a patriotic duty to resist foreign encroachment by refusing to sell out to BAT. The economic argument was valid, he acknowledged, wherever nationalistic advertising caused customers to differentiate Nanyang brands from BAT ones, but such advertising was effective only in South China and Southeast Asia. It was not effective, he maintained, in the Yangtze River regions, North China, and throughout the rest of the country where smokers chose their cigarettes on the basis of quality, price, and accessibility and were relatively unconcerned about a product's nationality. If nationalistic advertising had served any useful purpose outside South China, then why, he asked his family, had the several other Chinese companies which had employed it in Shanghai and the North been unable to succeed there? "Without goods of high quality at reasonable prices, do as you please [with advertising], but no matter how hard you plead, our countrymen will not hear." The other Chiens failed to appreciate these regional variations, Chien Chao-nan charged, because they lived in a parochial world confined to the Cantonese urban communities of Hong Kong, Canton, Singapore, and Bangkok. From their limited vantage point, the national goods slogan might seem to be a magic solution to Nanyang's financial problems, but they had not seen first-hand, as he had, the relatively unenthusiastic reception that the national goods movement had been given along the Yangtze and in North China.[38]

Similarly, Chien Chao-nan cited his brothers' parochialism as the explanation for their distrust of BAT's foreign directors. Lacking much exposure to Westerners, the other Chiens were guilty, he told them, of bigotry. They were afraid to work with BAT executives because they had not overcome old-fashioned stereotypes of racial differences between Westerners and Chinese. Claiming to be more cosmopolitan, Chien Chao-nan found Chien Yü-chieh's racism particularly offensive and chided him for it, accusing him of being "a hidebound bigot" (yü-fu). As an indication that Westerners were not so power-hungry and determined to grasp control as his brothers supposed, he informed them that BAT had agreed to a special concession whereby all authority in Nanyang would remain in the old management's hands after the proposed BAT-Nanyang merger, with the president (Chien Chao-nan) at the top, the other Chiens in positions as directors, and Westerners specifically excluded from the board of directors.[39]

His family's third objection to the merger—patriotism—Chien Chao-nan considered a product of pride, fuzzy thinking, and self-righteousness. His brothers' idea that a business deal with BAT would lead to a violation of their own patriotic principles seemed to him "extremely unclear." He suspected that their so-called patriotism was primarily an expression of hurt pride. It had originated in a history of defeats and humiliations at the hands of foreigners in South China since the Opium War, and now it was coloring his family's attitudes toward business, making their blood boil in a fiery, peculiarly Cantonese way (cheng ch'i). These feelings of shame for the failures of one's predecessors to stand up to foreigners in the past seemed to Chien Chao-nan irrelevant to business considerations in the present. Moreover, he felt that his family members' obsession with past humiliations caused them to follow imaginary social norms that had no basis in reality. To pursue patriotic impulses based on a nebulous notion of the people's "larger interests" was, he said (invoking the Buddhist idiom), as illusory as "mistaking flowers for the reflection of flowers in a mirror." In other words, he felt that the Chiens were wrong to assume that Chinese outside the company would support Nanyang's fight against BAT merely because the Chiens considered it a patriotic crusade. Allowing this emotional identification with the public to determine their business decisions would be financially ruinous for Nanyang, he predicted. "If you [oppose BAT] as an expression of your pride and it turns out that BAT defeats Nanyang, do you think that our countrymen will, for the sake of their pride, reimburse us for our losses?"[40]

All these objections to the merger were based, according to Chien Chao-nan, on irrelevant philosophizing in which the Chiens, as businessmen, could not afford to indulge. The stark financial fact of the matter, he claimed, was that Nanyang either had to merge with BAT or be crushed by it. Challenging his brothers to face this fact, he asked them, "Is the little world which we have created for ourselves indestructible?" He believed that their world was not indestructible because it might soon collapse under competitive pressure from BAT. In the predatory world of big business where they now found themselves, he insisted that the very existence of their company was at stake. As long as they remained unaffiliated with BAT, he warned, they were vulnerable to an attack that might destroy their company at any time. "If the foreign devil continues to be our enemy, then every day we shall have to look [anxiously] left and right. There is no way to struggle with ferocious tigers. Whenever careless, we shall be swallowed." He viewed the proposed merger as the only way out of this jungle, and he hoped that after the merger BAT might give Nanyang a sense of security and no sense of inferiority. In his folksy Cantonese metaphor, amalgamation would amount to a "family alliance" in which the Chiens would be "like children who have an amah [BAT] whose job it is to keep them in a secure position."[41] In this way, he suggested, the Chien family's control would be preserved and BAT would play a subordinate role in the family, serving merely as an amah, protecting the Chiens and looking after their welfare but remaining their social inferior.

In addition to sheltering Nanyang from BAT competition and offering the Chiens security, the merger would, Chien Chao-nan predicted, be very profitable. In preliminary negotiations Chien Chao-nan had asked for ¥10 million as an initial bonus, and BAT had agreed to as much as ¥2 million, an amount larger than Nanyang's total paid-up capital at the time. Moreover, although the anticipated volume of Nanyang's trade would be specified in the contract, the amount would be adjustable and would not limit Nanyang's growth. Best of all from Chien Chao-nan's viewpoint, Nanyang's share in a monopoly and access to BAT's manufacturing, marketing, and purchasing systems would greatly increase Nanyang's profits. According to his calculations, these prospective profits, not fuzzy philosophizing, should be the most important determinant of the Chiens' decision.

If his family members needed some philosophical basis for this decision, then he urged them to consider the fundamental importance of finances in promoting their family, their business, their country, or any other institution they valued. "The principle thing which has enabled people to bring prosperity to the world and its societies and to generate power and wealth in the conflict among nations and to glorify their countries," he maintained, "is money." Hard cash, not patriotic passion, was Chien Chao-nan's antidote for Nanyang's and for China's ills. They were undercapitalized in a harshly competitive world where he defined the fittest as the richest. "The rich are respected, the poor despised," he told his more idealistic younger brother. "Life is like that and there is no way to help it."[42]

The fundamental differences between Chien Chao-nan and the other Chiens grew out of their attitudes toward regional and national loyalties in China. Chien Chao-nan's family opposed the merger on the basis of an orientation that may be characterized as nationalistic regionalism. Undoubtedly their regionalism was, as Chien Chao-nan charged, partly a product of their provincial backgrounds. They were willing to confine Nanyang to regional markets in the South rather than merge with BAT to reach a nationwide market partly because they had seldom ventured outside the Cantonese communities of South China and Southeast Asia. Wedded to these regions, they had seen less of the rest of China than Chien Chao-nan had. On the other hand, however parochial their orientation, the key to the Chiens' regionalism lay, paradoxically, in their nationalism. Disappointed to find that the national goods movement was less effective in the Lower Yangtze and North China, their tendency was to limit themselves to the regions that were most responsive to their nationalistic advertising rather than surrender their independence to a foreign rival. The result was that their nationalism merged with and reinforced their regionalism.

Chien Chao-nan, by contrast, viewed regional and national loyalties with greater equanimity. He approved of using nationalistic advertising to differentiate Nanyang's product from that of BAT, and, in fact, he personally created innovative advertising and devised imaginative promotional techniques to accomplish this purpose.[43] Unlike the other members of his family, however, he was not committed to any type of regionalism

or nationalism as a principle on which to base Nanyang's business strategy. In his estimation, his brothers' zealous commitment to the nation
had blinded them to realities that required a change in business strategy.
Unless they went ahead with the merger, he felt that BAT would drive
their company into extinction.

Convinced that he was right and they were wrong, Chien Chao-nan
began negotiations with BAT despite his brothers' and other relatives'
opposition. The first day of bargaining with James Thomas, March 16,
1917, went well and seemed to him to vindicate his position. The talks
were marred, he complained to Chien Yü-chieh, only by the failure of the
rest of the family to cable an expression of their interest in the merger and
support for his participation in the negotiations. Without it, he was
unable to negotiate forcefully and decisively. "You, my brothers, have
been businessmen for a long time. Don't you understand the fundamen-
tals of business? If you are so cautious, how can you expect to have success? Or perhaps you are afraid that I am not shrewd and will be
duped?"[44] His sarcastic questions reflected not only impatience with his
brothers but also his confidence in negotiating with James Thomas.

As Chien Chao-nan became more hostile toward his family, he became
more attracted to James Thomas and other Americans in BAT. Whereas
the other Chiens expressed the belief that Westerners were grasping and
untrustworthy, Chien Chao-nan characterized Thomas as sincere
(ch'eng-tu). To illustrate the sincerity of Thomas and his American assistant, Thomas Cobbs, Chien Chao-nan described for his younger brother
the BAT representatives' willingness to take him into their confidence
and privately disclose to him ways in which BAT could beguile the Chinese public. For example, he was pleased with Cobbs' candid comments
on Nanyang's nationalistic advertising. Cobbs acknowledged that it had
been effective and expressed the hope that the Chiens would continue to
use it after merging with BAT. How would Nanyang retain its image as a
nationalistic company after selling out to its foreign rival? With no difficulty whatsoever, Cobbs explained, as long as the BAT-Nanyang
merger was kept a secret. To spur sales, he felt that Nanyang should continue its vigorous attacks on BAT in newspaper advertising and other
media after merging so that consumers would think that the two companies were still battling for the market while in fact they were secretly
cooperating. As part of this strategy, he suggested that he and Chien
Chao-nan should keep their personal friendship a secret and should
avoid being seen socializing together.[45]

Such revelations gave Chien Chao-nan the feeling that he had won the
Americans' confidence, and he turned their trust to his advantage by trying to deceive them on several major issues. In answer to Thomas' questions about the Chien family's support for a merger, for example, Chien

Chao-nan admitted that there were problems, but he made it seem that a sufficiently large bonus—¥10 million—would eliminate dissent within the family. In answer to Thomas' inquiries about Nanyang's profits, Chien Chao-nan provided inflated figures. He said that the company's profits had been ¥300,000 in 1914, for example, when it had barely cleared half that much. When Thomas pressed him and asked to examine Nanyang's financial records, he stalled and sent instructions to Chien Yü-chieh to make up an entirely new set of books falsifying the information in them to conform with what Chien Chao-nan had told Thomas.[46]

In June 1917, at the very moment when Chien Chao-nan felt his influence with Thomas and Cobbs was reaching its peak, Nanyang's negotiations with the Peking government collapsed. Though disappointing for the rest of the Chien family, the elimination of the Chinese government as a possible partner delighted Chien Chao-nan. "You have been patriotic," he wrote with relish to Chien Yü-chieh after the government's debacle, "but now do we cooperate with the government or with the foreign devil?"[47] Believing the answer was obvious, Chien Chao-nan demanded in early July that his relatives approve his plans for a merger, and at that point he discovered the depth of their opposition to collaboration with BAT.

Rather than merge with BAT, the other Chiens were prepared to sever relations with Chien Chao-nan not only as a business partner but as a member of the family. The youngest brother, Chien Ying-fu, wrote to the middle brother, Chien Yü-chieh, that the merger had to be blocked at all costs. If Chien Chao-nan obstinately insisted on it, "We should not hesitate to break off our relationship with him . . . and take matters into our own hands." To show that this was more than pure talk, Chien Ying-fu disclosed that he had already looked into the possibility of relocating the family's business in Singapore. It would be better to abandon one brother and the established company in China, Chien Ying-fu reasoned, than to surrender to BAT and permit it to destroy the entire family and business organization.[48] Chien K'ung-chao, a cousin who managed Nanyang's branch in Thailand, was equally adamant in his opposition to the proposed merger. To show Chien Chao-nan that they were serious about the issue of control, he wrote to Chien Yü-chieh, they might take the absurd position that they would sell everything to BAT but not merge with it. If Chien Chao-nan tried to force his family into merging, Chien K'ung-chao was as willing as Chien Ying-fu to break up the family. At that point, "Each man must make up his own mind."[49] Chien Ching-shan, the Chien brothers' uncle who also helped manage Nanyang's business in Thailand, echoed Chien K'ung-chao's sentiments and rounded out the Chien family's opposition to a merger.[50] Feeling ran so deeply on this issue that the Chiens thus were willing to ostracize Chien Chao-nan before they would agree to merge with BAT.

While Chien Chao-nan ran into opposition from Nanyang's directors, James Thomas had similar problems enlisting support for the merger within BAT. Thomas complained to Chien Chao-nan that while he, Cobbs, and the other Americans in BAT favored acceptance of Chien Chao-nan's proposal, British directors in the company opposed it. If authorized, Thomas said that he would gladly pay the ￥10 million that Chien Chao-nan was asking, but he could not as long as British executives in the company disapproved.[51]

By mid-July 1917, the negotiations for a merger began to break down. Neither James Thomas nor Chien Chao-nan was able to muster the necessary support to carry out the merger both men wanted. Discouraged, Chien Chao-nan informed his family that he was dropping the negotiations, and he disavowed all responsibility for the consequences—which he was sure would be dire—of Nanyang's failure to offer terms acceptable to BAT. Any further discussion of the matter he left to a subordinate. He did not, however, cease to admire BAT or abandon his hopes that Nanyang would merge with it. A few months later, in October 1917, Chien Chao-nan expressed his regret that Nanyang had not been prepared to "strike while the iron was hot" and pointed out to Chien Yü-chieh that BAT's attitude toward the negotiations had begun to cool. On October 21, 1917, Thomas formally terminated the talks.[52]

Even after the negotiations officially ended, Thomas and Chien Chao-nan tried to keep the idea of a merger alive. Thomas suggested that Chien Chao-nan might help him persuade the directors of BAT to agree to the merger by going to the United States with him. Chien Chao-nan inferred from the invitation that Thomas hoped the two of them could circumvent British opposition to the merger within BAT by approaching James Duke in person and persuading him to make a decision favorable to them. Still without encouragement from his family, Chien Chao-nan decided that, if invited by Duke, he would go, but he received no invitation.[53] He and Thomas, both lacking the support of their colleagues, held no additional talks in 1917, and their hopes for a cooperative Sino-American enterprise remained, at least for the moment, unfulfilled.

In 1917 the outcome was a victory for Chien Chao-nan's relatives, who refused to permit Nanyang to be absorbed as Duke's other rivals had been. As a result of their intransigence, they immediately suffered reprisals in the market place, for by October 1917 Chien Chao-nan noted, "The foreign devil in its war [against Nanyang] is becoming more violent every day."[54] But the other Chiens viewed this intensified pressure as a challenge and questioned whether Chien Chao-nan was responding to it as vigorously as he should. Chien Chao-nan's nephew, Chien Ch'in-shih, for example, was incensed at the deference Chien Chao-nan paid BAT. Visiting Shanghai and North China to observe Nanyang's operations there, he reported back to Chien Yü-chieh in Hong Kong that "an indignant fire" had burned within him when he became aware of Chien Chao-nan's lack of fighting spirit. Nanyang's brands were less popular in Shanghai and North China, according to his assessment, not because Northern Chinese were less patriotic than Cantonese but because Chien Chao-nan and Nanyang's other executives had truckled to BAT and had failed to push nationalistic advertising as aggressively as Nanyang had in the South. He recommended that the Chiens become even more deeply committed to a nationalistic business strategy at the end of 1917 than they had been before Chien Chao-nan began his talks with BAT a year earlier, and the other Chiens who resided south of Shanghai shared his view.[55] Taking this hard line, they were unequivocal in opposing a merger with BAT.

The family's rejection of the proposed BAT-Nanyang merger left Nanyang's growing need for capital still unmet. Their search for capital had led them into talks first with the Peking government in 1916 and then with BAT in 1917, but after deciding not to collaborate with either of these, they still needed funds to finance Nanyang's expansion. So they turned to a third alternative: the formation of a joint stock company.

Nanyang's Formation of a Joint Stock Company

Like most Chinese businessmen in the early twentieth century, the Chiens were not enthusiastic about forming a joint stock company. The concept of limited liability was introduced in China at least as early as the 1850s and was endorsed by the Chinese government as early as 1904, but the joint stock company did not become a popular type of industrial organization among late nineteenth or early twentieth century Chinese. Some refrained from adopting it because they did not fully understand it; Chang Chien, for example, the distinguished scholar-industrialist who became Minister of Agriculture and Finance in the Republican government, did not grasp the basic difference between a partnership and a limited liability company. Others were reluctant to organize joint stock companies because of deficiencies in China's financial environment; they were afraid to raise capital by selling stock within the joint stock arrangement as long as China lacked a well-developed market for the transfer of equity shares.[56] Still others, perceiving the same problem from a somewhat different perspective, felt that the joint stock company, as a Western financial institution, was alien to Chinese social and cultural values and therefore either would fail to attract Chinese capital or would attract capital from investors who would abuse their authority as shareholders. It was the latter danger that most disturbed Chien Chao-nan.

Setting himself against the rest of the family as he had on previous proposals for Nanyang's financial reorganization, Chien Chao-nan vehemently opposed converting Nanyang into a joint stock company. It would be foolish for the Chiens to sell Nanyang stock on the open market, Chien Chao-nan cautioned, because they would attract few shareholders and no loyal ones. Persuading any Chinese from outside the Chien family to buy stock, he maintained, would be "as difficult as poling a boat up a mountain." And those who did invest in Nanyang would demand short-term profits, interfere with plans for long-term expansion, meddle in managerial affairs, and generally undermine the Chiens' authority. As he put it, such self-seeking investors would, "on the pretext that they are stockholders, grasp for authority and form cliques out of pure selfishness and without respect for the interests of the majority nine times out of ten. My brothers have lived in Chinese society no small number of years, how can you not know."[57] If Chinese society had accepted a stronger rule of law as in the West, then greater trust between Chinese management and stockholders might have been possible on a legal basis, and this Western financial institution might have served entrepreneurs' needs more effectively in China, but as long as Chinese stockholders were restrained neither by law nor by deference to management, Chien Chao-nan felt that they were bound to have a destructive influence on his company.[58]

Those major Chinese enterprises that were organized as joint stock companies seemed to Chien Chao-nan to prove that the management of a firm could not hope to command the respect of its Chinese stockholders. Of these, three were cited by Chien Chao-nan as examples of companies suffering from their stockholders' excessive demands at the time—the Shanghai Commercial Press which was capitalized at ¥2 million, the Chung-hua Book Store which was capitalized at ¥1 million, and the Sincere Company which owned a chain of department stores.[59] Moreover, the case of Mu Hsiang-yueh (1876-1943), a contemporary of Chien Chao-nan and a well-known industrialist of the time, further illustrated his point. Suspicious stockholders secretly hired workers to spy on Mu in the cotton textiles factory where he was manager, and they even appointed a vice-manager without consulting him. Ultimately stockholders hounded him into resigning from the company.[60]

Chien Chao-nan feared that reorganization of Nanyang into a joint stock company would pose similar threats to his family's and his own authority. For this reason, rather than impose this Western-style joint stock structure on Nanyang, he preferred to retain the Chinese safeguards that had protected his authority in the past—familism and regionalism. The Chiens had consistently relied upon these institutions since the founding of Nanyang by restricting its ownership to members of the family as much as possible and by never selling to anyone else except fellow Cantonese from their native place (t'ung-hsiang). According to Chien

Chao-nan, this familism and regionalism had permitted the development of rapport between managers and owners and had enabled Nanyang to strive toward long-term goals without fear of interference from short-sighted stockholders, but if stock were now sold publicly, then this basis for trust between management and owners would disappear. And if the Chiens tried to preserve a sense of regionalism by restricting distribution of their stock to Cantonese in South China and Southeast Asia, they would fail, he felt, because he doubted that their fellow Cantonese would buy more than half of the ¥10 million in Nanyang stock that his brothers proposed to sell; the other half would then have to be sold to non-Cantonese.[61]

Although Chien Chao-nan had eagerly and at times creatively incorporated into Nanyang's operation Western business methods in producing, distributing, and advertising and had adapted them successfully to the Chinese commercial environment, he was afraid to abandon Chinese financial institutions and impose a Western one. To strip his company and his authority of protection from traditional institutions, float stock in the open market, and open Nanyang to anyone who might buy a share would reduce Nanyang, he feared, to a business organized on the basis of mutual distrust. He predicted that if such a suicidal change were made, within a year "the foreign devil will be triumphantly drinking champagne and we shall be vomiting blood."[62]

The other executives in Nanyang were not much more enthusiastic about forming a joint stock company than Chien Chao-nan was. They felt a little more confident than he did that ¥10 million could be raised by selling stock to Chinese stockholders. Chien K'ung-chao observed, for example, that political instability might, paradoxically, make stock easier to sell because the change in goverment and China's other political difficulties in 1917 had outraged many wealthy men and aroused their nationalistic feelings; consequently, some of them might be more inclined than ever to invest in Chinese companies for patriotic reasons.[63] Nonetheless, even if believing that a joint stock form of organization would attract capital, neither Chien K'ung-chao nor the other Chiens resorted to it until their talks with the Peking government collapsed in the summer of 1917. As late as the early autumn of that year, Chien Chao-nan continued to oppose the idea, but when his talks with BAT ended in October, he joined his brothers in considering whether it might be feasible.

All the Chiens agreed that a joint stock company would only be practicable if they could find ways to prohibit unruly or untrustworthy investors from buying shares in the company. Chien Ying-fu suggested that they might prevent indiscriminate distribution of stock by selling it to selected (presumably Cantonese) merchant associations rather than opening it to the public. Chien Chao-nan proposed that the Chiens control distribution even more closely, screening the interested parties and selling only to those "who have contributed to the efficacy of the company in the past and will help it grow in the future." After hearing these various proposals, all the major stockholders in Nanyang gathered in Shanghai in March 1918 and agreed that no stock would be sold outside the family without unanimous consent of the five Chiens who comprised the board of directors, and none would be sold to foreigners (including representatives of BAT) under any circumstances.[64]

This decision protected the family but apparently did not give Chien Chao-nan as much security as he felt he needed. In the spring of 1918 he took steps to guarantee that no other individual inside or outside the family would be able to subvert his authority or oust him from the presidency of the company. He first acquired an option on ￥800,000 worth of stock for an indefinite period of time, and a week later he brought his branch of the family together and confirmed the revaluation of it and certain other stock. Under the new arrangement, ￥7,500 worth of stock he had bought in 1912 from the only investor outside the family at that time became revalued at ￥150,000, twenty times its original price. Added to his past and recent holdings, this addition gave him a total of ￥762,500 in stock as compared to ￥367,500 for Chien Yü-chieh, ￥122,500 for Chien Ying-fu, and ￥122,500 for their nephews, Chien Han-ch'ing and Chien Ping-jen (sons of their deceased brother, Chien Chien-ch'üan). The Chiens all accepted this distribution of the stock and gave it a familial sanction by swearing to honor it "in the presence of our honorable mother." The other branch of the Chien family, led by Chien K'ung-chao, had previously held as much stock as Chien Chao-nan's branch, but after these financial maneuvers, Chien Chao-nan became Nanyang's leading stockholder and did not relinquish this position until his death in 1923.[65]

Chien Chao-nan reinforced his growing financial control by taking as his title "permanent president" (yung-yuan tsung-li) and by securing his hold on the top position in Nanyang's corporate pyramid more firmly after the reorganization than before. This new formal definition of his position gave him additional protection from his opposition in that it became virtually impossible for anyone to force him to resign. He even reserved the right to dictate in his will the successor to whom he wished to bequeath his office. Among other privileges Chien Chao-nan secured at the same time was a salary of ￥12,000 and an expense account of ￥15,000 per year. He justified these unprecedented steps to buttress his authority as responses to the reorganization of the company and to the problems raised by the new joint stock approach.[66] They served this purpose, and they probably served another one as well. In the aftermath of his family's rebellion against him over his proposal to merge with BAT, it seems likely that he was taking this opportunity to reassert his leadership and control over the company and the Chien family.

Chien Chao-nan's success at elevating himself above the other Chiens was reflected in their handling of the company's registration with the Chinese government as a joint stock company in the summer of 1918. Their petition to the Ministry of Agriculture and Commerce in Peking clearly indicated that, even though they wanted to distribute stock outside the family, they fully expected control over the company to remain in the family's, and primarily Chien Chao-nan's, hands. In it they named the initial members of the board of directors and comptrollers of the company, all of whom were members of the family and were scheduled to serve three-year terms; prospective shareholders from outside the family would be allowed no say in these appointments. This and similar provisions in Nanyang's bylaws caused officials in the ministry to doubt whether there was any substance to the Chiens' professed willingness to "cooperate with the people" who might buy stock.

If the officials were disturbed by the Chien family's arrogation of power, they were aghast at Chien Chao-nan's prerogatives. They censured one provision which stated that the company's bylaws could never be changed in such a way that the company's president might suffer. They were also critical of Chien Chao-nan's title and powers as "permanent president." They reminded the Chiens that, according to Chinese law, no one man in a joint stock company was supposed to be made invulnerable to changes in a company or empowered to make decisions by himself. For these reasons, the ministry rejected the petition.[67]

The two representatives sent to Peking to secure approval for the petition, Chien Yü-chieh and Chien K'ung-chao, responded to these rulings in a way that revealed their determination to retain control for the family. They ignored the substance of the officials' criticisms and merely revised their strategy for arranging to have Nanyang registered. They did not change their concept of the company itself; they simply deleted the names of the members of the board of directors and the term "permanent president" from the bylaws. To protect themselves from any future investigation of the unchanged features of the company, they employed a time-tested technique for evading bureaucrats while at the same time ensuring protection from official reprisals: they enclosed with their petition a lengthy history of the company "for reference." Within this history, they buried information indicating the extent of the Chien family's and Chien Chao-nan's powers. In their private correspondence, they acknowledged that the aim of this revision was to make the bylaws flexible so that the members of the family could later "stretch and shrink" them as they wished. Writing from North China, Chien Yü-chieh and Chien K'ung-chao assured their kinsmen in the South that "in the future, we shall all discuss and define the powers of the president and board of directors by ourselves, and for the purposes of registration we do not need to indicate in the bylaws what they are."[68] The Chiens realized that retaining these powers contravened the ministry's ruling and the law, but they expected the document to be deposited unread in the ministry's files and took comfort in the prospect of invoking it to prove, if later questioned, that they had accurately reported to the ministry the nature of Nanyang's corporate structure. Thus, to placate the ministry, the Chiens revised their description of Nanyang—or gave the impression that they were revising it—but did not change their conception of how the company would be organized and run.

From the Chiens' viewpoint, this rather meaningless adjustment of the bylaws was simple; guiding their petition through the maze of bureaucrats responsible for validating their registration was a more difficult task. Winning approval from the Ministry of Agriculture and Commerce was only the beginning. They then tried to convince the minister (*tsung*) and vice-minister (*tz'u-chang*) to exert pressure on lower ranking officials, expedite the petition, and save it from being smothered under other paper work on the desk of some procrastinating bureaucrat. Failing in this attempt to work from the top down, they "fixed" (*t'o*) things directly with a department head (*ssu-chang*) and a section chief (*k'o-chang*). Only then, after proper lubrication, did the bureaucratic machinery at the Ministry of Agriculture and Commerce efficiently produce Nanyang's registration. In July 1918, a few days after the last draft of the Chiens' petition was submitted, Nanyang was registered with the Peking government as a joint stock limited liability company capitalized at ¥5 million (54 percent paid). At this time, all of the ¥2.7 million worth of paid-up stock was owned by the Chien family.[69]

With Nanyang's reorganization and legal registration as a joint stock company in 1918, all the Chiens except Chao-nan were relieved to have ended their long search for capital and pleased to have found a means of broadening their capital base without resorting to a merger with BAT. Although they would have preferred the sponsorship and financing they had sought from the government in 1916 and 1917, at least they now had a corporate structure that conformed with their nationalistic business strategy and kept them aligned with their countrymen and free of foreign financial backing in their battle against BAT.

Chien Chao-nan, by contrast, remained unconvinced that the joint stock structure would attract sufficient capital to solve Nanyang's financial problems and perplexed that his family clung so tenaciously to nationalism as the basis for both the strategy and the structure of their business. Nationalistic advertising, he conceded, may have served Nanyang's purposes while it was a small firm in Southeast Asia and South China.

Now, however, as it grew into a big business with hopes for developing a still larger nationwide commercial empire, he advocated a more complex strategy—one which allowed Nanyang not only to disseminate nationalistic advertising but also to upgrade its goods, train new personnel, establish new branches, build new factories, install new machinery, form new dealing agreements with marketing *hang*, secure freight rebates with transport *hang*, and negotiate tax rebates with governments. In coping with these complicated problems, he felt that nationalism was no panacea. The key instead was capital—more capital than he believed China's money markets would yield in response to Nanyang's issuance of stock. Still urging his brothers to recognize the limitations of their nationalism and the folly of defying the bigger and better financed BAT, he wrote to them in January 1919, "Money is all powerful in this world, and everyone is caught in its web . . . If you have money, nothing else matters, and it will always bring you fame. Otherwise, no matter how patriotic you are, if you lack money with which to accomplish things, people will think nothing of you."[70]

Thus, the debate between Chien Chao-nan and his family that had begun in 1916 was still underway in early 1919. During these four years of arguing over the proper means of financing their company's expansion, Chien Chao-nan and his family had defined two different approaches to business based on two different visions of China's future. His family anticipated an intensification of Chinese nationalism in the future which would vindicate their approach, and Chien Chao-nan feared an intensification of competitive pressure from BAT which would vindicate his. Paradoxically, both sides' expectations were vindicated by the events that followed, for both Chinese nationalism and competitive pressure from BAT reached unprecedented heights during the May Fourth Movement of 1919.

5.

Commercial Campaigns in the May Fourth Movement

THE MAY FOURTH incident set off a wave of strikes and boycotts that swept through China's cities and heralded the beginning of a new era in Chinese history. The incident began on the morning of May 4, 1919, when about 3,000 students gathered to protest the news that diplomats at the Paris Peace Conference had decided Japan would be awarded the former German concession area in the Chinese province of Shantung as part of the peace treaty with Germany following World War I. Angry with Chinese officials in the Peking government as well as foreign diplomats in Paris for condoning this seizure of Chinese territory, the students held a demonstration which marked the first of a series of mass protests in China's major cities over the next two months that ultimately reached a successful conclusion when the Chinese delegation did not join in the signing of the Treaty of Versailles on June 28.[1]

Of these mass protests, perhaps the most effective was the May Fourth boycott against foreign goods in general and Japanese goods in particular. Led by students and merchants, Chinese consumers participated much more fully and effectively in it than they had in the anti-American boycott of 1905 and 1906 or the anti-Japanese boycotts of 1908 and 1915, and this boycott, like the earlier ones, stimulated Chinese entrepreneurs to establish new companies that produced and advertised national goods as substitutes for foreign ones.[2] As in previous boycotts, BAT and Nanyang were among the most visible of the firms that became involved in the boycott, for they distributed advertising on a massive scale. In fact, their May Fourth promotional campaigns dwarfed all previous ones. They published almost two hundred expensive newspaper advertisements over the thirty-five day period between mid-May and mid-June 1919, an average of between five and six every day.[3] The exact number of Chinese touched by this flood of publicity cannot be measured precisely because studies of newspaper circulation at the time did not take into account the common practice of renting or reselling newspapers which enabled many people to read a single newspaper, but the cigarette companies' May Fourth advertising unquestionably reached a large segment of the literate population of Canton, Hong Kong, Shanghai, Hankow, Peking, and Tientsin through newspapers and became known to a still wider audience through handbills, posters, and word-of-mouth.[4] Scholars in China who specialize in the compilation of newspapers have probably not exaggerated in calling the cigarette companies' series of campaigns during the May Fourth movement in 1919 "a great war of words" (ta-cheng lun) on a scale unprecedented in Chinese history.[5]

Since the Chiens had benefited from earlier boycotts, one might have expected them to benefit even more from this one, but, ironically, it hurt rather than helped them. They spent ¥1.5 million on advertising during the summer quarter of 1919—almost as much as BAT's advertising for the whole year—but their sales still slumped and their reputation suffered during the May Fourth protests and for several months thereafter; as late as March 1920 Nanyang had still not recovered.[6] Meanwhile BAT, after suffering briefly from boycotts against its brands and strikes in its factories during May and June, enjoyed a booming business throughout the latter half of 1919; by the early autumn of 1919, the demand for its goods rose to the point where it began hiring more workers and even added a new night shift in its Shanghai factories.[7]

The fact that Nanyang did not benefit from the May Fourth boycott at the expense of BAT raises a historical question about the role of nationalistic boycotts in the two big businesses' rivalry: if nationalistic boycotts had worked to the advantage of Chinese tobacco companies in the past, why did this one work to Nanyang's disadvantage during the May Fourth protests in 1919? The process whereby this boycott turned against Nanyang will be analyzed here with illustrations from the three cities in which the cigarette companies' May Fourth commercial campaigns became the most intense: Canton, where attacks against Nanyang were led by members of the gentry; Shanghai, where its chief critics were businessmen; and Peking, where it was assailed by politicians.

The Chiens and the Gentry in Cantonese Society

Like other Chinese companies that became targets of the May Fourth boycott, Nanyang began to suffer as soon as it was shown to have affiliations with Japan, the country whose claim in Shantung had led to the May Fourth incident. In Canton Nanyang's affiliation with Japan was publicized immediately after the May Fourth incident when Chiang Shu-ying, a local BAT agent, arranged for the publication of documents in

Canton's newspapers showing that Chien Chao-nan was a Japanese citizen with a Japanese name who had used his citizenship to file suits in Japanese courts.[8] Under ordinary circumstances this action need not have posed a threat to the Chiens in Canton. After all, four years earlier when Nanyang had first entered the China market at Canton, BAT had tried to stir up nationalistic feeling against Nanyang for maintaining ties with Japan but had failed to stop the growth of the Chiens' business. As Cantonese, the Chiens had been able to overcome BAT criticism and establish their credibility by drawing upon their knowledge of the city and forming liaisons with their fellow Cantonese in merchant associations and other organizations in their home region. In surviving these attacks in 1915 and using Canton as a beachhead from which to advance northward thereafter, they had appeared invulnerable to criticism in their hometown. Unfortunately for the Chiens, however, the circumstances in Canton were different in May 1919 than in 1915. During the May Fourth campaigns, they had to cope not only with new anti-Japanese hostility created by the May Fourth incident but also with critics representing BAT whose attacks were potentially far more damaging than any earlier ones. For in Chiang Shu-ying, the BAT agent who publicized Chien Chao-nan's citizenship following May 4, they faced an opponent who spoke not merely as a BAT agent but as a member of one of Canton's most prominent gentry families. Because of their gentry background, Chiang Shu-ying and his father, Chiang K'ung-yin, a salaried employee of BAT in Canton at the time, had a different basis for exerting influence than the Chinese BAT merchant-agents with whom the Chiens generally competed in China.

Chiang K'ung-yin's claim to gentry status was perfectly genuine. In fact, by traditional standards, he had reached the pinnacle of social status and moral influence among members of the gentry. Born in Nanhai (the Chiens' home county) in 1865 (five years before Chien Chao-nan), he had climbed to the top of the traditional ladder of success by passing the imperial examinations at every level, earning the metropolitan degree (*chin-shih*), and serving in the imperial institute for distinguished scholars in Peking, the Hanlin Academy. Then, following the traditional pattern, he had capped this illustrious academic career by putting his training to use as a political activist. Returning to Canton, he had become the single most influential member of the gentry in Kwangtung provincial politics during the decade between the Boxer Uprising of 1900 and the fall of the Ch'ing dynasty in 1911. According to the assessment in Edward Rhoads' recent study of the province in these years,

> Chiang K'ung-yin . . . was undoubtedly the most active member of the gentry in Kwangtung during the post-Boxer decade. He had taken part in practically every aspect of the merchants'

nationalist movement as well as the gentry's self-government movement. He had been a leader, though a controversial one, in the anti-American boycott. A member of both the Merchants' Self-Government Society and the Macao Boundary Delimitation Society, he had spoken out successively on the West River Patrol, the *Tatsu Maru* case, the *Fatshan* incident, and the Macao question. He had helped organize the provincial assembly and, though he declined election, he had been elected to the assembly. He had also been a leader of the campaign to abolish licensed gambling. Finally, when Governor-General Chang Ming-ch'i inaugurated the rural pacification and reconstruction program in the Canton area in May 1911, he had chosen Chiang K'ung-yin to head it up.[9]

Nor did Chiang K'ung-yin's political influence end with the fall of the dynasty. Like many other members of the gentry, he showed remarkable staying power in politics following the revolution of 1911 and secured a place for himself at the nexus of urban elite society in South China, shuttling from one group to another, mediating among them, and thus becoming the person best able to link gentry and scholars, students and revolutionaries, merchants and newspapers editors. Through this array of personal contacts he maintained his influence in politics and his prominence in Cantonese society. Known popularly by the nickname of Chiang the Shrimp (Chiang Hsia) because of his hunched posture, he retained sufficient prestige throughout the Republican period to attract the attention of President Yuan Shih-k'ai, Generalissimo Chiang Kai-shek, and other political leaders even after he officially ended his political career in the mid-1910s.[10]

Recognizing that these credentials would enable Chiang K'ung-yin to wield an influence that might be damaging to Nanyang, the Chiens began to spar publicly with him soon after he left politics to join BAT in 1917. Why, they challenged him to explain, had a man of his background taken a job that reduced him to the status of a lowly comprador groveling for a job with a foreign company? Chiang replied that he had chosen to work for BAT precisely because he was able to do so without reducing his status at all. He did not deny that compradors were disesteemed in Chinese society, but he argued that his appointment made him entirely different from compradors. Whereas compradors and other Chinese had been forced to appeal to foreigners for employment and had been assigned tasks on the seamy side of business such as guaranteeing payments and arranging squeeze, BAT had shown its deference to him, he said, by seeking him out and by appointing him to a more exalted position. He received a salary from BAT as compradors did and performed functions comparable to theirs, but according to Chiang, the primary distinction between compradors and himself lay in his social prestige. The Chinese people might have reason to distrust compradors, Chiang implied, but since he was not becoming a comprador, he expected them to respect him and defer to him in his new post as in his previous ones, and with his background, he had justification for anticipating that many Cantonese would.[11]

Rejecting Chiang's self-defense, the Chiens questioned his claim to a high social status that supposedly elevated him above other people. They said that he needed to do more than merely invoke his *chin-shih* degree and his membership in the gentry to win the respect of Chinese people because in recent years new standards had emerged for judging whether a person was worthy of respect. As they were well aware, since the abolition of the imperial examination system in 1905, Chinese had climbed a variety of new ladders of success in pursuit of social status not only as members of the gentry but in the Chiens' case as businessmen, and in other cases as students and teachers in Western-style schools or as soldiers, lawyers, engineers, doctors, and professionals in other fields. The members of these emerging elite groups had advanced in Chinese society and had begun to earn respect not by passing the imperial examinations as Chiang had but by demonstrating that they were patriotically and usefully serving the nation. By this latter criterion, the Chiens charged, Chiang K'ung-yin had ceased to be respectable. Whatever social status he may have achieved earlier by traditional standards, he forfeited it when he demeaned himself and betrayed the nation by going on the payroll of a foreign company. His willingness to work for BAT exposed him as a member of the "evil gentry" (*lieh-shen*) who had become a "tiger's accomplice" (*hu-chang*)—one helping the tiger-like BAT to prey on innocent victims such as the Chiens.[12]

Chiang responded to these charges by agreeing that a person should be judged according to his service to the nation, but he turned the question of patriotism back on his accusers, proclaiming that he worked for BAT because it served the interests of the Chinese people. To give the concept of patriotism meaning and to discern who was the real "tiger's accomplice," the fundamental question, he asserted, was how much love one showed for the masses (*ai ta-ch'ün*) as opposed to how much one merely pretended to contribute to their welfare. He justified working for BAT by citing its expenditures in Kwangtung province and China as a whole. He claimed that in Kwangtung BAT distributed cigarettes to small retailers purely out of an altruistic desire to serve the common man (*hsiang-jen*) and bought tobacco leaf in the Kwangtung countryside, which benefited peasants, purchasing 70 to 80 percent of the more than ¥2 million worth of tobacco exported from Kwangtung's tobacco growing region (around Nan-hsiung) in 1918. Throughout China as a whole, he asserted, the company's contributions were more varied: an additional ¥10 million spent procuring tobacco leaf grown in China; employment of almost 20,000 Chinese in BAT factories; distribution of between ¥8 million and ¥9 million worth of dividend-bearing shares in the company to Chinese stockholders; donations of "several 10,000's" of *yuan* for flood relief in the North. BAT's premise, implicit in these policies, he concluded, was that it "must put a bit of the money taken from society back into society." He lauded this approach and contrasted it with the disingenuous patriotism of his critics. "In actual fact, they use foreign tobacco, foreign packaging, and foreign tins . . . Hiding behind masks and chanting 'Love China! Love China! Buy native goods! Buy native goods!' (*ai-kuo, ai-kuo, t'u-huo, t'u-huo*) . . . they are deceiving others as well as themselves. They are the real tiger's accomplices."[13] Although Chiang K'ung-yin never named his critics, he was undoubtedly referring to the Chiens.

Exchanges such as these between the Chiens and Chiang K'ung-yin in Canton's newspapers immediately prior to 1919 established the context for the May Fourth battles between Nanyang and BAT. Within this context, when Chiang Shu-ying published evidence of Chien Chao-nan's Japanese citizenship in Canton's newspapers early in May 1919, he inflamed a rivalry that had been growing not only between two companies but between two families. Coming from very different social backgrounds, each family seems to have sought to wage commercial warfare against the other by exerting its social influence and demonstrating that it and its company commanded more respect in local society than the other.

Immediately after the May Fourth incident and Chiang Shu-ying's publication of the documents showing Chien Chao-nan's Japanese citizenship, the Chiens deliberately set out to establish their credibility by securing endorsements from influential groups in Cantonese society. On May 6 they invited members of seventeen Cantonese civic associations (*she-t'uan*)—eight charitable halls (*shan-t'ang*), four hospitals, four merchant and labor organizations, and one academic group—to travel the ninety miles from Canton to Hong Kong and see for themselves that Nanyang's Hong Kong factory was Chinese, not Japanese. After the tour, a statement was issued, signed by the civic associations, which said that the Chiens used Chinese or American raw materials and Chinese labor, that all the rumors about Nanyang being a Japanese company were untrue, and that any objections to Chien Chao-nan's Japanese citizenship were petty or even self-defeating. After all, if the Chinese government found fault with double citizenship in this case, then other Overseas Chinese (who were largely Cantonese) might well feel that their double citizenship was in jeopardy and might withdraw moral support and capital from the national goods campaign in China—a loss that would be financially disastrous for the Chinese economy and for Cantonese people out-side China as well as within it. Accordingly, the petition signed by the representatives of Cantonese civic groups urged that the Chiens be exon-erated from all charges against them and that Nanyang be designated without qualification as a "national goods company."[14] Moreover, repre-sentatives of hospitals and other charitable institutions among these civic associations argued that the Chiens' company was even better than other Chinese businesses. It was more than a Chinese "capitalist enterprise" (*tzu-pen ying-yeh*); its philanthropic contributions to aid Cantonese vic-tims of recent disasters—floods, famine, warfare—and its donations to orphanages and hospitals made it a kind of "public welfare agency" (*kung-kung fu-wu*) as well. According to these administrators (whose charitable institutions the Chiens had helped finance), Nanyang deserved not only to be vindicated from criticism but to be held up as a model for all Chinese capitalists genuinely interested in the salvation of the nation and the welfare of the Chinese people.[15]

Unfortunately for Nanyang, three men refused to sign the endorsement and released a separate statement. The so-called "investigation," they said, had amounted to no more than a perfunctory tour of the Hong Kong factory lasting "a matter of minutes." The visitors from Canton were shown no records of who owned stock in Nanyang and no docu-mentary proof of where Nanyang's foreign tobacco and other imported raw materials had originated. They saw tobacco marked with white chalk indicating that it had come from the regions mentioned in the majority report, but none of the visitors had the expertise to tell from the leaves whether these markings were accurate. They were told that Nan-yang had imported from Japan bamboo mouthpieces for its cigarettes until 1917 and cigarette paper right up to the May Fourth incident but that recently Nanyang had begun to make the tips in Fatshan and now intended to begin to import the paper from the United States. Although the tour was brief and uninformative, the majority of Nanyang's guests from Canton were satisfied with the visit, according to these dissenters, because they were concerned less with asking serious questions than with enjoying the treatment that Nanyang lavished upon them—accommoda-tions at Hong Kong's posh Chung-hua Hotel, transportation by taxi, guided tours of Hong Kong's scenic spots, banqueting at a leading restau-rant, and partying in a night club. Not until the guests had become in-ebriated, reported the dissenters, did Chien Ch'in-shih, Nanyang's director of public relations, come forward with a report of the day's "in-vestigation," prepared beforehand, for everyone at the party to sign.[16] Though only three of the Chiens' forty visitors withheld their endorse-ments, their critique suggesting that Nanyang had suborned its visitors was publicized as widely as the "investigation."

Other commentators criticized the Chiens' supporters for leaving the impression that the Chiens' philanthropy made them selfless men of virtue giving to good causes for purely altruistic reasons. In reality, said the critics, the Chiens made donations to charitable causes merely as a way of "buying people's hearts . . . If one is doing good things with one hand and waving his trademark with the other, then the good things are not being done out of altruism." Nanyang's critics conceded that the Chiens had made donations to charities but said that the figures quoted in the endorsements were highly inflated. In actual practice, the critics surmised, Nanyang took high profits from its Chinese customers and did not return as much to them as would have been proper. Nanyang's critics also complained that the Chiens' philanthropy was irrelevant to the central questions of whether Chien Chao-nan held Japanese citizenship and whether Nanyang was a Japanese company. If the philanthropy had come from a company that was authentically Chinese, Nanyang's critics claimed, sarcastically, then they would have "appreciated it, praised it with tears [of gratitude], even worshipped it, and, of course never criticized it." But if it were traceable to Japan, they argued, then it would not be so highly esteemed because the Chinese people were in no mood to welcome charity from the Japanese.[17]

A variety of other groups lent their names to the controversy. Perhaps the most prestigious group solicited by the Chiens consisted of scholars from Canton Teachers College (later known as Chung-shan or Sun Yat-sen University). The scholars accepted the Chiens' invitation to visit Nanyang's factory in Hong Kong, and after their tour, they publicly commended the company for its "vast scope" and "sophisticated research" and for relying on Chinese rather than foreign capital, raw materials, technicians, and laborers.[18] This endorsement lent some academic respectability to the Chiens, who lacked formal academic training and credentials, and gave them a counterweight against the superior status of the Chiang family.

Nanyang's critics also exhibited special credentials which, they said, made them privy to inside information and, therefore, more credible as partisans in the debate over Nanyang's national status. A critic claiming to be a former Nanyang employee, for example, said that the experience of working for the company had made him better qualified to pass judgment than the scholars who visited Nanyang's factory. "The scholars [from Canton Teachers College] may be knowledgeable about Chinese society, but they were deceived by Nanyang," he concluded, for the real truth was the exact opposite to what the learned men had reported: Chien Chao-nan was Japanese and Nanyang did import raw materials and finished products from Japan. This source introduced no tangible evidence to support his allegations. Similarly, Cantonese students and merchants residing in Japan claimed to be authoritative commentators because they, too, had access to evidence not available to others—the material appearing in Japanese sources. This evidence, they said, made the case against Nanyang iron-clad. Their references to Japanese newspapers, popular magazines, and business guides, however, merely reaffirmed the known fact that Chien Chao-nan was a Japanese citizen and included no proof that Nanyang used Japanese raw materials, employed Japanese technicians, or issued stock to Japanese financial backers.[19]

Throughout most of the May Fourth battles between Nanyang and its critics in Canton, the Chiens were more visible than the Chiangs because the Nanyang owners felt compelled to recruit Cantonese leaders to tour their Hong Kong factory whereas the Chiangs did not have to reveal publicly the extent of their participation in the May Fourth campaigns against Nanyang. In light of the familial rivalry already underway in 1917 and 1918 and in light of the expense of publishing items in Chinese newspapers and other media, however, it seems very unlikely that the May Fourth criticism of Nanyang all originated spontaneously in Canton. Instead, the Chiangs probably drew upon their extensive contacts and inspired criticism of Nanyang (that is, they paid for it and wrote it or recruited others to write it) in the same way that the Chiens demonstrably inspired their allies to defend themselves and their company.

While the Chiangs appear to have been discreet in this regard, at points the family rivalry broke into the open as when Chien Ch'in-shih of Nanyang published a private letter from Chiang Shu-ying of BAT and thus set off an angry exchange of letters between them in newspapers. Chien Ch'in-shih used Chiang Shu-ying's letter without his permission to substantiate the Chiens' contention that the May Fourth criticisms of Nanyang were part of a BAT plot to crush its Chinese competition. Construing an ambiguous passage in the letter to mean that Chiang Shu-ying intended to buy out Chien Chao-nan, Chien Ch'in-shih published it along with an anti-BAT polemic which was signed by 12,000 people—supposedly all of Nanyang's staff and workers. This polemic cited Chiang Shu-ying's letter as an example of the pressure that BAT had exerted on the Chiens to give up their business. In the past, it said, "The Tobacco King of Canton and Hong Kong [with ties to] a certain other country" (in this context probably an allusion to Chiang K'ung-yin) had tried to overwhelm Chien Chao-nan by offering to buy out Nanyang for as much as ¥25 million, but Nanyang's president was determined to save China's face, prevent the drain of profits from China to the West, and promote Chinese industry. Therefore, according to this polemic, Chien Chao-nan had valiantly resisted BAT's offers, had only pretended to be interested, and had kept Nanyang solvent by turning it into a Chinese-owned joint stock company. Frustrated, BAT had sought to intensify the pressure on the Chiens through slanderous campaigns following May Fourth, which tried to convince the public that Nanyang's cigarettes were "evil goods" (lieh-huo) tainted by foreign influence, as opposed to "national goods" (kuo-huo) of Chinese origin. Promoters of this campaign had paid local gangs to remove Nanyang's cigarettes forcibly from cigarette stands and tea shops and had hired local agitators to deliver harangues against Nanyang in theaters, ferry boats, and other public places in Canton. Although BAT had not openly identified itself with this campaign, the polemic concluded that no one else would have been able to afford the "endless sums of money" that were spent on it. In sum, according to the polemic, BAT had used a combination of coercion and enticement (wei-po li-yu) to oppose Nanyang.[20]

In reply, Chiang Shu-ying fired back an open letter rebutting Chien Ch'in-shih in Canton's newspapers. In saying that BAT had offered the Chiens ¥25 million for Nanyang, Chien Ch'in-shih had told "a big lie," he retorted; one would have to be "crazy" to pay so much for a firm that was not worth ¥3 million. Besides, in a financial transaction BAT did not have the power to force Nanyang to do anything because "Nanyang is not a child and BAT is not an office of the government." As for the May Fourth campaigns, he agreed that there had been oppression, but complained that Nanyang, not BAT, had been the oppressor. "Concerning the question of whether BAT has oppressed (ch'i-ya) your company, it seems to me that precisely the opposite is true." The oppression had taken many forms, he said, ranging from the publication of lies about BAT to harassment and beatings of Nanyang's critics and even including threats against the lives of members of the Chiang family—all ordered by the Nanyang owners. By resorting to these dark and sinister methods, the Chiens may have expected to drag the Chiangs down to their low level of business ethics, but he said that his family had unwaveringly continued to engage in "proper competition." In short, Chien Ch'in-shih's allegations of BAT oppression amounted to nothing more than a diversionary tactic according to Chiang Shu-ying. By stirring up xenophobic attitudes (p'ai-wai hsin) among Cantonese against BAT, Chien Ch'in-shih was trying to distract attention away from the real May Fourth issue, namely Chien Chao-nan's and Nanyang's foreign entanglements with Japan.[21]

It is not possible to confirm whether the Chiens and Chiangs actually used tactics as vicious as the ones they attributed to each other in these impassioned exchanges, but it is apparent (by comparing their allegations with the findings in Chapter 4) that each side distorted the other's past record and that the commercial rivalry between the two families and between the two companies deepened during the spring and summer of 1919. In the intensely nationalistic atmosphere generated by the May Fourth movement in Canton, both families became more defensive about their affiliations with foreigners and each tried to represent itself as more patriotic and less hypocritical than the other.

Whether or not BAT won this advertising war, Nanyang clearly fared less well than it had during its conflict with BAT over Nanyang's Japanese connections in 1915. This time, despite a heavy investment in elaborate campaigns to win endorsements from Cantonese social groups, the Chiens were not able to suppress the anti-Nanyang campaign so quickly in Canton, nor were they able to prevent it from spreading to Shanghai.

The Chiens' Rivals and Allies in Shanghai

Chien Chao-nan had more at stake in Shanghai than Canton, for within the past two years he had moved his residence to Shanghai, had reorganized Nanyang's Shanghai office, had adopted the currency of the Shanghai Bank of Communications as the standard for Nanyang, and had expanded Nanyang's Shanghai factory, raising its productive capacity to four million cigarettes per day.[22] Under his leadership, though Nanyang's headquarters formally remained in Hong Kong, the center of its operations had begun to shift to Shanghai. And if Chien Chao-nan expected to establish a good reputation for himself and his company in Shanghai he could ill afford large-scale anti-Nanyang commercial campaigns like the ones in Canton. Like it or not, however, he faced campaigns in Shanghai similar, though not identical, to the ones in Canton. In Shanghai, as in Canton, he was condemned or endorsed in newspapers and other advertisements by Chinese from a variety of social groups, but in Shanghai by contrast with Canton, his most prominent critics and supporters were members of the city's commercial and financial elite rather than members of the gentry.

BAT seems to have been one of his critics in Shanghai, but, as in Canton, the extent to which it was responsible for the campaigns is difficult to say. In Shanghai, as in Canton, the Chiens charged that BAT was the moving force behind criticism of Nanyang, but they could not prove it.[23] BAT remained even more shadowy in Shanghai than in Canton because BAT agents and compradors were not identified by name with the Shanghai campaign as the Chiangs were in Canton. Nonetheless, in light of the tactics used by BAT on earlier occasions, the Chiens were probably justified in their suspicion that some participants in the campaign against them benefited from BAT assistance in preparing newspaper articles critical of Nanyang or at least in covering the high cost of publishing these articles.

Whether or not Nanyang's larger Western rival engineered and financed much of the campaign against Nanyang in Shanghai, smaller Chinese-owned rivals unquestionably participated in it. In the cigarette industry, as in other businesses, smaller Chinese companies exploited the May Fourth boycott at Nanyang's expense in much the same way Nanyang and other small Chinese companies had exploited the anti-American boycott of 1905 at BAT's expense; that is, they identified their larger rival with the foreign country being boycotted and suggested that they, by comparison, were more patriotic because of their independence from foreigners. For Nanyang the most troublesome of these Chinese competitors was Ch'en Ts'ai-pao, the manager of a new Chinese tobacco firm, the Hsing-yeh Tobacco Company of Shanghai. Ch'en had once worked for Nanyang and had helped found Nanyang's Shanghai branch in 1917, but in 1919 he showed little sympathy for his former employers. According to the Chiens, Ch'en had begun to abuse them even before the May Fourth boycott, counterfeiting their brands, "stirring up people in the cigarette industry against us, sabotaging our displays, and doing everything else to subvert our firm out of jealousy." Already antagonistic toward the Chiens, he seized upon the May Fourth boycott, they said, and tried to turn it against Nanyang by hiring young people to impersonate patriotic students and sending them out in this guise to distribute handbills and paste up posters in Shanghai and Hangchow saying Nanyang was Japanese. For one Chinese to manipulate a patriotic movement against his fellow Chinese in this way was unconscionable, the Chiens said, and his willful exploitation of the situation revealed him to be a creature so despicable that "no jackal or tiger would touch his flesh."[24]

Although the Chiens had more specific information on Ch'en's role in the anti-Nanyang campaign than on BAT participation in Shanghai, they concentrated most of their countercampaigns on their larger Western rival rather than on smaller Chinese ones. They characterized BAT as a ruthless monopoly that victimized people throughout Chinese society, striking down merchants who tried to compete with it, reaching into the countryside to extract tobacco and "oppress peasants and make them suffer indescribable bitterness," and siphoning off ¥70 million to ¥80 million in profits (according to the Chiens' estimate) from the Chinese economy. In all these ways BAT had oppressed the Chinese people by depriving them of their economic rights, the Chiens maintained, and it was these economic rights that they had sought to regain for the sake of the nation. The charges against them for unpatriotic tendencies were absurdly ironic, they said, because they had established a reputation as leaders of the national goods movement—the most patriotic Chinese industrialists of all. To reinforce this point, they secured endorsements from commercial organizations, student groups, and hundreds of fellow industrialists and other prominent people in Shanghai. In Shanghai as in Canton these countercampaigns defended the Chiens' record of service to the nation and attacked their rivals in the tobacco industry for opportunistically exploiting the May Fourth boycott at Nanyang's expense.[25]

After fighting back throughout the month of May, the expense and the adverse effects of May Fourth commercial campaigns on Nanyang's business forced Chien Chao-nan to concede defeat on the issue of his Japanese citizenship. He refused to accept his critics' contention that he had ever ceased to be Chinese, but in response to the May Fourth campaigns, he renounced his Japanese citizenship. "I am a Cantonese," he began. "If Cantonese are natives of China, then I am certainly a Chinese citizen." Since he had always regarded China as his "ancestral homeland," he felt that he had a legal right to retain his Chinese citizenship after becoming a citizen of Japan, even as other Overseas Chinese had a legal right to retain theirs after becoming citizens of countries of Southeast Asia, Europe, or America. Although he still possessed this legal right and although double citizenship would still be useful for business purposes, he was now willing to terminate his Japanese citizenship, he said, because he had become aware of the Chinese public's antipathy for Japan and wished to bow to the will of the people.[26]

On the same day that he renounced his Japanese citizenship, May 27, 1919, Chien Chao-nan inaugurated a new kind of commercial campaign by announcing that Nanyang would immediately issue ¥10 million in stock for sale on the open market—the first stock in Nanyang to be made available to non-Cantonese investors outside the Chien family. "In our humble opinion," he proclaimed, "limiting economic opportunities to one family is not as good as opening them to the people of the whole country."[27] In subsequent publicity, as in this announcement, the Chiens tried to give the impression that they welcomed investment by Chinese regardless of familial or regional background.

Selling stock at ¥20 per share, they urged all their countrymen to invest in Nanyang not only to gain a share of the profits but to join the Chiens' heroic fight against a foreign rival. "In view of repeated harassment by foreigners," read one plea for stockholders, the Chiens "feel a company run by one family cannot endure and must be open to all of our countrymen. Our policy of united and collective action offers all compatriots opportunities to invest so we can advance together."[28] Another advertisement called for prospective stockholders "to rally under one [Chinese] banner on the commercial battlefield" and give Nanyang "the strength of the masses (*chung-jen*) in order to expand national production and stop the leak [of profits drained from China by foreigners]."[29]

Though publicly proposing to "open Nanyang to the whole country," the Chiens privately relied as much as possible on fellow Cantonese to carry out the plans for financial expansion. On June 28, 1919, for example, the Chiens acknowledged the necessity of moving Nanyang's headquarters from Hong Kong to Shanghai in order to gain access to China's largest money markets, but at the same meeting they appointed two Cantonese to Nanyang's board of directors specifically to serve as liaisons between the Chiens and Shanghai's financiers. These two men, Ch'en Ping-ch'ien and Lao Ching-hsiu—the first members of Nanyang's board ever to come from outside the Chien family—were eminently qualified to serve as a bridge between the Chiens and the financial world of Shanghai. On the one hand, both Ch'en and Lao were Cantonese and so had regional ties with the Chiens. On the other hand, both men had lived in Shanghai for many years and had established extensive connections there in their positions as compradors for foreign firms—Ch'en with BAT among others—and as members of the boards of directors of various Chinese businesses mostly in real estate and insurance. In appointing Ch'en and Lao, the Chiens established a preference for choosing Cantonese residents of Shanghai as members of their board of directors that was reflected in subsequent appointments. Seven of the eight directors (including Ch'en and Lao) appointed from outside the family in 1919 were Cantonese and six of the eight were residents of Shanghai.[30]

With financiers from Shanghai on Nanyang's board of directors, the company immediately received the backing of Shanghai's financial elite. Among its sponsors, perhaps the most useful in helping attract risk-capital were key members of the wealthy and powerful "Chekiang financial clique" in Shanghai: Chu Pao-san, a successful comprador and businessman who had come to Shanghai from Chekiang province in the early 1860s at age sixteen; Yü Hsia-ch'ing (Yü Ho-teh), a comprador and the leading Chinese investor in steamship lines; and Ku Hsing-i, an industrialist with investments primarily in food processing companies and cotton mills. These men were successors to a long line of fellow provincials from Chekiang who had managed Shanghai's money markets for several generations, and they exerted influence not only as their predecessors had but also through an influential institution created in the early twentieth century, the Shanghai General Chamber of Commerce, in which Chu, Yü, and Ku each served as president. In addition to the Chekiang financiers, many others sponsored Nanyang and promoted the sale of its stock in Shanghai: leading figures from Kiangsu province such as Chang Chien, the scholar-industrialist who was renowned for the development of his native place in nearby Nan-t'ung; Cantonese residing in Shanghai like Yang Hsiao-ch'uan, the commissioner of foreign affairs for Kiangsu province at the time who joined Nanyang's board of directors later in 1919; and hundreds of other members of Shanghai's financial, industrial, and political elite.[31]

With this kind of support, Chien Chao-nan finally seemed to have weathered the May Fourth storms in Shanghai. His announcement on May 27, 1919, renouncing his Japanese citizenship and "opening Nanyang to the people" by issuing stock in it seemed to undercut his critics; and his transfer of Nanyang to Shanghai seemed to eliminate any doubt that it was a Chinese company based on Chinese soil. By coincidence, on June 28, 1919, the same day the Chiens approved the contract reorganizing Nanyang and transferring its headquarters to Shanghai, the May Fourth protests came to an end with the signing of the Treaty of Versailles.[32] With these two events, by late June 1919, Nanyang's May Fourth woes seemed to be over. After enduring two months of commercial campaigns that had besmirched his reputation, overburdened his advertising budget, and reduced his sales, Chien Chao-nan was relieved to have a respite from pressures exerted upon him since the May Fourth incident. Unfortunately for him, however, he soon discovered that he was not free from the charges brought against him during the preceding months. In mid-July the controversy over his citizenship resumed, this time moving from the commercial battlefields of Shanghai, Canton, and other cities to the halls of government in Peking.

The Chiens' Involvement in Peking Politics

On July 17, 1919, Ho Hsun-yeh, a recently elected member of parliament in Peking, demanded that the government force Nanyang out of business by rescinding its registration as a Chinese company. The company should have been closed down as soon as its president's Japanese citizenship had been revealed, Ho insisted, and it had no right to continue to operate until Chien Chao-nan was legally recognized as a Chinese citizen. A week later another member of parliament, a military representative from Shansi province named Chou Wei-fan, broadened and intensified the attack on Nanyang. He demanded cancellation of Nanyang's registration and rejection of the petition which Chien Chao-nan had made to reaffirm his Chinese citizenship. Chou maintained that Chien Chao-nan had been reluctant to terminate his Japanese citizenship because by Japanese law anyone rescinding naturalized citizenship was subject to confiscation of two thirds of his property by the Japanese government, and he doubted that Chien Chao-nan sincerely wanted to rescind it at all. Whatever Chien Chao-nan's reasons for holding Japanese citizenship, according to Chou he had not only deceived his countrymen but had also violated the law. To stop him and others like him from acting as dummy fronts for foreign capital and thus facilitating foreign penetration of China's economy, Chou called upon the government to punish Chien Chao-nan and his company harshly by canceling Nanyang's registration, withdrawing its license, and cutting off its trade. Closing down Nanyang and making an example of it would serve as an appropriate warning to other men of money, he argued, and it might well deter them from becoming "treacherous merchants" (*chien-shang*).[33]

While Ho and Chou called for the Ministry of Agriculture and Commerce to get tough with Chien Chao-nan, others from outside the government lobbied against him too. For example, Wang Shu-kung, representing the Committee for the Investigation and Promotion of National Goods, presented a report to the Ministry of Agriculture and Commerce which was more accurate and less bombastic than Chou's but which called for the same action: revocation of Nanyang's license and termination of Nanyang's business, a ban which was to be enforced by officials at all levels of government. In addition (according to Chien Yü-chieh), Huang Ch'u-chiu, a "vagrant merchant" and professional lobbyist, was paid ¥400,000 by BAT to "fix things" and spent ¥200,000 of it "bribing high and low officials to carry out the plot." Under this combination of pressures, at the beginning of August 1919 the cabinet of the Peking government ordered the Ministry of Agriculture and Commerce to cancel Nanyang's license and declared that Nanyang should be treated as though it were Japanese. On August 9 the ministry issued the cancellation.[34]

In response to this decision, protests from Nanyang's supporters poured into Peking. Many were sent by the same groups that had defended Nanyang in May and June 1919—chambers of commerce and other commercial associations, charitable institutions, labor organizations, schools—but some also came from a new source, Overseas Chinese. The Association of Overseas Chinese in Singapore, for example, immediately sent a telegram to the cabinet of the Peking government (*kuo-wu yuan*) protesting its ruling. The Overseas Chinese complained that the decision on Chien Chao-nan had ominous legal implications for all Overseas Chinese, for the Chinese government had never objected before to double citizenship. If this new precedent were permitted to stand, they predicted that all Overseas Chinese would be so fearful that "their blood would run cold" and, as a consequence, they would sever all ties with China and withdraw their investments from the Chinese economy.[35]

The contention by Overseas Chinese that the decision against Chien Chao-nan was a deviation from established norms is understandable against the background of China's nationality laws. In the eighteenth and nineteenth centuries, the Ch'ing government had prohibited emigration from China under penalty of death and, when emigration had occurred anyway, the government had assumed that all Chinese, wherever they lived, owed perpetual allegiance to China.[36] The first nationality law comparable to those in the West was instituted in 1909, but it did not reverse the doctrine of perpetual allegiance to China. On the contrary, it implied acceptance of this doctrine in its omission of the principle that the acquisition of a new nationality automatically extinguished a previous one—a principle generally accepted in the West—and, by this omission, it seemed to leave open the possibility of double citizenship.[37] The law was revised slightly in 1914, but it still clearly stated that the loss of Chinese nationality as a result of voluntary acquisition of foreign nationality was "limited to persons . . . in whose cases the consent of the Ministry of the Interior (*Nei-wu pu*) had been given."[38]

In general, during China's decade of experience with nationality laws since 1909, the Chinese government had seemed disposed to interpret and enforce these laws in favor of Overseas Chinese. In the late 1910s, for example, the Peking government had passed legislation giving Overseas Chinese additional privileges as citizens, such as eligibility to run for elective office in China. Such policies had seemed to reinforce the idea that double citizenship was permissible and, as the Association of Overseas Chinese in Singapore put it, that "our government has never discriminated against us."[39] But the judgment on Chien Chao-nan's case implied that the government was for the first time beginning to discriminate against them, they feared, by forbidding Chinese to hold double citizenship.

Besides protesting directly to the cabinet of the Peking government, Overseas Chinese also took Chien Chao-nan's case to a potential ally, T'ang Chi-yao, the military governor of Yunnan province who had expanded his sphere of influence in Southwest China without interference from Peking. "Our government has been bribed by thieves to sabotage merchants and the people and cause their downfall," the Overseas Chinese from Singapore complained to T'ang. They expressed the belief that Nanyang's challenge to BAT's domination of the market had prompted the foreign firm to plot against its Chinese rival. They charged that foreigners had spent "millions of *yuan*" bribing officials in the Ministry of Agriculture and Commerce and politicians in parliament—corrupt men who accommodated the foreigners because they were "traitorous thieves" like their cronies in the Anfu Clique, a group of warlords notorious for their cordial relations with foreign capitalists. Through these maneuvers, said the Overseas Chinese, the foreigners were responsible for the unlawful cancellation of Chien Chao-nan's citizenship and Nanyang's registration. To make amends, the proper action would be to punish and to censure officials in the Ministry of Agriculture and Commerce. Some Overseas Chinese pleaded with T'ang Chi-yao to protest the decision in order to protect Overseas Chinese and to avoid losing their investments in China. Others alluded to the Sung patriot Yueh Fei's efforts to recover North China from the barbarians and urged T'ang to take up their cause in order to "save the country," assuring him that once all Overseas Chinese became aware of this misconduct in Peking, their hatred for the Northerners in power would unite them behind him.[40]

Judging from the timing, the combination of these protests from Overseas Chinese and other pressures exerted on Nanyang's behalf probably hastened, if it did not directly bring about, action by the Peking government on Chien Chao-nan's request for repatriation (although the Ministry of the Interior claimed to have withheld its decision on Chien Chao-nan's citizenship merely because the Japanese consul had failed to supply confirmation of the termination of Chien Chao-nan's Japanese citizenship). On September 13, 1919, the Ministry of the Interior finally restored Chien Chao-nan's Chinese citizenship and ordered the Ministry of Agriculture and Commerce to reinstate Nanyang's license.[41] Five months after the May Fourth incident and one month after the cancellation of Nanyang's license the Chiens had regained legal status as a Chinese company owned entirely by Chinese stockholders.

The Chiens welcomed the government's decision, but for them legal exoneration was not enough. They felt the need to be judged favorably not only by Chinese legal authorities but also by Chinese society. To achieve this, they appealed to the group in Peking that had instigated the May Fourth incident and shaped popular attitudes more profoundly than any other during the spring of 1919, the student leaders at Peking National University (Peita). The Chiens did not simply try to buy these students off—an approach that dedicated May Fourth idealists would have found crass and insulting. Instead Chien Chao-nan devised and invested in educational programs designed to attract the support of the academic community in general and student leaders of the May Fourth movement in particular.

Chien Chao-nan appealed to these students and persuaded them to become associated with his company primarily through individual Nanyang scholarships, lucrative grants which covered all expenses for long-term study at the best universities of Europe and the United States.[42] In 1919 the Chiens awarded all of the first five of these scholarships to students who had become famous as leaders of the May Fourth movement at Peita: Lo Chia-lun, who had drafted the May Fourth Manifesto on May 4, 1919; K'ang Pai-ch'ing, who had been among the first students elected to Peita's student union at its founding on May 5; Tuan Hsi-p'eng, who had been elected chairman of the student union of all China at the height of the movement a few weeks later; Meng Shou-ch'un, who in early 1919 had helped found *New Tide* (*Hsin-ch'ao*), an influential monthly magazine which prepared the way for May Fourth; and Wang I-hsi, who, like all these first winners, later became a distinguished academician.[43] In this way, the Chiens linked the name of their company to the young people who had become famous as leaders of the May Fourth movement. Over the next four years, forty other scholarships were awarded (ten per year between 1919 and 1923), and the winners all attended Western universities entirely at Nanyang's expense (or, more precisely, with Nanyang covering two thirds of the costs and Chien Chao-nan himself covering the remaining third).[44] In the process, Nanyang's scholarship program soon became widely known and respected. Even Liang Ch'i-ch'ao, the brilliant scholar, journalist, and political figure, then in his mid-forties, seriously contemplated soliciting Chien Chao-nan for one of the scholarships. Presumably because Nanyang's program was so successful, BAT imitated it in the mid-1920s, conducting examinations in South China and awarding the winners scholarships for study at Hong Kong University.[45]

Though publicly presenting themselves as patrons of scholarships, the Nanyang directors revealed in private discussions among themselves that they did not donate the money purely out of a belief in the virtues of higher education. When they first agreed to set aside ￥200,000 for educational purposes in 1919, they did so with the stipulation that it be used "to bring the best return on [the stockholders'] money [and] to advance the reputation of the company as a promoter of education."[46] Similarly, after weighing the question of whether to allocate more for individual scholarships or for universities, they decided in favor of scholarships on the basis of business considerations and without mentioning China's educational needs. The board of directors preferred to spend the money on scholarships because these "brought effective results faster than donations to universities and thus enhanced the company's reputation more."[47] In other words, the student scholarships program was developed not merely for academic or philanthropic reasons but to coopt the names and reputations of young May Fourth leaders for advertising purposes and to turn May Fourth sentiment to Nanyang's commercial advantage.

Thus in Peking as in Canton, Shanghai, and other cities, the Chiens came under fire for their Japanese connections, and they retaliated by publicizing their own patriotism. Around Peking as in most of China's major markets, despite the Chiens' heavy expenditures on advertising, Nanyang's sales continued to fall. For the Chiens, the nadir came in the autumn of 1919 after the Peking government canceled their license. At that time newspapers reported that Nanyang's cigarettes virtually disappeared from the Yangtze River Valley and North China.[48]

The Ambiguity of Nationalistic Boycotts

The Chiens' painful experience with the May Fourth Movement in 1919 demonstrated that nationalistic boycotts did not invariably serve all owners of Chinese businesses at the expense of their foreign competitors. In this case, it had the opposite effect, benefiting BAT at the expense of Nanyang. Nor was Nanyang the only Chinese company victimized in this way. The May Fourth boycott dealt a severe blow, for example, to three of China's largest Chinese-owned department stores—the Sun (*Hsing-ta*) Company, the Sincere (*Ta-hsin hsien-shih*) Company, and the Chen Kwong (*Chen-kuang*) Company—and it drove at least ten Chinese companies in the Shanghai foreign concession area into bankruptcy. And on earlier occasions as well as in 1919, other Chinese businessmen besides the Chiens had felt driven by nationalistic boycotts and demonstrations to change the nationality of their stockholders and make their companies appear free of foreign capital.[49]

All of these examples suggest that for a Chinese businessman with foreign affiliations, a nationalistic boycott could turn into a double-edged sword. On the one hand, if he could convince Chinese consumers that he was untainted by foreign connections and more "Chinese" than his rivals, then a nationalistic boycott might help him break into a market that had been dominated by a foreign firm (as a boycott had helped Nanyang enter the market at Canton in 1915). On the other hand, if his own foreign affiliations made him seem sympathetic to the country being boycotted, then nationalistic sentiment might be turned against him in commercial campaigns led, for example, by an influential Chinese employee of a foreign company (like Chiang K'ung-yin in Canton), an ambitious Chinese manager of a rival Chinese firm (like Ch'en Ts'ai-pao in Shanghai) or a zealous—and corrupt?—Chinese politician (like Ho Hsun-yeh or Chou Wei-fan in Peking). When Chinese businessmen from competing firms within a highly competitive consumer industry clashed, the one whose commercial campaigns more effectively publicized the patriotism of one or exposed the hypocrisy of the other might well enable one to gain ground at the expense of his rival. But if a Chinese businessman's connections were with a country that had become highly unpopular, then even if he recognized the need to project a favorable image of himself and willingly spent large sums on publicity to counteract his critics (as Chien Chao-nan did in 1919), he might still be unable to prevent consumers' attitudes from shifting against him, tarnishing his reputation, and reducing his company's sales.

By the end of 1919, if not before, the Chiens had learned about the ambiguities of nationalistic boycotts. After benefiting generally from boycotts prior to 1919 and suffering from the one in 1919, they became well aware that a boycott could hinder as well as help their business. By carrying out elaborate and expensive countercampaigns, they at least salvaged enough of their sales and reputation during the May Fourth boycott so that they were not forced to close down (as they had been forced to do during business slumps a decade earlier), but for them 1919 was unquestionably a long, agonizing year. Not until early 1920, after the May Fourth commercial campaigns ceased to affect their business, did Nanyang's sales finally begin to grow again.

6.

The Postwar Golden Age

IN DISCUSSING the development of Chinese-owned industry during the years following World War I, historians have generally compared it with the preceding wartime years, and the postwar period has almost invariably suffered by comparison. The wartime period, it is said, was a golden age (*huang-chin shih-tai*) during which Chinese-owned industry flourished while its Western competition was preoccupied in Europe. The postwar period, by contrast, marked the beginning of dark ages for Chinese-owned industry as Western firms returned to the market and cut off the growth of their Chinese-owned rivals from 1920 on. This golden age thesis is useful as a characterization of the period during World War I, for many Chinese-owned firms (including Nanyang) developed rapidly in the mid- and late 1910s. But was their growth suddenly stunted during the postwar period because of the resumption of Sino-foreign rivalries? In some industries this might have been the case, but not in the cigarette industry, for Nanyang's sales rose by 27 percent (from ¥25 million to almost ¥32 million) between 1920 and 1923, and its profits reached the highest level in its history, averaging more than ¥4 million annually during these years.[1]

Why did Nanyang prosper when Chinese industry was supposedly suffering? One explanation may be that Nanyang was not atypical and that Chinese industry suffered less in these years than historians have generally thought. Recent studies by Albert Feuerwerker and Marie-Claire Bergère suggest that Chinese-owned textile manufacturers and other industrial enterprises prospered in the early 1920s as Nanyang did. The golden age extended into the postwar period, according to Feuerwerker and Bergère, because a recession in the West prevented Western companies from resuming full trading operations in China and because the restoration of transoceanic shipping after 1919 permitted Chinese industrialists to import capital goods and technology from abroad that had been lacking in China during the war.[2] This revised version of the golden age chronology, though more persuasive than the original thesis, still fails to account for Nanyang's growth because Nanyang did not benefit from any curtailment of foreign competition in China during the postwar years. On the contrary, as this chapter will show, BAT grew even faster than Nanyang did, raising sales by at least 70 percent between 1920 and 1923 and making more than twice as much in annual profits as it had during the war.

If neither the original nor the revised version of the golden age thesis fully accounts for the rise in Nanyang's and BAT's sales and profits during the early 1920s, then how is the postwar growth of the firms to be explained? To provide an explanation, it is necessary to shift the focus away from the impact on China of events abroad (such as World War I and its consequences, which are emphasized by advocates of the golden age thesis)[3] and concentrate instead on developments in the history of the cigarette industry within China: negotiations between BAT and the British and Chinese governments; expansion of BAT's and Nanyang's operations; Nanyang's adoption of specialized professional management; and the heightening of competition between the two firms. By examining each of these developments, it is possible to suggest why the postwar period was a golden age not only for the Chinese big business in the cigarette industry but for the Western one as well.

BAT's Negotiations with Governments

The end to World War I might have appeared to open the way for Western firms to export new capital goods and technology to China, but James Duke and other BAT executives complained about two political obstacles that militated against investing there: Americans in BAT objected to a new policy of the British government requiring that a British subject be appointed managing director of every British-licensed company in China, and all BAT's Western executives were concerned about increases in likin and other local taxes in China. As BAT expanded in the postwar period, its management negotiated with the British and Chinese governments to have these policies changed.

James Duke was particularly disturbed by the British policy because it threatened to break a critical link in the American chain of command he had always maintained between himself and BAT's branch in China. As Duke was well aware, the British order in council, promulgated in 1919, applied to BAT because the company had held a British license since its founding in 1902.[4] But Duke had always appointed an American to serve as managing director of the China branch, C. E. Fiske from 1902 until 1905 and James Thomas since 1905. And in the early 1920s, as always,

Duke was determined to retain an American in that post, for only then could he be confident that it would be held by a close associate of his who would follow his orders, emulate his example, and remain unwaveringly loyal to him as Thomas had.

Duke doggedly fought the order in council, but he found the British government determined to enforce it. First, he complained bitterly to the British ambassador in the United States, but to no avail. Next, he tried to circumvent the order by changing the legal status of BAT's operations in China from a branch to a subsidiary; accordingly, what had been a BAT branch since 1902 became incorporated in Shanghai as the British-American Tobacco Company (China), Limited, in 1919. The parent company retained 97.5 percent of the stock in this new subsidiary (and seven individuals whose names and nationalities were not disclosed held the remaining shares), but legally the operation in China ceased to be a branch of BAT.[5] As an almost wholly-owned subsidiary, BAT (China), Ltd., was probably set up not only to help Duke get around the order in council but also to permit the parent company to escape taxation in London.[6] Unfortunately for Duke, this maneuver didn't work either, for the British government maintained that the order in council applied to BAT whether its operations in China constituted a branch or a subsidiary. In either case, according to British officials, BAT had to appoint a British subject to serve as its managing director in China.

Unable to persuade the British government that BAT should be exempted from the order, Duke then went so far as to move his headquarters for China from Shanghai to Hong Kong, a British colony outside China's boundaries, rather than replace his American managing director with a British one.[7] If forced to withdraw BAT's headquarters from China, it is not surprising that Duke chose to go from Shanghai to Hong Kong, for, as Paul Cohen has pointed out, the two places had so much in common during the late nineteenth and early twentieth centuries that "physical movement between them was natural" along a kind of "Hong Kong-Shanghai corridor."[8] Nonetheless, the 850 mile move down the coast caused the American managing director of BAT's subsidiary considerable inconvenience, for it prevented him from supervising the company's operations within China as he and his predecessors had in the past. There is evidence that the acting managing director in 1920, Thomas Cobbs, an American, dared to violate the order in council by venturing from Hong Kong into China at least once, but he wrote to James Thomas at the time that he did so "against advice of some of our Legal friends."[9]

After stationing the American managing director of BAT (China), Ltd., outside China for three years, Duke and other Americans in BAT's management finally agreed to compromise with the British government in 1922. They were still unwilling to subordinate the American managing director to a British one, but they restructured BAT's management in China so that there was no longer one man at the top. Instead, ultimate authority over BAT's operations in China was vested in six or seven directors—some of them British, some of them American, all of them stationed at the subsidiary's Shanghai headquarters, and no single one of them possessing authority greater than the others.[10]

Though the Americans in BAT might have felt that the reorganization went too far in depriving them of authority, BAT's management was fortunate that the British government considered it to be sufficient; after all, the company still had not consented to appoint a British managing director as the order in council stipulated. But BAT succeeded in convincing British officials in Shanghai to accept its solution in 1922 largely because of the persuasive powers of one of its British directors, Archibald Rose. A British diplomat in China for two decades before joining BAT, Rose frequently presented BAT's case to his former colleagues in the diplomatic corps in the 1920s; and in the spring of 1922, it was Rose (as one of the American BAT directors wrote to James Thomas) who succeeded in finally "getting the Order in Council squashed."[11]

From the standpoint of the company's history in China, the most important consequence of these negotiations was that BAT's postwar expansion was carried out by a top management that was less American and more British than it had ever been before. James Thomas, James Duke, and other Americans were loathe to relinquish American authority in China, but in the early 1920s they had to face up to their loss of it. Perhaps because he recognized that he would never again have the degree of authority he had once held, James Thomas went on leave from BAT just after the order in council went into effect in 1920, and he resigned from BAT altogether just after it replaced the position of American managing director with an Anglo-American board of directors in 1923.[12] Unlike Thomas, James Duke remained active in BAT throughout most of the postwar period, but he too resigned because of growing British influence within the company in 1923.

Though Thomas, Duke, and other American BAT executives suffered from the British order in council, their suffering does not seem to have impaired the effectiveness of BAT's management in China during the postwar period. Whether BAT's American managing director resided in or outside China, BAT's managerial staff within the country not only continued to conduct business as usual but also took new initiatives to raise sales and profits for the company. Of these initiatives, one of the most important was a lobbying effort to persuade the Peking government to renegotiate cigarette taxes.

BAT's management and the Peking government sought to reach a settlement in the postwar period that would end the company's disputes with the Peking government's Wine and Tobacco Monopoly. Since the founding of the monopoly, BAT's management had repeatedly come into conflict with it. Between 1915 and 1917, for example, BAT had refused to pay any taxes (besides import and transshipping duties) until permitted to help administer the collection of tobacco taxes. Denied this privilege, James Thomas, Wu T'ing-sheng, and others in BAT had begun lobbying for the appointment of an American adviser in the monopoly, undoubtedly hoping to exert indirect control over the levying of tobacco taxes.[13] In 1919 they had succeeded in persuading the monopoly to award the post not to a Japanese (as Chinese officials in the monopoly threatened to do) but to a former member of the American consular service in China, C. L. L. Williams. Ultimately, BAT's management had been disappointed in Williams, for he had little power and even less inclination to use it on behalf of the company.[14] Nonetheless, BAT's success at convincing the government to appoint an American indicates that it had possessed a measure of strength as a lobbyist during World War I.

After the war BAT continued to resist taxation by bringing suit against the government. In 1919 and 1920, for example, when Chinese officials in the Wine and Tobacco Monopoly charged the company with tax evasion and impounded its goods, BAT's lawyers marshaled support from the British and American governments and forced the monopoly to back down.[15]

After demonstrating its effectiveness as a lobbyist and legal adversary in instances such as these, BAT's management was in a strong bargaining position in the postwar period. Chinese officials admitted that they were disappointed in the monopoly's total revenue between 1915 and 1921, which had totaled about $48 million per year, only one third as much as they had anticipated and one thirtieth of that obtained from wine and tobacco taxes collected at the time in the United States.[16] To raise more revenue, they were prepared to make extraordinary concessions to BAT.

The concession the Peking government agreed to grant BAT in exchange for tax payment was one the company (and other Western businesses in China) had long sought: exemption from likin and other local taxes. On August 3, 1921, BAT and the Peking government signed a private agreement granting BAT this exemption and giving the government, in return, assurance from BAT that it would pay a factory tax of ¥2 on every case of 50,000 cigarettes it manufactured in China (in addition to paying the usual import duties at 5 percent of value on goods shipped from abroad and transit taxes at 2.5 percent on goods transshipped from one Chinese port to another).[17]

When the agreement between BAT and the Peking government was put into effect, Chinese radicals strenuously objected to the government's willingness to exempt BAT from paying likin and other local taxes. To the critics, such a concession represented a sellout to foreigners at the expense of the provinces and the Chinese people. One critic of the agreement was none other than Mao Tse-tung, who denounced it in "The Cigarette Tax," one of his first published essays. Using the earthy language that eventually became one of his hallmarks, Mao wrote at the time: "If one of our foreign masters farts, it is a lovely perfume [for the Peking government] . . . [If] . . . our foreign masters [in BAT] want to bring in cigarettes, the Council of Ministers [in Peking] thereupon 'instructs the several provinces to stop levying taxes on cigarettes.' Again, I ask my 400 million brethren to ponder a little: Isn't it true that the Chinese Government is a counting-house of our foreign masters?"[18]

Though vivid, protests by Mao and others did not prevent the government's agreement with BAT from going into effect and yielding considerable benefits to the Western company during the postwar golden age. As the government and BAT realistically anticipated, the agreement was unenforceable in several southern provinces—Kwangtung, Kwangsi, Kweichow, Yunnan, and Hunan—where Peking admitted that it exerted no authority and where Sun Yat-sen's provisional government and various warlords held administrative control at the regional and local levels; tax passes issued in Peking, BAT and the Peking government acknowledged, would not "be duly respected in those provinces." But elsewhere in China they expected the passes to be honored, and in the early 1920s, their expectations were almost completely fulfilled.[19] Partially compensated by Peking, the provinces cooperated to the full satisfaction of BAT and did not begin to rebel against BAT's exemption from likin until 1923. Only then, near the end of the postwar golden age, did provincial governments, led by Chekiang, unilaterally cancel the arrangement, overcome legal protests from Peking and BAT, and reclaim their right to tax BAT at the provincial level. And in the mid-1920s, even after provincial taxes went back into effect, BAT's total tax burden remained so much lighter in China than in the United States that the company was able to distribute the same cigarette in both countries, charge a price 40 percent lower for it in China than in the United States, and still make higher profits on the one sold in China.[20]

Thus, BAT's management wrested concessions from the Peking government and improved its fiscal arrangements in China despite the restrictions that the British government had placed on its American managing director. As a result, though likin and other local taxes might have proliferated in postwar China,[21] BAT's goods were free from these taxes in most of China's provinces. The success of BAT's management at winning fiscal concessions from the Peking government partially explains why the company's sales and profits grew during the postwar period. But perhaps a more important explanation for the company's postwar golden age lies in the expansion of its operations.

Expansion of BAT's Sino-Western Operations

BAT's expansion during the postwar period was built upon the foundation it had laid earlier in China. Between its founding in 1902 and the early 1920s it had never stopped growing. Even when World War I had curtailed Western shipping and had supposedly diverted the attention of Western companies away from East Asia, BAT had steadily expanded its operations in China. For example, BAT had built a large factory in Shanghai in 1917—its second in that city—and had raised its total production in China from 7.3 billion cigarettes in 1916 to 16.5 billion in 1919. It had promoted the cultivation of bright tobacco with extraordinary success during the war, and by 1919 the yield of flue-cured bright tobacco had amounted to 45.5 million pounds, 90 times more than in 1915. And it had boosted annual sales to approximately 22.5 billion cigarettes in the last years of the war, allowing Nanyang no wartime respite from competition.[22]

Yet, the postwar golden age was more than a continuation of this pattern of sustained growth, for BAT's postwar policies regarding investment and recruitment sharply accelerated its rate of growth. These policies were similar to those used by James Duke and James Thomas during the company's first years in China. As in the first decade of the twentieth century, BAT's management exported new personnel, capital, and technology from the West, and once again it recruited Chinese to staff its distributing, advertising, manufacturing, and purchasing systems and to make them work on a large scale.

Large amounts of new Western capital, personnel, and technology were assigned to China. In fact, the capital of BAT (China), Ltd., was so large in the postwar period that it seems unlikely all of it could possibly have been available for investment in China at the time. The total was $231 million—fourteen times more than the paid-up capital of BAT's China branch during the war, and 3.7 times more than the total paid-up capital of BAT's London headquarters in 1919. However much of this capital was available in China, BAT (China), Ltd., invested large sums in its distributing, manufacturing, advertising, and purchasing systems. To supervise distribution, it added new Western salesmen, bringing its total in China to between five and six hundred. To raise production, it built a large new cigarette factory in Tientsin between 1921 and 1922, the company's first in North China. To improve tobacco purchasing, it paid ¥1.5 million for redrying machinery, warehouses, and loading docks within its principal reception stations at Wei-hsien in Shantung province plus additional unknown sums for new reception stations it built at Hsuchang in Honan province and at Fengyang in Anhwei. To enhance its advertising, it enlarged its printworks to include (in the judgment of journalist Carl Crow) "the largest color-printing plant in the world and one of the finest."[23] And to strengthen its advertising still further, it invested $150,000 in imported photographic equipment which gave BAT, according to its English cameraman William H. Jansen, "one of the most modern and best motion picture studios that at the time could be found anywhere in the world outside the United States."[24]

These investments represented substantial additions to BAT's physical plant in China, but still more impressive was the expansion of Chinese participation in the company's operations. In recruiting Chinese as in making investments, BAT's management followed precedents it had established earlier in China, but it relied on more Chinese and sought to integrate them more closely into the company's marketing, advertising, manufacturing, and purchasing systems than it had before.

In marketing, for example, BAT's management extended its reach and tightened its hold on distribution by investing heavily in Cheng Po-chao's Yung-t'ai-ho Company. Previously (as suggested in Chapter 2), Cheng had been more of a "capitalist" than a "comprador" in the sense that he had had greater financial independence than most compradors, but during the postwar period BAT acquired a controlling interest in his company by buying 51 percent of its stock in 1921.[25] By thus drawing him more fully into BAT's financial and administrative structure, BAT's management attempted to foreclose the possibility that he might break away from the foreign firm and go independent (as Chinese compradors had been known to do in the nineteenth century). Such a strategy proved to be effective, for Cheng never did go independent or compete with BAT. When he showed an interest in inaugurating his own enterprise in the early 1920s, he was permitted to open a small cigar factory under the name of his marketing agency, the Yung-t'ai-ho (Wing Tai Vo) Tobacco Corporation, Ltd., which employed 220 workers and was capitalized at ¥100,000, with BAT maintaining control over it by retaining a majority of its stock. But when Cheng wished to go one step farther in 1923 by investing in his own cigarette factory in Chungking, he succeeded only in persuading his old friend James Thomas (by then retired from BAT) to recommend the scheme to the company—not in winning the support of BAT's management. So, rather than defy BAT's management, Cheng refrained from establishing a new cigarette factory, continued to market BAT's cigarettes, and—now more a "comprador" than a "capitalist"— proceeded to accumulate a fortune worth between three and five million American dollars in the 1920s.[26] Thereafter (and throughout the history of BAT) neither Cheng nor any other Chinese BAT comprador or agent ever broke away to establish an independent cigarette company in competition with the Western one. If Cheng's case is any indication, the ones who contemplated such a venture had sufficient capital to try it but did not go through with it because they lacked the necessary experience at industrial management (as Cheng surely did) or because they were reluctant to give up the income and financial backing that they received as a result of their affiliations with BAT.

BAT made new investments in other Chinese distributing operations which were intended to bind them more closely to the company at about the same time. In 1920, for example, when two agents who had risen to the top of BAT's regional office in North China, Ts'ui Tsun-san and Kung Ho-ch'ien, showed initiative in forming a regional distributing agency known as the San-ho Tobacco Company, BAT gave them exclusive rights to distribute two popular brands, Rooster and Kingfisher, in southern Chihli and Shansi provinces. To reward Ts'ui and Kung for launching the new venture and to deepen their loyalty to the parent company, BAT sent the two men on a six-month expense-paid tour of BAT's headquarters in the United States and England. Upon their return, Ts'ui and Kung set up a business that was profitable for themselves as well as for the company.[27]

BAT seems to have invested most heavily in Chinese holding top positions, such as Cheng, Ts'ui, and Kung, but, unlike most businesses in China at the time, it also helped finance its smaller dealers by permitting them to take goods on credit for as long as thirty, sixty, or ninety days. According to the journalist Carl Crow, BAT's Chinese dealers commonly used but rarely abused this access to credit because they "made so much money that they were keen to keep their credit rating with the company."[28] Without some idea of the volume of business at each level of the distributing system, it is difficult to estimate the size of profits that were earned, but a contract Cheng Po-chao signed with BAT in 1919 specified the following profit rates: Cheng took responsibility for the cigarettes from BAT at a 2 percent profit to himself; he then worked through the Chinese directors of his sales divisions in twelve regions of China, who each made a 1 percent profit; they, in turn, conveyed the goods to "all villages, towns and districts" where they had connections with wholesalers; and finally the cigarettes were delivered to the retail market at a 4.5 percent profit, which was divided among the merchants, dealers, subdealers, and all others involved in handling cigarettes at the local level.[29] At the bottom of the distributing system, subdealers, shopkeepers, and hawkers often worked on a minute scale and earned far less profits than major Chinese distributors; their incomes were so low that a Chinese-speaking American BAT representative who surveyed local distribution of BAT's goods in several regions characterized them as "living from hand to mouth."[30]

Tile store front with BAT cigarettes on display in Hankow, 1926

Within this marketing hierarchy, BAT's Chinese agents distributed enormous quantities of cigarettes during the postwar golden age. Sales of the company's Ruby Queen brand, for example, which were handled exclusively by Cheng Po-chao, shot up during these years, and it became the second most popular cigarette in the world. Almost twice as many Ruby Queen (which was made in the United States and England as well as in China) and other BAT cigarettes were imported annually than in the last years of the war, an average of nearly 8.5 billion per year between 1920 and 1923.[31] If conservatively estimating that a new plant built by the company at Tientsin in the early 1920s added 6 billion to the 16.5 billion annually produced by existing factories in China, then it seems probable that the company's total sales of all brands of domestically produced and imported cigarettes amounted to about 31 billion per year during the early 1920s—an increase of 72.5 percent over the volume of BAT's sales in 1919.[32] For the year 1924, BAT's sales in China were valued at $57.9 million—a 140 percent increase since 1916 which suggests that the value of BAT's sales grew by at least 70 percent (which would be proportionate to the increase in the volume of the company's sales) during the period between 1920 and 1923. As BAT's sales rose, so did its profits, bringing the company ￥18 million (US $9.1 million) in gross profits and ￥15 million (US $7.6 million) in net profits during the first year after the war, 1920—more than twice its profits in 1916.[33]

Single hawker selling BAT cigarettes in Fan-ch'eng, 1906

Reviewing this sales record in 1923, one of BAT's Western directors remarked to James Thomas, "We have only demonstrated that it is possible to organize an efficient selling force of Natives."[34] Though this comment is misleading, not to mention patronizing, it contains a grain of truth. It is misleading in that BAT did not "organize" Chinese or make them into "efficient" salesmen so much as it delegated authority to Chinese, gave them financial backing, and left the organization of the distributing system to them. Nonetheless, the BAT director was right to attribute the company's success to the "natives"—the salaried Chinese compradors and nonsalaried Chinese agents and jobbers and dealers and merchants and hawkers at all levels of the marketing hierarchy throughout the country.

In sum, Chinese, not Westerners, continued, as in the past, to manage the markets and perform the functions of jobbing and retailing that made BAT's distributing system effective. At the highest levels, the company's management extended its reach more deeply into its selling organization by buying a controlling interest in Cheng's distributing agency and by furnishing credit to Chinese agents, but for wholesale marketing it continued to rely largely upon independent, nonsalaried, Chinese agents.

In advertising, as in distributing, BAT's management relied on Chinese to reach its customers. For example, BAT commissioned Hsi Ying, a painter known for portraits of Shanghai women, to do original designs, and it opened a school to train Chinese workers in lithography so that they could use the printing machinery BAT had imported during the postwar years. When the reams of colorful posters, scrolls, calendars, hangings, handbills, and cigarette cards came rolling off BAT's new presses, the company relied, as always, on Chinese to circulate these advertisements throughout the country.[35]

Such advertising matter was posted not only in treaty port cities but also in other cities and even small towns in these years. It reached the small, inland walled town in the Lower Yangtze region where Pearl Buck lived at the time, for example, and left commercial marks which she deeply resented—"Huge cigarette-posters plastering those dark and ancient bricks and defiling their sombre age with crass and glaring colors."[36] And similar BAT posters went up where Wang Li, a famous Chinese philologist, grew up, even though, as he later recalled, it was only "an out-of-the-way, small market town" (*hsiao ch'eng-shih*) in South China.[37]

Besides plastering posters on walls, BAT used billboards, which it erected at wharves, railroad stations, highway junctions, and other transportation nexuses. Unlike most other firms in China, BAT had its own billboards. So, while others rented space from English, French, Chinese, and Japanese advertising agencies and had difficulty coordinating advertising campaigns in one region or another because of nationalistic

jealousies (for example, Japanese advertising agencies in Manchuria refusing to rent to French companies), BAT controlled its own boards and
ran advertising campaigns throughout the country without difficulty.[38]

To make movies for advertising purposes, BAT's management also
recruited Chinese. The limitations of relying exclusively on its English
cameraman, William Jansen, were evident in the company's first film,
made in 1922. This was an animated cartoon which, in Jansen's words,
showed "a donkey wagging its ears and refusing to move until it smelled
the smoke of a cigarette its Chinese leader lit; the smoke spelling the
name of the cigarette in the air."[39] To make films that showed more
awareness of Chinese theatrical traditions and that used the Chinese language, BAT recruited a Chinese movie director, Kuan Hai-feng, and Chinese actors, actresses, and crews from Shanghai, Canton, and Hong
Kong.[40] The Chinese and Jansen then proceeded to use BAT's imported
cameras and other photographic equipment intensively, shooting more
footage than any other film maker in China by 1924—including, according to Jansen's inventory, "191 reels, 100 ft. of town topics, 59 one-reel
scenic pictures, 15 one-reel educationals, 6 two-reel comedies, and 4 fea-
ture length pictures."[41] The purpose of all these films, as Jansen readily
acknowledged, was the promotion of cigarettes sales, and with this purpose in mind, he saw to it that the names of various BAT cigarettes were
printed at the foot of every subtitle in every frame of every film. At some
theaters BAT eliminated any doubt about the connection between its
films and its product by selling packages of cigarettes at the door to serve
as the tickets for admission.[42]

To promote and maintain control over its films, BAT created a network of movie theaters through which to have them shown. It bought
some theaters outright and gained special access to others through its
cigarette distributor, Cheng Po-chao, who built and owned a large one
in Shanghai, and through James Thomas who, after officially retiring as
director of BAT in 1922, became associated with the Peacock Motion
Picture Corporation (K'ung-ch'iao tien-ying kung-ssu), a well-endowed
Sino-American firm which distributed movies in China.[43] In fact, according to a critique of BAT's motion picture operation by Chinese journalists in the mid-1920s, the company directly or indirectly controlled so
many of China's approximately 100 movie theaters at the time that it
determined the fate of the entire Chinese film industry because it was able
to dictate whether any film would or would not be widely shown.[44]

Thus, in advertising as in distributing, BAT recruited additional Chinese to staff its expanding operations but attempted at the same time to
maintain or tighten control over these aspects of the business. Whereas it
increased control over distributing by investing in Chinese distributing
agencies and financing local dealers, it maintained control over advertis-

"Pirate," a BAT brand, advertised in Tientsin, late 1910s

"It's Toasted" on a BAT billboard

ing by hiring its own artists, making its own movies, and creating its own networks of billboards and movie theaters. Using these techniques, BAT's management successfully integrated distributing and marketing more fully into the company's expanded organization during the postwar years, but when BAT's management tried a similar approach in manufacturing, it encountered resistance of a kind it did not face in other sectors of the business.

In manufacturing, as in distributing and advertising, BAT's management had no difficulty recruiting Chinese, but in this area of the business it failed to accommodate Chinese workers' most fundamental grievances. It had a total of 25,000 Chinese working in its factories during the postwar period—nearly twice as many as in 1915. To win support from these workers, BAT became one of the first businesses in China to provide certain services for its workers: dormitories, clinics, schools for the children of workers, a savings fund, a rice allowance. And by a comparative standard, BAT seems to have offered its workers better conditions on the job than did cotton and silk mills and other industrial plants in China at the time—hours were shorter, work lighter, and night shifts less frequent.[45]

Though undoubtedly of some benefit to workers, these measures did not eliminate the workers' two chief complaints: low wages and job insecurity. Hired on a contract basis and paid by the day, unskilled workers received between ¥0.20 and ¥0.50 per day in the 1910s and 1920s—low wages compared to the amounts paid workers in heavy industries or even in other light industries in China at the time.[46] If accused of insubordination or any infraction of the rules, they lost the cards that guaranteed their places on the assembly line and were abruptly fired the same day. It was difficult to regain cards (which were often obtainable only through family contacts or bribery of supervisors), and unskilled workers recall feeling obsessed with the fear of losing their jobs. Only the most highly skilled workers were paid by the month.[47]

Driven by their need for higher wages and job security, BAT workers vented their frustrations in strikes. They had gone on strike several times before 1919—generally for a few days at a time and in 1917 for as long as three weeks—but they had always returned to work without accomplishing their avowed objectives rather than run the risk of losing their jobs permanently. In perhaps the biggest BAT strike prior to the 1920s, 6,000 workers in Shanghai had struck in support of the May Fourth movement and had enjoyed the satisfaction of participating in patriotic demonstrations in June 1919 that came to a successful conclusion when the Chinese delegation did not sign the Treaty of Versailles, but this victory was purely political.[48] Not until the early 1920s did BAT's workers begin to win economic concessions from BAT's management.

BAT workers claimed their first economic victory over BAT's management at Shanghai in July 1921—the same month that the Chinese Communist Party was founded and the month that China's "first big wave of labor struggles" originated according to labor historian Jean Chesneaux. The strike was called, as earlier ones had been, because of a dispute between a factory supervisor and a group of workers, but this time in forming a strike organization, the workers received guidance from Li Ch'i-han, a militant Hunanese student, a labor organizer, and a member of the Chinese Communist Party. BAT workers later recalled that Li had first made contact with them through the Green Gang secret society (into which he had become accepted) and had soon become their "secret friend." With him as their adviser, they rented a meeting hall, elected representatives, and formulated demands calling for better wages, conditions, hours, and job protection for strike leaders who had been fired or feared being fired at the end of the strike.[49]

Thus organized, BAT workers refused to work until their demands were met. BAT's management initially took a firm stand, but after the strike continued for three weeks, William Morris, one of·BAT's American directors in Shanghai, grudgingly agreed to grant most of what they asked, including ￥1,800 in compensation for wages lost by workers while on strike. Gratified, strike leaders converted their provisional strike organization into a permanent labor union, the Tobacco Union (*Yen-ts'ao kung-hui*), in late 1921. Throughout the following year, the Tobacco Union met regularly in the British concession area in Shanghai near the offices of the Chinese Communist Labor Secretariat, led brief strikes whose outcomes its leaders considered victories, and strengthened its ranks by allying with the Pootung Textile Union (*P'u-tung fang-chih kung-hui*).[50]

In Hankow BAT's workers also founded a militant, Communist-backed union that succeeded in winning concessions from BAT's management. As in Shanghai, the workers' organization in Hankow originated during a strike, in this case a strike by 800 women workers in October 1922 against new packaging techniques which had reduced their incomes and against the abusive methods of foremen who had put the new procedures into effect. And, again as in Shanghai, the strikers became affiliated with a Communist-led organization, this time the Hupei Provincial Labor Federation. The Federation, led by a prominent Communist labor organizer who had been a delegate to China's first National Labor Congress earlier in 1922, Hsu Pai-hao, probably appealed to BAT workers because it had already shown its ability to fight for workers' interests in representing railroad and steel workers. With the help of the Federation, BAT workers formulated demands, and after two weeks on strike, sixteen workers' representatives and two union officials gained BAT's approval for the desired concessions, including BAT's promise that the foremen in question would be fired.[51]

Though yielding on these occasions, BAT's management was rid of the militant labor leaders by the end of 1922, for the new unions disappeared as rapidly as the broader wave of labor struggles receded. In June 1922, less than a year after the founding of the Tobacco Union in Shanghai, it was weakened by the loss of Li Ch'i-han, who was arrested during a police raid that closed the Chinese Communist Labor Secretariat (of which he was a member) and banned its publication, *The Labor Weekly* (*Lao-tung chou-k'an*). A few months later, in November 1922, the Tobacco Union was threatened more directly by the establishment of a rival, moderate union known as the Pootung Society for the Advancement of Morality in Industry (*P'u-tung kung-yeh chin-teh hui*). The Pootung Society came into existence during a three week strike by 4,000 workers against BAT, which the Tobacco Union started because it wished to show sympathy for striking textile workers and to seek better wages, hours, and vacations for BAT workers. The founder of the Pootung Society, Shao Ping-sheng, took as his motto, "A promise of rice for every bowl and oppression for no one" and guaranteed that those who joined his union would return to work sooner than if represented by the more militant Tobacco Union. Responding to his appeal, BAT workers followed Shao back to work at the end of the second week of the strike. As the rivalry between the two unions heightened, fights broke out in the streets around BAT factories between strikers and strikebreakers during which three women and two men were stabbed. Angered and frustrated by the new development that broke the strikers' solidarity and turned workers against each other, some members of the Tobacco Union marched to police headquarters and demanded that the chief of police arrest the strikebreakers. Others took matters into their own hands, broke into the luxurious home of BAT's chief comprador, and smashed the place to bits. This act of destruction cost the Tobacco Union dearly, for the police responded to it by ordering the arrest of the union's leaders, and BAT responded to it by declaring that forty-one of the union's members would never be reemployed and that all other workers who failed to return to work unconditionally by November 23, 1922, would be punished in the same way. Rather than permanently lose their jobs, most of BAT's workers, led by Shao Ping-sheng, complied with the BAT management's ultimatum, gave up their demands, and returned to the factories; 400 who did not return to work were fired, and the Tobacco Union's leaders went underground. Stripped of its leadership, the Tobacco Union then dissolved.[52]

BAT's management had to contend with the militant union in Hankow even more briefly than in Shanghai. On January 3, 1923, the union led a strike of about 3,000 workers to protest BAT management's failure to carry out the agreement signed a few months earlier when the union had first been founded. But this time BAT's management took a much stronger stand. Initially its directors refused to negotiate, protesting that fair bargaining was impossible because Chinese newspapers had poisoned the atmosphere with "distorted accounts" of the origins of the strike—particularly the newspaper *Ch'en-pao,* which had obvious ulterior motives, according to BAT, because its "Editor is connected with the cigarette firm of Nanyang Bros."[53] Not until two weeks later did BAT's directors agree to receive the union's leaders at all and then offered to comply only with the union's minimal demands. Disappointed but feeling pressure from workers to end the strike, the union leaders accepted BAT's terms, and their decision to return to work at the BAT factory in Hankow on January 8, 1923, marked the end of the first series of strikes that had produced victories for militant labor unions over BAT management.[54]

The role BAT's management played in subverting the militant unions is not entirely clear. Western BAT directors unquestionably contributed to the demise of the unions by bargaining more aggressively in 1922 and 1923 compared to 1921. According to the recollections of former BAT workers, however, BAT's management attacked the militant Shanghai union not only at the negotiating table but also through a conspiracy with the police and the leader of the moderate union, Shao Ping-sheng. The workers maintained that Archibald Rose, a BAT director with decades of earlier experience in China as a member of the British diplomatic corps, was at the center of the conspiracy. On the one hand, the workers said, Rose arranged for a local warlord to use his police force to intimidate the strikers and compel them to return to the factory, and, on the other hand, Rose recruited Shao to undercut the militant Tobacco Union by luring workers away from it and into the more moderate Pootung Society for the Advancement of Morality. Whether Rose manipulated warlords, police, and Shao in this way is difficult to prove, but he and other BAT directors indisputably tried to undermine militant unions by operating BAT's assembly line during strikes—using scabs (*kung-tsei*) as well as regular employees.[55] Whatever methods they may have used to maintain high production in these years, BAT's directors discovered that keeping factories running in Shanghai and Hankow was not so simple as in the past, for they could no longer take labor's cooperation for granted.

By contrast, BAT's management had less trouble getting what it wanted from China's countryside, for it was able to rely on Chinese to enable it to expand its tobacco purchasing system during the postwar period. During World War I, Westerners in BAT had been able to buy land for tobacco reception stations directly from the German administration, which controlled the land along the Kiaochow-Tsinan Railroad in the province where bright tobacco grew best, Shantung.[56] But after the Germans were defeated in World War I and had to withdraw from Shantung in 1919, then provisions in China's treaties with the West and Japan, prohibiting foreigners from buying land outside China's treaty ports, applied to this land. No longer legally eligible to buy land, Westerners in BAT delegated formal responsibility for land purchasing to the company's Chinese compradors.

The legal device through which BAT's management bought land in rural China in the name of its Chinese compradors was the Hung-an Land Company, formed in 1920. This company was ostensibly owned by nine Chinese, but in fact all nine were BAT compradors who merely acted as a blind for their Western employers. Although the true ownership of the Hung-an Land Company was an open secret, BAT compradors succeeded in arranging for it to be registered as a Chinese-owned company through personal contacts with government officials, which were provided by BAT's two most influential lobbyists in Peking, Wu T'ing-sheng, who had recently been made an honorary adviser to the Peking government's cabinet and an honorary member of its Commission for the Investigation of Financial Conditions, and Shen K'un-shan, a man characterized by Archibald Rose as "born into the world of the gentry and capable of wielding enormous social influence." Another BAT comprador, Jen Pai-yen, used equally illegal techniques to buy land in Honan province, first operating under the name of a fictitious Chinese family estate, the Yung-an t'ang, and later setting up the ostensibly Chinese-owned Hsuchang Tobacco Company as a front for BAT.[57]

Registered under the names of these Chinese dummy fronts, BAT induced peasants to convert land to the cultivation of bright tobacco with astonishing speed immediately after the war. The amount of land devoted to this crop increased from 39 acres in 1913 to 23,210 acres (153,500 *mou*) in 1920 and reached 30,300 acres (200,000 *mou*) by 1923 in Shantung province alone, not to mention the land converted in Honan, Anhwei, and other provinces.[58] At harvest time in the early 1920s, the landscape around Wei-hsien in Shantung, Hsuchang in Honan, and Feng-yang in Anhwei was dominated by the bright green plants and dotted with tens of thousands of recently built, small, white, thatch-roofed curing barns (*hung-fang*). The total yield of flue-cured bright tobacco in these three areas rose meteorically from 2.4 million pounds in 1916 to a peak of 54 million in 1920 and averaged more than 37 million pounds annually between 1919 and 1923.[59]

In the postwar years, the cultivation of bright tobacco spread so rapidly that BAT's friends as well as its enemies became alarmed. Its enemies complained that the new crop exhausted the fertility of the soil and displaced needed food crops, especially soybeans and grain. Officials in the American government expressed agreement with the latter charge, and even James Thomas admitted that the abandonment of food crops in favor of bright tobacco on such a large scale was deleterious to China's economy.[60]

Why did so many Chinese peasants begin to cultivate bright tobacco? The question has special significance because it has been cited and debated on both sides of a major controversy among historians and social scientists over the nature of the prewar Chinese peasant economy. One answer has been given by Ch'en Han-sheng (Chen Han Seng), a Chinese sociologist who has been identified as a leading advocate of "distribution theory" because of his contention that China's land and wealth became increasingly maldistributed and its peasants increasingly impoverished largely as a result of high rent extracted by landlords and usurious interest rates charged by moneylenders in the early twentieth century. A different answer has been suggested by Ramon Myers, an economist who has been identified with the "technologist" approach to China's agricultural history because he has denied Ch'en's assertion that sociopolitical relations caused acute peasant distress in the first half of the twentieth century, arguing that such misery occurred only during "severe climatic disturbances and wars;" and because he has concluded that the "key problem" for Chinese agriculture in this period was the absence of technological improvement.[61]

On the basis of field work conducted in North China in the mid-1930s, Ch'en Han-sheng presented his interpretation of why peasants cultivated bright tobacco in a monograph published in English, *Industrial Capital and Chinese Peasants: A Study in the Livelihood of Chinese Tobacco Cultivators* (1939). In it he blamed the introduction of American seed for contributing to the impoverishment of peasants. Peasants were initially induced to grow bright tobacco, he argued, not because it proved to be a more profitable crop in the long run but because they wished to take advantage of incentives offered by BAT—especially free seed, access to credit, and cash payments for their crops. Once the peasants began planting seeds, took out loans from Chinese BAT compradors and agents and from members of the local gentry, and made investments in expensive items such as fertilizer, iron pipes, and coal (the latter two to be used in curing sheds), then, according to Ch'en, they became irretrievably committed to growing the crop. When BAT ceased to subsidize the operation, they were caught in a vicious cost-price squeeze. As a result, peasants whom Ch'en interviewed in the mid-1930s complained that many among them had been reduced to desperation and some had been driven to suicide. Appalled by what he found, Ch'en concluded that bright tobacco growing had spread because of "the common and cooperative efforts of the foreign and Chinese capitalists, the compradors, and the local gentry" and that the participants in this "entente cordiale" had used their control over loans and prices to benefit themselves at the peasants' expense.[62]

Ch'en's explanation for why Chinese peasants grew bright tobacco has received special attention from Ramon Myers in his summary and critique of the exponents of distribution theory. Indeed, of the various bodies of data collected by distributionists on China's rural economy, Ch'en's research on tobacco cultivators is the only one Myers has directly confronted and explicitly questioned. Myers has accepted Ch'en's findings on the large investments that peasants made in tobacco growing, but he has drawn different inferences from the data. According to Myers, "High costs of production such as high land rents, interest rates, and fertilizer costs on tobacco land can also be interpreted to mean the peasant anticipated and earned a high income. Ch'en did not compare tobacco profits with that of other crops nor did he explain why so many peasants abandoned conventional crops to grow tobacco."[63] Myers thus implies that peasants chose to plant tobacco not merely because they were lured into it by BAT but because tobacco was more profitable than food crops. He does not deny that the peasants whom Ch'en interviewed were suffering, but he notes that *Industrial Capital and Chinese Peasants* (as well as Ch'en's other studies) was a "point in time analysis" based largely on data from a period of only two years (1933 and 1934); accordingly, it offers no conclusive evidence that tobacco-growing peasants suffered prior to 1933, leaving open the possibility that their suffering in the mid-1930s might have resulted from natural disasters, warfare, and the collapse of farm prices between 1930 and 1933 rather than prolonged exploitation by BAT agents, compradors, and other moneylenders throughout the 1920s and 1930s.

Whether "distribution theory" (exemplified by Ch'en) or the "technologist" approach (exemplified by Myers' critique of Ch'en) has greater validity is an historical question of immense importance which will not be settled here. But perhaps the specific point of disagreement concerning tobacco growing can be clarified: did peasants grow bright tobacco because it was more profitable in the long run than other crops?

The answer to this question—like many others over which "distributionists" and "technologists" have come into conflict—depends on whether labor costs are included in the computations. If labor costs are taken into account, then tobacco growing was less profitable than other crops. Contrary to Myers' allegation in the above quotation, Ch'en did compare tobacco profits with that of other crops and showed that in 1933 and 1934 growing tobacco was more profitable for some cultivators than growing wheat or kaoliang if labor costs were excluded but less profitable for all cultivators if labor costs were included. Moreover, other field research has corroborated Ch'en's conclusions. A careful survey of Weihsien by Presbyterian missionaries in the 1920s and a study of tobacco growing by Japanese from the research department of the South Manchuria Railway Company in the late 1930s both found that bright tobacco was less profitable than food crops (such as sweet potatoes, yellow soybeans, and millet) whether profits were measured by net profit per acre or net profit per man-work unit—as long as all labor costs were included. Only if the expense of family labor was not included among the costs of cultivating bright tobacco—a labor-intensive crop—was it as profitable as or more profitable than other crops.[64]

Thus, though Ch'en's own evidence, as Myers has observed, was largely based on research from a very limited time period, his findings and conclusions concerning the comparative unprofitability of bright tobacco growing have been substantiated by surveys conducted earlier and later than his. With the exception of the 1910s when BAT subsidized bright tobacco growing (a period for which data on its profitability are not available), it was never more profitable than food crops for peasants unless their labor costs are ignored.[65]

If comparatively unprofitable, why did peasants continue to grow it? They might, as Ch'en argues, have been locked into this crop because of their investments in equipment and their debts to BAT agents, compradors, and other creditors (like the notorious landlord and moneylender T'ien Chün-ch'uan, who reportedly took advantage of his position as a BAT agent to gain control over the markets for fertilizer, coal, iron pipes, and other items needed to grow and cure tobacco and amassed a fortune by requiring peasants to pay high rents and interest rates in Shantung during the 1920s and 1930s).[66] Or, as Myers implies, peasants might have discounted all or part of their labor costs (especially for labor supplied by family members) and might have perceived bright tobacco as the crop that brought them the highest possible income. Or, as the Japanese study of bright tobacco growing in the late 1930s suggests, peasants might have taken labor costs into account and continued to anticipate year after year that the ideal soil and climate, the nearness of railroads, and above all, the availability of cheap surplus labor would make bright tobacco more profitable than other crops.[67] If the last, then their expectations were not fulfilled.

Whatever their reasons for cultivating bright tobacco, peasants produced large amounts of it that contributed directly to BAT's postwar golden age by permitting the company to cut costs on raw materials. In the postwar period BAT not only paid far lower prices to Chinese peasants than for American imports—an average of $0.08 for Chinese-grown flue-cured bright tobacco compared to $0.44 per pound for imports from the United States—but also saved on transportation costs and lowered the danger of spoilage en route to its factories in China.[68] Well aware of the comparative advantages of using Chinese-grown leaf, BAT's directors were delighted to buy large quantities of it, all of which were used in China. After the company purchased 50 million out of the 54 million pounds marketed in China in 1920 at less than $0.10 a pound (while the price of imported American tobacco was about $0.43), a BAT director acknowledged the importance of Chinese tobacco growing to the company at the time. "This is very cheap," he exulted, "and will be our salvation in China."[69] Throughout the postwar golden age, BAT continued to import bright tobacco from the United States—where it had a direct purchasing system in Virginia and North Carolina with its own buyers, warehouses, and enormous stemmeries and handling plants—but in these years the company appears to have relied more on Chinese- than American-grown flue-cured bright tobacco, for from 1918 throughout the early 1920s, the amounts of this kind of tobacco grown domestically in China consistently surpassed the amounts imported.[70]

In summary BAT grew rapidly and enjoyed a postwar golden age because of its management's investments and efforts at integration and because of the performance of Chinese in all aspects of the business—distributing, advertising, manufacturing, and purchasing. Its management's investments made it one of the financially strongest and technologically best endowed companies in China at the time. And its management's efforts at integration not only coordinated its salaried Western executives, technicians, and salesmen but even tightened its hold on Chinese through a variety of arrangements: investments in Chinese cigarette distributors' agencies, loans to local Chinese cigarette dealers, salaries and career opportunities for Chinese movie directors and actors, concessions to Chinese strikers, protection for Chinese strikebreakers, and rural credit to Chinese tobacco growers. These innovations and BAT's successes in negotiations with the Peking government raised the company's sales and profits during the postwar period and permitted it to enjoy a golden age.

Richer, bigger, more fully integrated than during World War I, BAT became an even more formidable opponent for Nanyang in the early 1920s than it had been in the past. How could the smaller Chinese company hope to cope with its remarkably resilient and ever-growing Western rival? In the aftermath of the losses that Nanyang sustained during the May Fourth movement, Chien Chao-nan once again raised this question with his family.

Near Merger and Intensified Rivalry

The relentless growth of BAT during World War I and the effects of the May Fourth protests on Nanyang in 1919 left Chien Chao-nan more convinced than ever that his company would be driven to the wall and that he would be ruined unless he could end the competition by arranging a merger with BAT. Accordingly, at the end of 1919, he sailed off to the United States in hopes of renewing discussions with BAT's top manage-

ment about a possible merger. Upon arrival in New York, he first arranged for a Chinese student from the Massachusetts Institute of Technology, Chung K'o-ch'eng, to act as his interpreter and then approached
James Duke and the other American tycoons who directed BAT's world-
wide operations. Speaking through Chung, Chien Chao-nan urged his
American hosts to bear in mind the potential benefits of a BAT-Nanyang
merger, but in early 1920 he was still in no position to conduct serious
negotiations. As George G. Allen, Duke's closest associate and adviser,
remarked at the time, "It was the same old story with him: that he could
not do anything until he went back and talked it over with his brothers
and associates."[71] Nonetheless, the several personal calls he paid BAT
directors in New York and the banquet he gave for them on this visit
marked resumption of talks between Nanyang and BAT at the highest
levels of executive leadership in both companies.

Upon his return to China, Chien Chao-nan marshaled support for the
idea of a merger from new members of the Nanyang board of directors
who came from outside the family, and by the end of 1920, he felt confident that the deal could be put through. Ready, he thought, for a final
settlement, he met several times in Shanghai with a representative from
BAT's top management, A. G. Jeffress, who had come from London spe-
cially for this purpose. In February 1921 the two men reached an agreement which seemed to Chien Chao-nan irresistible. BAT agreed to com-
bine the stock in its China branch with Nanyang's stock (minus dividends
which each company would retain) and divide the stock between the two
merged companies with BAT taking three quarters and Nanyang one
quarter. If, after the merger, Nanyang expanded more rapidly than the
rest of the BAT-Nanyang amalgam of which it was a part, then the agreement left open the possibility that Nanyang could gain control of a
greater proportion of the total stock issued. Considering the vast difference between BAT's and Nanyang's net profits during the preceding year
(1920)— ¥15 million for BAT as compared with ¥4 million for Nanyang
—Chien Chao-nan was pleased with this arrangement and felt that it represented BAT's final and best offer. He also believed that the provision
for Nanyang's eventual expansion should effectively allay his family's
fears that the merger would limit Nanyang's growth.[72]

Chien Chao-nan's relatives in South China and Southeast Asia wrote
to him in Shanghai that they were interested in the new proposal, but
they cautioned Chien Chao-nan to proceed carefully so that Nanyang
would not "get stung" (*ch'ih kuei*). If Chien Chao-nan were patient, they
felt sure that the foreigners would eventually "beg us" to carry out the
deal on better terms for Nanyang. In resisting Chien Chao-nan's proposal, Chien Yü-chieh and others in South China repeated the same patriotic arguments against the merger that they had used in 1917. They

reminded Chien Chao-nan that Nanyang's commitment to the nation and reliance on "national goods" slogans had brought it success. They reiterated the point that allying with the foreign BAT would be viewed as "defecting to a former enemy for which [Chinese] society would spit on us." Rather than suffer this ignominious fate, they were prepared to "continue to battle [BAT] for many years to come."[73] Chien Chao-nan's arguments in favor of the merger were also reminiscent of the earlier debates within the family. Invoking threats of an impending crisis as he had before, he warned that if the Chiens missed this chance to "join hands" with BAT, then the "foreign devil" would become more fiercely competitive, would underprice Nanyang's brands, and would make it "difficult for us to survive."[74]

Though the debate was similar to that of 1917, Chien Chao-nan's arguments had been bolstered by Nanyang's unhappy experience during the May Fourth commercial campaigns of 1919 which had shown that unforeseen events could rapidly turn popular feeling against the Chiens' company. Perhaps more because the Chiens feared this danger than for any other reason, on February 21, 1921, Chien Yü-chieh and Nanyang's other principal stockholders in South China sent a telegram to Chien Chao-nan in Shanghai giving their "reluctant approval" to the proposed merger and expressing their hope that there would be room for discussions before any agreement was signed. Delighted with this decision, Chien Chao-nan immediately dispatched two cables that reflected his optimism, one urging that Chien Yü-chieh hurry to Shanghai for the closing and the other arranging to buy more Nanyang stock for himself.[75]

Chien Yü-chieh was less eager to close the deal. A week and a half later, despite his elder brother's repeated urging, Chien Yü-chieh still had not left from Hong Kong for Shanghai. By the time Chien Yü-chieh and other members of Nanyang's board of directors finally reached Shanghai, Chien Chao-nan had already made BAT an offer, and Jeffress, representing BAT, had accepted it. The contracts were prepared and the deal appeared to be settled. Then, at the last minute, Chien Yü-chieh and other members of the family who had come north to Shanghai refused to sign. All of Chien Chao-nan's efforts to coax and compel them to follow his lead were insufficient to overcome his family's long-standing doubts or to prevent their withdrawal from the negotiations.[76]

This abrupt, last-minute cancellation of the proposed merger greatly disappointed BAT directors, and they did not abandon hope for the merger without trying first to persuade and then coerce the Chinese family into going along with it. The effort to persuade the Chien family focused on Chien Yü-chieh. James Thomas, by now better acquainted with the Chiens than were others in BAT, singled out Chien Yü-chieh as the chief obstacle to a successful settlement and recommended that "someone live with him until he is won over." Thomas also suggested who that someone should be: Liang Shih-i, a Cantonese politician renowned for his manipulative abilities and a close friend of Thomas.[77] Liang had served as a scholar-official in the last years of the Ch'ing, and after the founding of the Republic, had become President Yuan Shih-k'ai's private secretary and leader of the Communications Clique in the Ministry of Posts and Communications in the Peking government. Taking advantage of these positions, Liang had created a network of bureaucratic and financial power which made him immensely influential with Chinese economic leaders. Nicknamed "the God of Wealth" by Peking newspapers, Liang seemed, in Thomas' words, "big enough to put the deal through."[78] Liang devised a plan whereby he would use his influence and BAT's money—a potent combinaton—to convince all Nanyang stockholders opposed to the merger to sell their stock in the Chinese company. Duke approved of the plan, and Thomas and Liang were confident that it would work, but they all underestimated the difficulty of persuading the Chiens to sell.[79]

When Liang Shih-i's efforts at persuasion failed, the directors of BAT then decided to try to coerce the Chiens into merging by attacking Nanyang in the marketplace. "In order to bring this [merger] about quickly," Thomas recommended in June 1921 to BAT directors in London that the company unequivocally "fight them [the Chiens and Nanyang] in the districts where they are selling a brand of cigarettes which shows them a good profit." To put Nanyang in its place, he was prepared to resort to the most aggressive merchandizing techniques. "My idea would be to . . . go into this market [anywhere in China that Nanyang is making a profit] and for a few days give away as many cigarettes as they were selling of a cheap brand which would muddy the waters sufficiently to get the desired results."[80]

BAT's board of directors in London agreed with Thomas that the situation called for drastic action. "We are going out after the competition," the London office informed Thomas at the end of 1921. If BAT previously had shown restraint, it now signaled its men in China to wage all-out commercial warfare. William Morris, an American BAT representative in China, was given "practically a free hand . . . to undercut them everywhere in China where they are showing any signs of strength." The parent company backed this decision, as it had previous ones, with solid financial support. "This [undercutting of Nanyang] will cost money," a BAT executive in London wrote to Thomas, but "the Company is prepared for it." Morris and other Western BAT representatives in China immediately put the directive from London into effect, and according to Thomas, began with new zeal to go "right after" Nanyang "in a systematic way."[81]

Even where BAT agents felt they had exerted maximum competitive pressure on Nanyang, they were exhorted to do more. In the South, for example, the region in China where Nanyang had marketed cigarettes the longest, the company instructed its agents to compete more keenly with Nanyang. In reply, a BAT representative in the region, V. L. Fairley, informed the company's headquarters that BAT representatives in his region could not wrest much more of the market away from Nanyang no matter what they did. During the May Fourth campaigns when Nanyang had been most vulnerable to criticism, BAT's marketing system had secured the cooperation of all dealers and retailers in South China "right down to the smallest restaurants and cigarette hawkers," he said, but even then the sales of none except BAT's most expensive brands had increased. With Nanyang so well entrenched, Fairley felt that the best BAT could do was to keep "plodding away . . . but any idea of bringing this [BAT domination of the South China market] about at short notice must be abandoned."[82]

Fairley's superiors, however, regarded such a gradual approach as inadequate. In 1920 James Thomas insisted that "the only thing to do is go in and fight back [against Nanyang in South China]. I am satisfied that you will win."[83] And during the same year, a delegation composed of two BAT executives from London, Lord Achison and Sir Arthur Churchman, and two American representatives from Shanghai, Thomas Cobbs and William Morris, visited Canton and discussed with their Cantonese employees the appropriate strategy to use against Nanyang. In consultation with Chiang K'ung-yin, the gentry leader and BAT employee whose family had led the May Fourth opposition against Nanyang in Canton, they raised the possibility of organizing a new tobacco company to compete directly with Nanyang in South China, but no such company was formed. In response to BAT's demands for keener competition, Chiang Shu-ying, Chiang K'ung-yin's son, urged the Western directors to make him a kind of roving agitator responsible for leading, in his words, "a general attack on the [Nanyang] company." Still convinced that Nanyang continued to maintain affiliations with Japan and still determined to discredit the Chiens for doing so, he wrote to William Morris of BAT in 1922 expressing his feeling that "in Manchuria, Yan[g]tse-kiang, Hon[g]kong, Canton and Malay States, the true condition of the N.B.T. [Nanyang Brothers Tobacco Company] must be exposed to the public, so that confidence in the company is undermined, injuring its finances thereby and creating a popular clamour for investigation of its organization, with the view to the annulment of its charter by the North and Southern Governments."[84] There is no evidence showing the BAT executives' reaction to this scheme, but in light of their other aggressive tactics and in light of their feeling that "things seem to be 'pop-ping' in the cigarette business in China [in the spring of 1922] . . . [and] in these times of severe competition the organization needs all of the strong men it can get into it," they might well have hired Chiang to carry out his proposal.[85]

Despite the intense competitive pressure BAT exerted on Nanyang in the early 1920s, Chien Yü-chieh and his compatriots did not sell. From their perspective, the rationale against merging they had formulated in 1917 (as discussed in Chapter 4) still obtained. If consenting to a merger, they feared their control over their business would be weakened and ultimately lost, their advertising would be rendered ineffective, their leadership in the national goods movement would be repudiated, their reputations would be tarnished, and their moral principles would be compromised. Because of personal convictions and social obligations as well as economic considerations, they steadfastly refused to reverse their decision. Chien Chao-nan continued to grope for a formula that would satisfy his family on the one hand and BAT on the other, but he could find no solution that would please everyone, and in the summer of 1922 even James Thomas—previously Chien Chao-nan's closest ally on the proposed merger—began to lose enthusiasm for the idea.[86] By then, any possibility of successfully negotiating an agreement between the two companies was lost, and the issue was never raised again.

Even after the talks between BAT and Nanyang had collapsed and BAT's directors had abandoned all hope for a merger, the Western company did not curtail its punishing commercial campaigns against Nanyang. In fact, some directors, representatives, and agents were so committed to the struggle with Nanyang that they were relieved to learn that the merger had not gone through. One director, for example, recognized that merging with Nanyang would probably have raised BAT's profits, but he could not help feeling pleased that the rivalry with Nanyang was continuing because it "certainly keeps our people alive in China and gives us something to fight against."[87]

This record of BAT's marketing strategy shows that it did not permit Nanyang to enjoy a golden age free from Western competition during the years following World War I (any more than during the war). In the early 1920s BAT not only developed its operation in China but also used its growing influence and financial and commercial power to try to subdue Nanyang first by absorbing it and, failing that, by attacking it in price wars and other commercial campaigns. And yet, as noted earlier, Nanyang still managed to raise its sales and profits in the early 1920s. Why, despite intense competitive pressure from BAT, did Nanyang grow in these years? The answer to this question lies in Nanyang's success at carrying out its own comprehensive program for expansion under the guidance of newly recruited professional managers.

Nanyang under Professional Management

Though unable to persuade the members of his family to merge Nanyang with BAT, Chien Chao-nan was able to convince them to make Nanyang more like BAT by hiring trained professional managers from outside the Chien family and expanding the company under the supervision of these specialists. The appointment of the new managers represented a radical departure from the Chiens' previous managerial policies. In the past they had kept the business family-managed and self-contained—a practice which seems to have been more common among Cantonese than Shanghainese businessmen and which continues to distinguish Cantonese-owned businesses from Shanghainese-owned businesses even in today's Hong Kong.[88] The Chiens had reluctantly admitted men from outside the family to Nanyang's board of directors in 1919 but even then had not placed them in managerial positions. In the early 1920s, however, Chien Chao-nan recognized that if Nanyang was to continue to grow and remain competitive with BAT, then a measure of family management had to be abandoned, for barring nonfamily members from all high managerial positions inevitably restricted Nanyang's size and limited its managerial expertise. Accordingly, he broke precedent by appointing professional managers from outside the family to high supervisory positions, and they, in turn, contributed directly to Nanyang's commercial successes during the postwar golden age by their innovations in finance, manufacturing, marketing, and purchasing.

In financing Nanyang's expansion, Chien Chao-nan turned for assistance to a man who was aware not only of Western financial institutions in general but also BAT financial techniques in particular, the former BAT comprador Ch'en Ping-ch'ien. Appointed in 1919 as one of the first members of Nanyang's board of directors from outside the Chien family, Ch'en rapidly rose to prominence within the firm. He demonstrated his commitment to Nanyang by buying more stock in it than anyone else (other than the Chiens) and formed a family tie with the Chiens by marrying his daughter to Chien Chao-nan's son. The Chiens, in turn, showed their faith in him by making him vice-president in charge of finance in 1921 (a position he held until 1929) and by appointing his brother and nephew to managerial positions in the company as well.[89]

Under Ch'en Ping-ch'ien's leadership, Nanyang financed its growth in the early 1920s largely through the sale of stock. During the Chiens' first public distributions of stock between 1919 and 1923, all of the ¥15 million worth of stock that it issued was purchased. The Chien family held 60 percent of it and the remainder was sold to stockholders outside the Chien family.[90] In addition, in 1922 and 1923, Ch'en Ping-ch'ien and Chien Chao-nan persuaded Nanyang's board of directors to create a vari-

1910s and early 1920s). Despite the furor, Ch'en Ch'i-chün continued to manage the factory, and he and other foreign-educated "returned students" who were appointed heads of each of its departments strived to make the management in this factory up-to-date by American standards so that it would be as "modern" in all respects as any BAT factory. The American-educated Chinese managers' effort was supported by the Chiens, who made a heavy commitment to American technology (even though it was more expensive than that which the Chiens had imported from Japan), almost tripling their total investment in Nanyang's manufacturing system between 1920 and 1922 by increasing it from ¥.59 million to ¥1.73 million.[99]

Though Ch'en Ch'i-chün alienated some older Nanyang managers, he appears to have been popular with other staff members and workers, and his popularity may help explain why the first big wave of labor struggles which hit BAT in this period scarcely touched Nanyang.[100] As strikes swept across China's industrial centers in the early 1920s, there were attempts to form labor organizations in Nanyang's Hong Kong and Shanghai factories, but no strong or militant unions took hold. In the company's Hong Kong factories, Chien Yü-chieh blocked formation of a union by taking a hard line. Union organizations extolling the "labor spirit" quickly recruited 400 members from the ranks of workers at Nanyang's Hong Kong factory in 1921, but Chien Yü-chieh prevented a union from developing by offering the workers insurance benefits comparable to the ones workers had been offered by the union and then firing the labor organizers and any other workers affiliated with the union. In Shanghai Chien Chao-nan was more accommodating and maintained remarkably harmonious relations with organized labor. Fortunately for him, he did not have to cope with the militant labor leaders who organized BAT workers in Shanghai at this time and was able to deal effectively with the moderate workers' organization that was formed in the Nanyang factories, the Shanghai Tobacco Union (*Shang-hai yen-ts'ao kung-hui*). This union led the only strike against Nanyang in the early 1920s, a brief one that was called on November 7, 1922, during a strike at BAT's factories in Shanghai. The Nanyang strikers demanded that the company's management recognize their union as spokesman for the workers, pay for the union's meeting hall, make monthly contributions to the union, give union leaders immunity from reprisals, raise wages for all workers, and provide more job security for piece workers. With some modifications, Chien Chao-nan acceded to all these demands, ending the strike within two hours after it had begun. At two o'clock 6,000 workers went on strike, at three o'clock their representatives met with Chien Chao-nan in his office, and by four o'clock everyone was back on the job. The workers' principal concession to Chien Chao-nan opened membership in the union to staff members as well as workers, a change that was made explicit in the new name of the organization, the Nanyang Staff and Workers Fraternal Association (*Nan-yang yen-ts'ao chih kung t'ung-chih hui*). Within a few weeks, a delegation from this organization informed the Nanyang workers in Hong Kong of their strike and settlement whereupon the Hong Kong workers demanded that the Chiens make a comparable arrangement with them. With Chien Chao-nan's approval, the Nanyang staff in Hong Kong then created an organization similar to the one in Shanghai called the Nanyang Tobacco Staff and Workers Club (*Nan-yang yen-ts'ao chih kung chü-lo-pu*), and on January 16, 1923, Chien Chao-nan and twenty representatives of the newly formed club signed an agreement almost identical to the one worked out in Shanghai.[101]

Nanyang's avoidance of conflicts with labor in this period of strikes at BAT and elsewhere in China may have been attributable to the popularity of Ch'en Ch'i-chün and other "returned students" as managers, or to the Chiens' manner of dealing with workers, or to the tendency of strikes to be held more against foreign- than Chinese-owned companies, but the lack of prolonged strikes against Nanyang did not mean that its 14,000 or 15,000 industrial workers, mostly women and children, were well paid or content.[102] Figures from the late 1920s give some idea of what wages would have been like at Nanyang's Shanghai factories in the early years of the decade: ¥0.80 per day for the average amount of piece work by a machine operator (virtually all of whom were men), ¥0.60 for the average amount of piece work by an unskilled adult (virtually all of whom were women), and ¥0.40 for the average amount of piece work by an unskilled child. The monthly wages of regularly employed workers in the company's various departments ranged from ¥15 to ¥40, and the monthly salaries of Nanyang's chemists and other technicians started at one hundred *yuan* and went as high as several hundred.[103]

The real value of unskilled workers' wages is difficult to measure in the context of the times, but a Chinese author's poignant characterization of one young woman who worked as a cigarette packer for Nanyang in the early 1920s suggests that her income kept her scarcely above the subsistence level. The writer, Yü Ta-fu, captured the awfulness of this Nanyang worker's daily life in a story that he wrote about her after he met and became acquainted with her in the dilapidated rooming house where they were both tenants. According to her description, she worked ten hours per day at a wage of ¥9 per month (¥0.03 per hour), which suggests that wages at Nanyang were considerably lower in the early 1920s than those quoted above for the late 1920s. Of this ¥9, she spent almost half on rice, a small portion on rent for a tiny unfurnished room, and all the rest on "squeeze" to her factory supervisor in order to retain her job.

Women packing cigarettes in Nanyang's cigarette factory

Women processing tobacco leaves in Nanyang's cigarette factory

She was not able to save anything. The meager wages combined with the close surveillance, harsh discipline, and ever present threat of dismissal at the factory embittered her toward Nanyang in general and her supervisor in particular. After Yü befriended her, she offered him complimentary cigarettes from the factory—her only luxury—but urged him never to spend money buying Nanyang brands because "I really hate the place in the worst way (hen-ssu)."[104] Whether all Nanyang workers were as impoverished and frustrated as this woman is difficult to say, but it is possible to infer from her budget and from surveys of industrial conditions conducted in the 1920s that workers in Nanyang's factories, like those in BAT factories, received low wages compared to the Chinese workers in other light industries or in heavy industries in China at the time.[105]

Judging from this evidence, industrial workers contributed to Nanyang's postwar golden age at minimal expense to the company's management. The Chiens improved working conditions in factories during the early 1920s—shortening hours, supplying free or nearly free housing in factory dormitories, and offering free or nearly free educational opportunities at company schools for workers' children—but this was a small price to pay for freedom from strikes in a period marked by intense labor struggles in BAT's factories and elsewhere in China.[106]

Pleased with the improvements in manufacturing and the absence of increases in labor costs that they achieved with the help of Ch'en Ch'i-chün, the Chiens sought help from another nonfamily member to expand their marketing system, the former BAT comprador, Wu T'ing-sheng. This time Wu did not secretly continue to work for BAT while pretending to side with Nanyang as he had earlier. When his original patron, James Thomas, took a leave of absence from BAT in 1920, Wu permanently severed his ties with the foreign company and took up full-time employment with Nanyang.[107]

Upon joining Nanyang, Wu immediately demonstrated his effectiveness at marketing. As a non-Cantonese (originally from Chekiang province) who had traveled widely, he concentrated on improving Nanyang's distribution in existing commercial networks outside the Chiens' home region of South China. His principal assignment was to regain ground for Nanyang in the early 1920s that he had helped BAT take from Nanyang in North China and the Lower and Middle Yangtze during the May Fourth Campaigns, and with his help, Nanyang rapidly reasserted itself in these regions.[108] In fact, Nanyang's recovery and expansion in the Lower Yangtze during 1920 was so swift that by the end of the year it prompted Thomas Cobbs, an American acting in James Thomas' stead as BAT's managing director, to travel from BAT's headquarters in Hong Kong to Shanghai to investigate. By doing so, Cobbs risked legal prosecution for violating the British order in council, discussed earlier, which prohibited him, as the American managing director of a British-registered firm, from entering China. But he went anyway because he felt that Nanyang's postwar marketing successes were so stunning that he had to investigate in person. Explaining his decision to make the illegal trip, he wrote to Thomas, "I want you to know that this competition [from Nanyang] is serious."[109]

In South China Nanyang had not suffered such severe losses during the May Fourth campaigns as in North China and along the Yangtze, and therefore, did not have so much lost ground to regain. According to studies BAT conducted in 1919, only members of the Cantonese elite had switched from Nanyang to BAT cigarettes as a result of the May Fourth controversy. The members of the Cantonese elite changed to BAT's brands, reported BAT's representative in South China, V. L. Fairley, because "the educated classes are not troubled with any imaginary patriotism, and the propaganda work we have been doing for some time, we believe, has fully convinced them that our opponents [the Chiens] are more or less imposters, so far as the native goods cry is concerned." But Fairley admitted that Cantonese consumers other than the elite had generally remained steadfastly committed to Nanyang brands throughout the May Fourth campaigns. "There is a very large section of the coolies and lower middle class who still believe that it is their duty to support what they consider are native products."[110] Aware of this brand loyalty during the May Fourth campaigns in South China, the Chiens built upon it in expanding during the 1920s.

Nanyang and BAT directors alike attributed Nanyang's marketing success in China during the early 1920s to the continuing growth of popular preferences for goods made by Chinese companies. The Chiens credited their company's resurgence after their May Fourth setbacks to the fact that "compatriots both overseas and at home used national goods and [Chinese] stockholders gave us their support."[111] Thomas Cobbs traced Nanyang's success to the same source, but, as managing director of BAT, he felt that the Chinese company was opportunistically exploiting "China for the Chinese" sentiment which was "becoming stronger throughout China today." To illustrate the point, he suggested that it was no accident that Nanyang's most popular brand was called Great Patriot (*Ta ai-kuo*).[112]

To exploit these popular attitudes, the Chiens did much more than choose patriotic names for their cigarettes. They also created a comprehensive advertising system, and in advertising, as in other aspects of their business, they sought to acquire expertise comparable to that of BAT and went outside their family to get it. In 1918, following the advice of P'an Ta-wei, an experienced Cantonese journalist who had earlier used newspapers to promote the revolution of 1911, the Chiens had bought their first two printing presses, but these were relatively unsophisticated and capable only of turning out simple items such as handbills. Not until the early 1920s did the Chiens gain access to facilities where more refined printing could be done at the Yung-fa Printing Company of Hong Kong. Initially they supplied half of the ¥200,000 at which this firm was capitalized, and subsequently they invested another ¥50,000 in it when it split its stock in 1923.[113]

With Yung-fa as its printer, Nanyang began distributing advertising matter that was as technically refined and as eye-catching as that of BAT. The Chinese cigarette company issued sets of cigarette cards, for example, showing characters from the Chinese novels *Romance of the Three Kingdoms* (*San-kuo chih yen-i*) and *Dream of the Red Chamber* (*Hung-lou meng*), printed in polychrome and appealing to collectors. In issuing set of cards drawn from Chinese literature, Nanyang was imitating BAT which had earlier issued an elegant set drawn from *Water Margin* (*Shui-hu chuan*), done in the style of traditional Chinese painting and graced with flowing calligraphy showing the names and nicknames of the novel's personae. In distributing one set of cigarette cards, however, Nanyang introduced a technique that even BAT had not tried by announcing that a complete set of its cards showing all the characters from *The Investiture of the Gods* (*Feng-shen yen-i*) could be traded in for valuable prizes—mostly Western imports such as bicycles and hot water bottles. The idea started a craze among collectors until rumors began to circulate suggesting that cards for eighteen of the characters were extremely rare and perhaps had never been printed. Eventually all the cards did appear, and in the meantime, Nanyang benefited from the attention its cards and prizes attracted.[114]

Additional investments in newspapers gave the Chiens a still larger publishing establishment. They set up their own newspapers in Hong Kong, Canton, and Shanghai in the early 1920s, each of which served, in Chien Yü-chieh's words, as a "mouthpiece" (*hou-she*) presenting a line dictated by Nanyang's advertising staff.[115] Publishing in these and other newspapers, Nanyang kept "pace with the 'B.A.T.' in the matter of newspaper advertising" according to a study of the Chinese press done in the early 1920s, which observed that the two cigarette companies stood out as "great advertisers" compared to other businesses in China.[116]

Newspapers and books produced by Nanyang's new publishing establishment projected a favorable image of the company by emphasizing above all the importance of its philanthropic activities. In the early 1920s the Chiens had more of these activities to publicize, for they not only maintained their reputation as donors of flood relief in Kwangtung province but began to institute and publicize welfare programs in other regions as well. During the disastrous drought of 1920 in North China, for example, they set up a "congee house" in Peking—a kind of soup kitchen that made free food available to the poor—and they dispatched men to work with the Buddhist Relief Society and other charitable groups throughout the region. To help finance the relief effort, they made a pledge to donate five *yuan* for every case of (50,000) cigarettes that the company sold in North China up to a maximum gift of ￥100,000.[117] By thus making the size of their "donation" contingent on the level of their sales, the Chiens enhanced the appeal of their cigarettes, for the arrangement left the impression that customers should buy Nanyang brands as a means of making an indirect contribution to a worthy uncommercial cause. In other regions of China—the Northwest and the Lower and Middle Yangtze—and in the countries of Southeast Asia as well, the Chiens followed a similar pattern, contributing to charities and service organizations and publicizing these contributions.[118]

These investments in publishing houses, newspapers, and philanthropy gave Nanyang direct access to a range of advertising comparable to that of BAT. Like all elements of Nanyang's operation, its advertising system remained smaller and less well financed than that of BAT but nonetheless became larger, better financed, and managed more by technically competent professionals than ever before.

Following a similar pattern in tobacco buying as in advertising and other areas of the business, the Chiens enlarged Nanyang's purchasing system and delegated greater responsibility for it to managers from outside the family. In China they made investments in the agricultural system they had started in 1918, which enabled Nanyang to become the largest purchaser of Chinese-grown bright tobacco next to BAT in the early 1920s.[119] The Chiens set up reception stations at the same three sites where BAT had reaped the most abundant harvest—Wei-hsien in Shantung province, Hsuchang in Honan, and Fengyang in Anhwei—and made their largest single investment in the same one of these as BAT had, Wei-hsien, where they built a tobacco leaf processing plant capitalized at ￥500,000 that occupied 20 *mou* of land near the railroad station in the town of Fangtze. For use at this and their other reception stations, they annually imported 1,000 pounds of bright tobacco seed from the United States in the early 1920s, and to persuade peasants in these areas to plant the seed, they imitated BAT's techniques for popularizing bright tobacco growing—distributing seed free of charge, lending fertilizer, and contracting to buy entire crops in advance regardless of the quality.[120]

The Chiens, like the directors of BAT, were eager to cut the cost of the tobacco they purchased by taking advantage of cheap agricultural labor in China, and they showed no signs of being more generous or scrupulous with Chinese peasants than BAT was. Their purchasing agents in the Nanyang reception station at Hsuchang, for example, consciously ex-ploited peasants by using "light" tobacco scales weighted to the buyer's advantage. Nanyang's accountants meticulously recorded in separate columns the amounts of tobacco for which the company paid peasants and the larger amounts which peasants actually delivered. The "excess poundage" (yü-pang), that is, the amounts the company received free through this fraudulent practice, averaged about 350,000 pounds of to-bacco each year, 12.6 percent of the total sold by peasants at Hsuchang between 1920 and 1928. In this way Nanyang cheated Hsuchang peasants out of an average of about ¥47,000 in each of these years.[121]

Though the Chiens thus took advantage—even unfair advantage—of cheap Chinese agricultural labor, their purchasing system in China was less productive in the early 1920s than they had anticipated. Disap-pointed by the results, they suspended operations at their leaf-drying plant in Anhwei province in 1921 and considered their reception station in Shantung a failure by 1923. Their difficulties in tobacco purchasing seem to have stemmed largely from the company's inability to hire and hold purchasing agents with the technical expertise needed to judge the quality of tobacco leaves. As mentioned earlier, Chinese "men of talent" (jen-ts'ai) whom the Chiens were able to recruit in this field tended to stay with Nanyang only briefly—just long enough to attract better offers from BAT. In 1923, hoping to reverse this pattern and recruit a former BAT man in this field (as they had recruited Ch'en Ping-ch'ien in finance and Wu T'ing-sheng in marketing), the Chiens offered the job of manag-ing Nanyang's purchasing system in Shantung province to James Thomas, but he politely turned them down.[122] Frustrated in their efforts to assemble a staff that could procure the quality and quantity of bright tobacco Nanyang needed in China, the Chiens continued to buy large amounts of it in the United States.

Chien Chao-nan tried to secure the technical and managerial expertise for Nanyang's purchasing system in the United States it lacked in China by traveling to the American South in 1918 and searching for an Ameri-can agent willing to serve as Nanyang's buyer in the major tobacco mar-kets of Virginia and North Carolina. On a tour of the bright belt, he found the man who seemed suited to the task, John Oglesby Winston "Captain Jack" Gravely (1862-1932) of Rocky Mount, North Carolina. Although Chien Chao-nan's world was geographically and culturally dis-tant from Gravely's, the two men discovered that they had much in com-mon. Like Chien Chao-nan, Gravely had dealt with Japanese tobacco men, selling to private Japanese buyers as early as 1895 and later to the Japanese Imperial Tobacco Monopoly; in all probability it was through these Japanese connections that Chien Chao-nan first heard about Gravely. In addition, both men had established reputations in their re-

spective countries for standing up to James Duke's multinational jugger-
naut and creating two of the few independent businesses that had managed to survive in competition with it. Even as Chien Chao-nan had
become known as a rival of BAT in East Asia, so Gravely had made a
name for himself throughout the American South as a staunch friend of
the "honest farmer" and as an unbending enemy of Duke's gigantic direct
purchasing system which had driven many independent American leaf
dealers out of business by the late 1910s.[123]

Drawn together by these mutual concerns, the Chinese industrialist
and the American tobacco man agreed in 1918 to form a new leaf agency,
the China-America Tobacco Company (*Chung-Mei yen-yeh kung-ssu*),
capitalized at $500,000. Chien Chao-nan invested $150,000 of Nanyang's
money in the company, making the Chiens its leading stockholders except for Gravely himself. This company established large storage plants
in New York, Kobe, Shanghai, and Hong Kong in 1919 and 1920 and became Nanyang's major supplier in the postwar period. To supplement
this source, the Chiens also struck a new bargain with their old supplier,
J. P. Taylor and Company, which had recently developed into one of the
world's largest leaf agencies, and they bought still more tobacco from
other American dealers. This group of suppliers gave Nanyang a network of active purchasing agents in the United States, and though this
network did not constitute a direct purchasing system as large as BAT's,
it contributed directly to Nanyang's expansion by supplying the company with a total of $8 million worth of American tobacco in 1920 and
comparable amounts throughout the postwar period.[124] By 1923 the large
amounts that were imported into China by Nanyang and the still larger
amounts imported by BAT helped make China the world's leading importer of American flue-cured bright tobacco next to the United Kingdom, a position it retained until the Japanese military invasion of China
in 1937.[125]

Thus, in tobacco purchasing as in financing, manufacturing, distributing, and advertising, Nanyang appointed a professional manager to
supervise the company's expansion during the postwar golden age. In this
sector of the business, Chien Chao-nan appointed an American, Gravely,
rather than a Chinese to the highest supervisory position, but the difference in Gravely's nationality was less important to Nanyang's growth in
these years than what he had in common with Chien Chao-nan's other
high-level appointees, Ch'en Ping-ch'ien in finance, Ch'en Ch'i-chün in
manufacturing, and Wu T'ing-sheng in marketing. As professional managers from outside the Chien family, each of these men was concerned
primarily with his own department and was eager to develop it—unin-
hibited, as the Chiens had been, about the danger that expansion might
cause the family to lose control over administration of the business. As

specialists with technical training or experience, each man was familiar with BAT's methods and was determined to make Nanyang more equivalent to BAT in his field. By the energy, ambition, and innovations they brought to Nanyang, these professional managers greatly stimulated Nanyang's growth. Driven by the professionals and led by Chien Chao-nan, Nanyang became more administratively specialized, structurally integrated, and generally more like Western big businesses.[126] In sum, Nanyang enjoyed a postwar golden age largely because it came closer in the early 1920s than ever before to battling BAT on the Western company's own terms.

The End of the Postwar Golden Age

After enjoying a postwar golden age, both of the big businesses in China's cigarette industry had to make major adjustments during the autumn of 1923, for at that time each lost its man at the top—BAT as a result of James Duke's resignation and Nanyang as a result of Chien Chao-nan's death. Throughout the history of BAT and Nanyang, these two men had been the only ones ever to serve as chairmen of the boards of their corporations, and their careers as international businessmen concerned with China had culminated in BAT's and Nanyang's best years to date during the postwar golden age. Their departures from the industry in 1923 raised the question of whether the two companies could continue to do as well without them.

Looking back on the lives of James Duke and Chien Chao-nan, the patterns of the two men's careers were strikingly similar. Both men had risen from impoverished childhoods. Both had received only the rudiments of formal education. Both had become merchants early in life. Both had capitalized on the cigarette as a product of the continuous-process machinery invented by James Bonsack in the 1880s. Both had laid the foundations for their businesses by establishing marketing bases in their home regions away from heavy competition (for Duke, the American South, and for Chien, Southeast Asia), and both had eventually dared to attack the established competition in big, urban, commercial capitals (for Duke, New York, and for Chien, Shanghai). Both had operated outside as well as within their home countries and had engaged in international business. Both had been aggressive and innovative commercial and industrial managers. Both had become upwardly mobile in countries that had myths of social mobility, both had risen in periods of rapid social change, and both had climbed ladders of material success, making their "rags-to-riches" stories genuine. Finally, in the end, both had bequeathed not only prosperous businesses but also legacies of effective management to their successors.[127]

Upon Duke's resignation, Sir Hugh Cunliffe-Owen, his successor as chairman of BAT's board of directors, tried to give the impression that he and other BAT executives regretted the loss of Duke's leadership, implying that they had no desire for change in the company's management. "There has been no fight for control at all," Cunliffe-Owen assured reporters after Duke's resignation was announced in 1923. "Mr. Duke has been a great friend of us all for many years, and we are all sorry he has retired . . . We have always asked him to stay on, but now he prefers to get the leisure to which he is so thoroughly entitled after his busy career."[128]

Cunliffe-Owen's words were kind, but behind this facade of Anglo-American harmony he and other English stockholders had been more anxious to oust Duke and elect an English chairman than Cunliffe-Owen led the press to believe. For several years BAT's English stockholders had gradually bought out their American counterparts, securing a majority of shares as early as 1915. Now, with the resignation of James Duke in 1923, the English finally wrested from the Americans managerial as well as financial control over the company.[129] As a result, BAT's managerial orientation began to change.

BAT's subsidiary in China was directly affected by the company's change in top management. As soon as Cunliffe-Owen became chairman of the board, he showed his high regard for BAT's business in China by immediately embarking on the long voyage from London to Shanghai to inspect BAT (China), Ltd., in person.[130] While there, he announced no radical changes in BAT's organization and gave no public indication that BAT would deviate from precedents set by Duke and Thomas, but soon major organizational changes were made.

The most significant change made in China after Cunliffe-Owen took over as BAT's chairman was its transformation from a centralized to a decentralized administrative organization. Whereas James Thomas had set up a single headquarters in Shanghai, had strived for national integration of the business, and had organized BAT's operation in China according to its functions (manufacturing, distributing, advertising, purchasing), now his successors managed all operations on a regional basis. Each region had its headquarters in a city where BAT had existing factories (Shanghai in the Lower Yangtze region, Hankow in the Middle Yangtze, Tientsin in North China, Mukden in the Northeast) or where it added new factories, as it did in 1925 in Tsingtao (near the bright tobacco fields of North China) and in 1929 in Hong Kong (for distribution in the South and Southwest).[131] BAT's regional headquarters in each of these cities had its own complete sales, advertising, accounting, and traffic offices and no longer needed to look to Shanghai for these administrative services. What James Thomas had once approached as a single administrative entity, his successors, as one American BAT executive put it in the late 1920s, "divided into self-containing units, each almost wholly independent of the others."[132]

In addition, responsibility for supervision seems to have passed from one nationality to another. In Shanghai BAT retained American as well as English directors, but in the regional headquarters and below, the staff became more English than American in manufacturing, marketing, and all other aspects of the business except tobacco purchasing. Moreover, as administrative authority passed into the hands of the English, the entire organization seems to have become even more dependent on Chinese than it had been under American executive leadership. The transfer of responsibilities from the Americans to the English and from Westerners to Chinese during the 1920s was well summarized by an American BAT executive, James Hutchison, who had previously served the company in China during the 1910s and then, after several years in the United States, returned to survey its operations in China once again in the late 1920s. During the 1920s "the company had gone completely British," he concluded, and "the foreigner was moving into the background." To illustrate this point, he noted a dramatic decline during the 1920s in the ratio of Western supervisors to Chinese workers at BAT's factory in Mukden: " . . . eight or ten English and Americans supervising a thousand or more Chinese. This was all new to me." And in marketing, the shift was even more pronounced. In BAT's office at Mukden, the company's regional headquarters for Manchuria, there were "eleven English, three Russian stenographers, and about three hundred Chinese. No Americans. The foreign sales staff was less than half of what it was fifteen years ago and, I soon learned, the [Western] men had practically given up travelling . . . The Chinese staff, mostly clerks, had increased ten times over." Hutchison described the operations at Mukden in greatest detail, but he saw evidence of "this same miniature evolution" in the company in Shanghai as well.[133]

As an American, Hutchison was quick to condemn the changes that were made in BAT's subsidiary during the 1920s, but he and other Americans in BAT might well have underestimated the virtues of the reorganization under English management. "The company had gradually grown, as the pioneer days came to an end," he concluded, and it had been transformed in the 1920s "into a ponderous, unwieldy, old-fashioned English accounting machine."[134] But Hutchison's judgment was perhaps distorted by Anglophobia or nostalgia for "pioneer days" and should not obscure a more plausible rationale for BAT's shift to a more regional organization and heavier dependence on Chinese. In the context of the time, this approach was an appropriate means of coping with floods and warfare—the twin scourges of the 1920s—that resulted in regional fragmentation in China. Under such circumstances each of

BAT's autonomous regional organizations was able to function effectively for long periods even when floods, warfare, and other disturbances cut transportation off between it and other regions.

Thus, the change in executive leadership at the top of BAT's multinational corporation had important consequences for its subsidiary in China, but these consequences were not necessarily detrimental to its operations after 1923. Though James Duke and other Americans might well have been dismayed by the transfer of managerial control to Hugh Cunliffe-Owen and other English executives, the new management's ouster of the old did not weaken the company's subsidiary in China in any discernible way. In fact, the new management's decision to decentralize its organization in China seems to have been appropriate in light of the circumstances at the time. Unimpeded and possibly helped by changes in management, BAT's subsidiary was able to extend its postwar golden age beyond Duke's resignation in 1923, achieving a high volume of sales in 1924.[135]

Nanyang, by contrast, was not so fortunate. After Chien Chao-nan's death in 1923, Nanyang's golden age abruptly ended. Whereas the Chinese company had earned profits averaging ¥4 million per year between 1920 and 1923, its profits plummeted to a mere ¥.479 million in 1924—less than 12 percent of its annual average during the four preceding years. This sudden downturn may be partially explained by general business conditions, which brought the postwar golden age to an end not only for Nanyang but also for many other Chinese-owned businesses in consumer industries. The Chiens cited several of these conditions in explaining to their stockholders why Nanyang's annual dividend was cut from an average of about 17 percent for the early 1920s to 4.4 percent for 1924: higher provincial taxes (as the provinces ceased to defer to Peking's policy on cigarette taxes and restored their own collection systems); natural disasters (especially floods in the tobacco growing regions of North China); increases in the cost of imported tobacco leaf as a result of a depreciation of currency (caused by the European recession which finally caught up with the Chinese economy in 1924); spoilage of large shipments of tobacco from the United States; and civil war and other sporadic fighting among warlords, which obstructed transportation, diverted the flow of goods and currency, interrupted planting and harvesting, and generally disrupted business.[136] Undoubtedly all these developments hurt Nanyang's business as the Chiens said, but there was another important reason for Nanyang's decline, which they omitted from their explanation: the loss of Chien Chao-nan's leadership.

At the time of Chien Chao-nan's death in 1923, hundreds of eulogists reminded his survivors of the contributions he had made to Nanyang's success. Chinese chambers of commerce, companies, unions, newspapers, tobacco leaf agencies, cigarette distributors, and even Chien Chao-nan's adversary in BAT, Chiang K'ung-yin, praised him for transforming Nanyang from an obscure small firm into an influential big business. Chinese crusaders in the national goods movement, patriotic Overseas Chinese, and well-known Cantonese leaders like K'ang Yu-wei and T'ang Shao-i lauded him for making Nanyang competitive with BAT and recovering economic rights from foreigners. People of several nationalities in Southeast Asia and Japan remembered him for his successes as an international businessman. And representatives of Chinese hospitals, relief agencies, orphanages, schools, and universities expressed admiration for his social concern and gratitude for his generosity.[137]

The members of Chien Chao-nan's family did not need these reminders to be made aware of his prominence in the history of the business. In recognition of his genius for finding commercial value in patriotic movements, philanthropic gestures, and other seemingly uncommercial events and activities, the Chiens held a peculiarly apt memorial service in honor of his death: every guest was given a badge bearing Chien Chao-nan's picture and a package of Nanyang cigarettes manufactured especially for the occasion.[138]

Though the Chiens recognized Chien Chao-nan's managerial accomplishments, they proved unable to maintain the managerial standards he had set at Nanyang. His successor as chairman of the board, Chien Yü-chieh, seems to have been a conscientious manager but was not as effective as his elder brother had been. When Nanyang's profits plunged during the first year after Chien Chao-nan's death, for example, Chien Yü-chieh expressed his feeling that he had been worth less to the company than his predecessor by cutting his own salary for the year, reducing it from ¥15,000 with an expense account of ¥20,000 (which Chien Chao-nan had received) to ¥12,000 with an expense account of ¥12,000; following this example, Nanyang's new vice-president, Chien Shih-ch'ing, took a comparable cut in pay at the same time.[139] Such a gesture was undoubtedly intended to raise morale, but it was no substitute for the close supervision over decision-making that Chien Chao-nan had exercised within the company, and it did not prevent administrative malfeasance from causing Nanyang's performance and profits to fall below the levels that had been achieved during the postwar golden age.

No longer under Chien Chao-nan's watchful eye, members of the Chien family took it upon themselves to make decisions without consulting Chien Yü-chieh or Nanyang's professional managers, and, as a result, they committed some costly blunders. Chien Chao-nan's son, Chien Jih-hua, for example, rashly signed a contract with suppliers in the United States for $20 million worth of tobacco when this was far more than Nanyang needed. "The tobacco came incessantly and the warehouses filled to overflowing," another of Chien Chao-nan's sons, Chien Jih-lin, later recalled, and even though the demand for cigarettes at the time did not justify this expenditure, "we carried out the contract and paid the money at the appointed time, and this almost caused the business to collapse."[140] The Chiens tried to avoid such blunders by delegating more responsibility for purchasing to "Captain Jack" Gravely in the mid-1920s, commissioning him to represent Nanyang exclusively and doubling their investment in his China-America Tobacco Company (bringing their total investment to $300,000), but the arrangement proved to be no panacea, for Gravely needed supervision too. When he wrote to ask advice from the Chiens, they responded by complaining that he was causing bothersome paper work and by insisting that he should take full responsibility for spoilage of tobacco in transit, fluctuations of currency exchange rates, and other purchasing problems. Unable to establish a good working relationship with Gravely, the Chiens never satisfactorily solved the problems of purchasing and importing American-grown tobacco in the 1920s, and Nanyang suffered from the rising costs of raw materials throughout the remainder of the decade.[141]

While Chien Jih-hua and others mismanaged tobacco purchasing, Chien Ying-fu, the younger of Chien Chao-nan's two brothers at Nanyang, carried out financial maneuvers that were even more damaging to the company and embarrassing for the family. Just after Chien Chao-nan's death in 1923, for example, with no authorization from Nanyang, Chien Ying-fu lent a friend in Malaya $100,000 in Straits currency, and when the friend's business failed, Nanyang had to cover the debt. Two years later he tried another speculative venture, this time an investment in a sulphur-mining operation which collapsed a year after its founding, leaving Nanyang a total bill of ¥157,000. While recklessly investing these large sums on unpromising business ventures, Chien Ying-fu took still larger sums from the company for his own personal use. According to the later recollections of Nanyang's staff and workers, Chien Ying-fu embezzled hundreds of thousands of *yuan* from the company to finance his extravagant life style which included indulging his taste for imported food and drink, taking no fewer than ten concubines, and traveling by train in a private "flower car" (generally reserved for officials and other worthies with good connections). Unable to ignore rumors of Chien Ying-fu's malfeasance, Chien Yü-chieh conducted an audit which revealed that Chien Ying-fu had in fact stolen from Nanyang. Embarrassed, Chien Yü-chieh and Chien Chao-nan's widow, P'an Hsing-nung, tried to dispose of the matter by quietly helping Chien Ying-fu repay the amount out of their own pockets.[142]

Covering up for Chien Ying-fu may have saved the Chien family from a scandal, but, perhaps partly because of Chien Ying-fu's example, corruption soon spread throughout the company. Near the end of his life, Chien Chao-nan had chastised Chien Yü-chieh for retaining managers of questionable integrity, and after Chien Chao-nan died, Chien Yü-chieh became still more indulgent toward his colleagues. Though apparently not corrupt himself, Chien Yü-chieh permitted illicit dealings in every department of the company. The department responsible for purchasing Chinese-grown tobacco, for example, became notorious for the amount of squeeze that was pocketed by those who provided the links in the long chain of marketing transactions that stretched from the tobacco fields of Shantung to the factories of Shanghai. Equally open to bribes, Nanyang's advertising department commonly accepted kickbacks from newspapers in exchange for commitments to buy advertising space.[143] And the company's marketing system was the most riddled with corruption of all. In branch offices throughout the country, Nanyang's distributing agents stole from the company by reporting lower profits than had actually been earned and keeping the difference. According to P'an Hsu-lun, an accountant who had earned a Ph.D. at Columbia University while on a Nanyang scholarship, such thefts were commonplace during Chien Yü-chieh's presidency because the company's system of surveillance and control over its branches almost completely broke down. After conducting a careful audit of Nanyang's books in 1929, P'an concluded,

> The company's business has spread throughout the country . . . but the eyes and ears of the company do not work perfectly. If only receiving a monthly report, how can we know the true story? . . . The stealing has gone on for years and the losses, when exposed, will amount to many 10,000s [of *yuan*] which are gone forever. Furthermore . . . the branch offices have not collected bills for the company properly, and the company's sluggish management has lost opportunities again and again. All of these things are because the main office of the company has not closely supervised the branch offices.[144]

As a result of this "sluggish management," Nanyang's administrative costs escalated sharply between 1925 and 1929, rising from 20 to 30 percent of the wholesale price of each cigarette produced.[145]

Thus, even as innovative management had helped Nanyang to enjoy a postwar golden age, so mismanagement helped bring the golden age to an end. Bereft of Chien Chao-nan's leadership, his successors were not able in the 1920s or any time thereafter to supervise Nanyang's specialized departments, curb corruption, and earn profits as successfully as he did in the early 1920s. Acknowledged as the leader of the Chiens' family as well as their business, Chien Chao-nan proved to be irreplaceable.

And yet, the death of Chien Chao-nan did not cause Nanyang to collapse or force the other Chiens to capitulate to BAT. His successors continued to compete with their Western rival, but they adopted a somewhat different business strategy than Nanyang had used under Chien Chao-nan's leadership. Whereas he had concentrated on the problems of business management—accumulation of capital, improvement of technology, procurement of raw materials and the like—his successors were more concerned, as they had been all along, about identifying Nanyang with patriotic movements and allying it with a potentially strong central government. As events transpired in China between 1924 and 1930, these concerns were crucial to the outcome of their rivalry with BAT, for the fates of both companies came to depend more than ever before on changes in the relationship between business and politics.

7.

Business and Politics

BETWEEN 1923 and 1930 the emergence of two Chinese political parties, the Kuomintang and the Chinese Communist Party, changed the relationship between business and politics in China[1] and impinged on the rivalry between Nanyang and BAT. During the postwar golden age, as noted in Chapter 6, the two big businesses had been politically unbridled—freed from paying likin by the Peking government's decision to exempt them from this local tax in 1921 and spared difficulties with all but a few brief boycotts, strikes, and other politically inspired protests. But between the time that Communists began to join the Kuomintang (as individuals, not as a party) in 1923 and the nominal unification of China under the Kuomintang government in the late 1920s, members of the Kuomintang and Chinese Communist Party led mass movements and instituted governmental policies that the directors of BAT and Nanyang could not afford to ignore.

The management of each company reacted to these political initiatives in markedly different ways. Between 1924 and 1926 the Chiens were gradually converted from ambivalence to enthusiasm for the Kuomintang while BAT remained unwaveringly hostile toward it. And yet, paradoxically, after the Kuomintang had penetrated the Lower Yangtze River Valley and taken Shanghai in 1927 (with the support of the Chiens and over the resistance of BAT), the Chinese company's sales and profits fell and the Western company's sales (and presumably profits) rose. Operating under Kuomintang rule in Shanghai during the last years of the decade, BAT recorded the highest sales in its history in China while Nanyang plunged deeply into the red, losing ¥2.25 million in 1928 and ¥3.2 million in 1929[2] and bringing the era of serious rivalry in the cigarette industry to an end.

This paradox raises questions about the relationship between business and politics in the rivalry between Nanyang and BAT. Before 1927 why did the Chinese businessmen in the tobacco industry become supporters of the Kuomintang and why did BAT oppose it? And after the Kuomintang began governing Shanghai in 1927, why did the rivalry between Nanyang and BAT take a sudden turn against Nanyang, the Kuomintang's past supporter, and in favor of BAT, the Kuomintang's past opponent? To answer these questions, it is necessary to analyze the two big businesses' perspectives on four important political developments of the late 1920s: Kuomintang and Communist leadership of the labor movement following the reorganization of the Kuomintang in 1924; the May Thirtieth Movement of 1925; the Northern Expedition of 1926 and 1927; and the establishment of the Kuomintang's government in the Lower Yangtze region between 1927 and 1930.

Nanyang's Early Ambivalence toward the Kuomintang: The Strike of 1924

The Chiens' decision to join the Kuomintang and campaign on its behalf was made gradually between 1924 and 1926. At the time of the party's reorganization in 1924, these Chinese businessmen were by no means fully committed to the Kuomintang. At best, their attitude toward it was then characterized by ambivalence. On the one hand, they were attracted by moderates in the Kuomintang, especially nationalistic labor leaders who had organized strikes against foreign-owned firms and had spared Chinese-owned ones. On the other hand, they were dubious about militants in the party, especially those labor organizers who opposed all capitalists, Chinese as well as foreign. Before 1923 these two groups of labor leaders had been readily distinguishable, for they had generally belonged to entirely separate parties, the moderates to the Kuomintang and the militants to the Chinese Communist Party. And while the line of demarcation between the two parties had been clear, the Chiens had shown their support for the Kuomintang moderates by donating ¥20,000 worth of rations to the Kuomintang Army in 1922. After the Kuomintang began to admit Communists in 1923, however, the distinction between moderates and militants became less obvious. Ostensibly, moderates and militants joined forces within the Kuomintang and agreed on a common policy, but in fact they continued to compete for the leadership of the labor movement.[3]

The tension between the two groups of competing labor leaders within the Kuomintang left the Chiens uncertain about the party's position on labor but still determined to court Kuomintang moderates. So in early 1924 they established a tangible link between themselves and the party by appointing two Kuomintang members who fully identified with management vis-à-vis labor, Kuang Kung-yao and Li Yuan, as supervisors in

Nanyang's Shanghai factory.[4] Unfortunately for the Chiens, this experiment backfired. Hired to help Nanyang avert strikes, Kuang and Li were accused of starting one; recruited to improve relations between Nanyang and the Kuomintang, their role in the strike ultimately drove the Chinese business and the Chinese political party apart.

Labor leaders who called upon workers to strike in September 1924 left no doubt that their principal grievance was the conduct of the Kuomintang members who worked as supervisors in Nanyang's Shanghai factory, Kuang and Li. The strikers charged that Kuang and Li had consistently acted as "henchmen" for management, indiscriminately punishing, frequently cashiering, and generally oppressing (*k'o-ya*) workers. These abuses had been tolerated, the strike leaders said, until September 8, 1924, when Kuang and Li tried to deprive the workers of leadership by arbitrarily firing two women workers who were officers in the Nanyang Tobacco Staff and Workers Club, the labor organization which had been created with Chien Chao-nan's approval two years earlier. Piqued by this incident, more than 7,000 workers followed their leaders off the job on September 9 and refused to return until Kuang and Li were dismissed, the two women workers were reinstated, harsh rules governing workers' conduct were abolished, strike pay was given, and wages, annual bonuses, and monthly rice allowances were improved.[5]

Though the Chiens might have forestalled the strike by discharging Kuang and Li, they chose instead to stand firmly behind the two supervisors, and as a result, the strike ensued. The factory was closed for two days, September 9 and 10, 1924, and when reopened it became the site of ugly confrontations between strikers and strikebreakers. On one side the company's men, according to strikers, included more than thirty mercenaries (*wu-shih*) and gangsters (*liu-mang*) and a hundred Shantungese thugs (*Shan-tung ta-shou*), who attacked people on the picket line, threatened them with weapons, and even dragged some into the factory against their will. On the other side, the striking workers organized teams to patrol the streets and tried to bar strikebreakers from entering the factory. Though both sides resorted to violence, this was an uneven battle, according to later recollections of the strikers, for they could not stop cars and trucks from carrying strikebreakers into the factory.[6]

Troubled by the violence and eager to attract workers back to the factory, the Chiens presented their case to the Kuomintang branch in Shanghai and, as they had hoped, received a sympathetic hearing. When the Chiens insisted that they and Kuang and Li had no choice but to fire workers who broke the factory's rules, the Kuomintang's newspaper, the *Republican Daily* (*Min-kuo jih-pao*), obligingly published their interpretation of the causes of the strike. And when the Chiens sought to end the strike by arbitration, Kuomintang members in labor, student, and civic organizations willingly served as mediators between management and labor.[7] In taking these actions, the moderates in the Kuomintang tried to position themselves between management and labor, publishing both the Chiens' and the strikers' statements in the party newspaper and acting as mediators between the two sides. But, despite their efforts, the strike continued.

If heartened by the Kuomintang moderates' willingness to mediate the dispute, the Chiens were infuriated by the response from militants. Unlike the moderates, the militants felt that anyone who showed a measure of sympathy for the Chiens thereby became an enemy of the strikers. Expressing this view in a series of articles published during the fall of 1924 in the principal Communist journal of the time, *The Guide Weekly* (*Hsiang-tao chou-pao*),[8] the militants charged that not only the Chiens but all of the moderate elite groups in Shanghai's capitalist society had betrayed Nanyang's workers. Though moderates might have claimed to promote the interests of both management and labor, the militants contended that anyone taking such an equivocal position was invariably drawn to the side of management by the magnetic power of the capitalists' money. Commercial newspapers, they said, made a pretense of being objective but neglected to cover the strike for fear of offending the Chiens and losing Nanyang's lucrative advertising accounts.[9] Police, they complained, brutally broke up picket lines for the sake of "maintaining order" but carefully refrained from disciplining unruly strikebreakers rather than antagonize the capitalists on whom they depended for regular exactions, unofficial as well as official. Lawyers, they pointed out, eagerly clamored for the chance to represent the Chiens as strike mediators but spurned striking workers who could not afford to pay high legal fees.

According to the militants, moderate Kuomintang members were most hypocritical of all, for they, too, rushed to the assistance of Nanyang's management rather than risk losing the Chiens' patronage. Publishing the Chiens' news releases, mediating between management and labor at the Chiens' request, retaining Kuang and Li within the party despite their mistreatment of workers, the Kuomintang had, according to the militants, joined in the capitalist conspiracy against the workers. The militants traced such promanagement policies to the Kuomintang's leaders in Shanghai, particularly Yeh Ch'u-ts'ang, a member of the Kuomintang's Central Executive Committee, the editor-in-chief of the *Republican Daily*, and the director of the Department of Youth and Women in the party's Shanghai headquarters.[10] And they demanded that Yeh and others in the party's Shanghai branch cease to equivocate and take a firm stand on behalf of the strikers.

To the Chiens' dismay, the militants succeeded in prodding the Kuomintang leaders into taking more vigorous action against Nanyang's management. The militants urged the party to act decisively on three issues. It should, they recommended, expel Kuang and Li from the party, raise funds for Nanyang's strikers, and initiate a boycott against Nanyang's goods. Stung by the militants' criticisms, the leaders of the Kuomintang's branch in Shanghai partially acquiesced, petitioning the party's Central Executive Committee in Canton to adopt the first two proposals but not the most provocative one, the boycott. Upon receiving this petition, Liao Chung-k'ai, chairman of the Kuomintang's Bureau of Workers in Canton and a leftist leader in the party, decided to go even farther than the Kuomintang's Shanghai branch had suggested, adopting all three of the militants' proposals including the boycott. The boycott appealed to Liao as a labor leader because he was able to use it to express his sympathy for Nanyang's workers and his solidarity with militants (as opposed to moderates) in the party. At the same time, as governor of Kwangtung province, he was able to use it to settle a grievance against the Chiens, for they had recently refused to lend money to his government.[11]

With the approval of Liao Chung-k'ai, government officials and labor leaders affiliated with the Kuomintang promoted a boycott against Nanyang's goods between October 1924 and January 1925. The boycott leaders intercepted at least 10 million of the company's cigarettes (valued at ¥50,000) between Hong Kong and Canton and confiscated them on the grounds that their packing cases lacked revenue stamps. The confiscations benefited the strikers not only by publicizing their cause but also by raising revenue, for Liao's government auctioned off the cigarettes and sent the proceeds to the Nanyang workers' strike fund in Shanghai. The boycott culminated in a rally at Shanghai in January 1925 organized by Communist leaders of the Nanyang strike, Hsiang Ching-yü and Yang Yin, and attended by a delegation from Canton and Hong Kong who came to Shanghai for the occasion. The latter group boosted the strikers' hopes for victory, for it included representatives of the Mechanics Union, the Co-operative Union, the Federation of Labor, and other organizations that had recently triumphed over management in the Hong Kong seamen's strike of 1922 and other labor struggles in South China. As the strikers and labor leaders converged on Nanyang's factory in Shanghai, Liao Chung-k'ai and Yang Yin wrote to the Chiens proposing that they accept arbitration and accede to the strikers' demands for reemployment of all strikers, recognition of their union, compensation for their lost income, and general improvement of wages and conditions in the factory.[12]

The boycott angered the Chiens but did not bring them to terms with the strikers. Bristling with indignation, they lodged a formal protest against the confiscations with Liao Chung-k'ai's government in Canton,

pointing out that revenue stamps were required only on individual cigarette packages, not large packing cases, and complaining, therefore, that
their cigarettes had been seized illegally. But they refused to negotiate
with Liao and Yang and flatly rejected the labor leaders' bid for arbitration. They saw no need even to acknowledge the strikers' organization.
For months they had been dealing with a more moderate labor organization, the Nanyang Tobacco Company Factory Friendly Workers and
Staff Fraternal Association (*Nan-yang yen-ts'ao kung-ssu chih-tsao
ch'ang kung-yu chih kung t'ung-chih hui*), which had replaced the striking Nanyang Club in the factory during the second week of the strike and
had solidified its position and gained legitimacy in December by displac-
ing the striking Nanyang Club in the Federation of Labor Organizations
(*Kung-t'uan lien-ho hui*), a group representing over thirty other labor
organizations. Accepting this arrangement and abandoning the strike, all
except 407 of the original 7,000 strikers had returned to work by January
1925 according to the Chiens' reckoning, and the remaining strikers were
neither needed nor wanted by supervisors in the factory. Noting that
1924 had been a bad year for business, the Chiens peremptorily dismissed
the strikers' demands that they improve pay and conditions for workers.
In fact, the Chiens did the opposite, reducing workers' wages, bonuses,
and vacations during the strike.[13] This uncompromising approach left
Liao, Yang, and other leaders with little room to maneuver, and the
strike dissolved in January 1925.

Though the Chiens endured the strike without yielding concessions,
they discovered in the process that they had enemies in the Kuomintang.
The militants' leadership of the strike against their Shanghai factory and
Liao Chung-k'ai's sanction for the boycott on their goods in South China
showed that Communists and other leftists within the Kuomintang were
prepared to mobilize workers against Nanyang even though it was
owned by Chinese rather than foreigners. And the Kuomintang moderates' endorsements of some of the militants' calls for action against Nanyang suggested that even the moderates were not dependable allies for
the Chiens. Accordingly, it is not surprising that the Chiens remained
aloof from the party's affairs in late 1924 and early 1925. Not until mid-
1925 did they become more favorably impressed with the party's possibilities when they witnessed its members' participation in the May Thirtieth movement.

The May Thirtieth Movement of 1925

During the May Thirtieth movement, the members of the reorganized
Kuomintang redeemed themselves in the Chiens' eyes by attacking foreign capitalists, sparing or aiding Chinese ones, and thus redressing Sino-
foreign rivalries—including the one between Nanyang and BAT. Like the

Nanyang strike, the May Thirtieth movement originated in a dispute between a factory supervisor and workers in Shanghai, but this time the supervisor was a Japanese employed by a Japanese-owned textile mill who shot and killed a Chinese worker. Again as in the Nanyang strike, militants called for a mass protest, but this time it was mounted against foreigners rather than Chinese and on a far greater scale than the strike and boycott against Nanyang had been. On May 30, 1925, three thousand demonstrators marched on the International Settlement in Shanghai to protest the slaying of the Chinese worker and other recent conflicts between Chinese labor and foreign management. As the protesters approached, a British officer in the International Settlement's foreign-controlled police force whose unit was supposed to quiet the crowed ordered his men to fire on it, killing ten people—some of whom were BAT workers—and injuring more than fifty others. As the news of the incident spread, anti-British strikes and boycotts were organized and attracted a large following. Throughout the summer of 1925 people from all strata of Chinese urban society in a dozen cities joined in these demonstrations, which have become known collectively as the May Thirtieth movement. The significance of the movement as a whole is still not fully understood and deserves further research, but it clearly had a profound effect on the Chiens, for it created opportunities not only to raise Nanyang's sales at the expense of BAT but also to become allied with leaders of the reorganized Kuomintang.[14]

During the summer of 1925, militant labor leaders, especially Communists, led May Thirtieth strikes against BAT as ardently as they had led the strike against Nanyang a few months earlier. Like many of the militant leaders of the May Thirtieth strike in Shanghai, the two leaders of the strike against BAT, Yang Chih-hua and Chang Tso-chen, were young Communists affiliated with the Shanghai General Union (*Shang-hai tsung kung-hui*).[15] Looking back on their May Thirtieth strike, former BAT women workers interviewed in China in the late 1950s remembered Yang with special affection. Wife of the brilliant Communist intellectual Ch'ü Ch'iu-pai, she was a prominent leader of the May Thirtieth movement and later the head of the women workers' division of the National Labor Organization in the People's Republic of China. The workers also had vivid memories of Chang, an eloquent advocate of their cause who was later killed by the Kuomintang during the anti-Communist purge of 1927. Under the guidance of these young Communists, all 15,000 of BAT's workers in Shanghai struck in early June 1925—providing almost 10 percent of the 160,000 workers who went on strike in Shanghai altogether—and, under the direction of another Communist, Hsu Pai-hao, workers in BAT's Hankow factory held a May Thirtieth strike at the same time.[16]

Naturally, the Chiens felt that their complaints during the Nanyang strike about the impropriety and illegality of leftist agitators interfering with the internal affairs of a business did not apply here. On the contrary, now that the labor leaders had their sights set on BAT rather than Nanyang, the Chiens unhesitatingly offered material aid to the labor movement, contributing ¥100,000 to the fund for Shanghai workers' strike pay, an amount representing about 12 percent of the total collected for strike pay from all sources (including labor unions) during the May Thirtieth movement.[17] Sustained by this financial aid, the May Thirtieth strikes kept BAT's workers in Shanghai off the job for 124 days, leaving the company's largest factories in China idle between June 4 and October 5. This was not only the longest of the May Thirtieth strikes—rivaled only by those against the Japanese-owned Naigai Wata Corporation and the British-owned Ewo Mills—but was the single longest strike in the entire history of the Chinese labor movement in Shanghai.[18]

While the Chiens solidly backed these strikes, they showed still greater enthusiasm for the May Thirtieth boycott. The Kuomintang government's major contribution to the boycott was a new fiscal policy put into effect in South China which discouraged consumption of foreign goods by requiring foreign cigarette companies to pay a 40 percent surtax that was not levied on Chinese companies. Pursuing the idea that Chinese should buy Chinese rather than foreign products, the Chiens used elaborate tactics to persuade consumers to switch from BAT to Nanyang brands not only in South China but throughout the country. They published and distributed handbills, posters, and newspaper advertisements identifying and railing against BAT brands and suggesting Nanyang brands as substitutes at equivalent prices. They appropriated BAT advertising sites, tore down BAT posters, and replaced them with advertising for Nanyang.[19] And in one of their most subtle campaigns, they sent recently recruited young agents to infiltrate the ranks of student leaders of the boycott. Once accepted by their peers, these pseudo-students supplied leaders of the boycott with BAT brand names and trademarks and guided boycott teams to BAT storage depots so these would be sealed.[20]

This May Thirtieth drive by Nanyang against BAT reached all of China's major markets. In Shanghai it was directed by Lao Ching-hsiu, a Cantonese resident of the city, who, just prior to the May Thirtieth movement, had been charged by his fellow members of Nanyang's board of directors "to make social connections with all of the organizations in the upper, middle, and lower echelons in Shanghai and to campaign there" on behalf of the company.[21] In other cities throughout North China, the Lower and Middle Yangtze regions, and South China, British consular officials reported efforts by Nanyang's agents to sabotage BAT's distributing and advertising systems and criticized the Chinese company for using the "most disreputable" competitive techniques (in Peking), "unscrupulous and unfair" promotional methods (in Nanking), and a "campaign of slander" (in Canton). On the basis of these reports, the British commercial attaché in Peking estimated that the total number of May Thirtieth "agitators" on Nanyang's payroll was in the thousands.[22]

The Chiens promoted the May Thirtieth strikes and boycott against BAT in concert not only with militant labor leaders and the Kuomintang government but also with other members of the Chinese elite. For example, the Chinese Newspapers Association resolved not to print BAT advertisements, and as a result, Chinese newspapers belonging to this organization carried no BAT advertising for more than a year. Another group of Chinese journalists attacked BAT for the movies that it had used as advertising. Just prior to the May Thirtieth movement, Chinese film critics had denounced BAT's movies for depicting Chinese as drunkards, prostitutes, gamblers, criminals, beggars, and "slaves of Westerners"— an inferior race "unfit to look at"—and at the height of the May Thirtieth movement the Chinese members of the BAT movie division's board of directors endorsed this criticism and resigned from the board in protest against the insulting characterizations of Chinese people in the company's latest releases. These complaints convinced BAT's directors that the company's movies had become self-defeating as an advertising device. So, even though the company's moviemakers had filmed more footage than anyone else in China in 1924, BAT's movie division was abolished and its cameras and photographic equipment were sold in 1925. Deprived of the use of newspaper advertising and films, BAT lost still more of its commercial glitter during the May Thirtieth movement when Chinese owners of *Shen-pao*, one of Shanghai's oldest and best known newspapers, turned off the lights in the cigarette company's big neon sign on top of *Shen-pao*'s building.[23]

While Chinese newspapermen disrupted BAT's advertising system, Chinese shipping magnates disrupted its marketing system. Yü Hsia-ch'ing, chairman of the board of the Ningpo Steamship Company, the largest private shipping company in China, cut off much of BAT's riverine distribution by refusing to transport BAT goods during the May Thirtieth boycott. Yü's role in the May Thirtieth movement as president of the Shanghai General Chamber of Commerce (*Shang-hai tsung shang-hui*) and as an intermediary between Chinese and foreigners during the movement will remain shadowy until his career has been fully researched, but whatever other actions he might have taken, in this instance he supported the May Thirtieth boycott by refusing to carry BAT's goods—perhaps because he was profiting from the boycott against his foreign competitors in the shipping business in the same way as the Chiens and other Chinese manufacturers profited from the boycott

BAT neon sign in Shanghai, the single largest and most expensive advertising
device in China, 1930

against their foreign competitors in consumer goods industries. Yü's leading Chinese competitor, the China Merchants' Steam Navigation Company, also joined the boycott and refused to touch BAT's goods.[24]

At lower levels in BAT's distributing system, other Chinese participants in the boycott prevented delivery and sale of BAT's goods. Boycott
teams intercepted BAT's own steam launches during the boycott, for
example, and confiscated or destroyed the cargo—in one instance setting
BAT's cigarettes ablaze using oil that had been taken from Shell Oil's
subsidiary, Asiatic Petroleum Company, another foreign firm being boycotted. At a still lower level in the marketing system, participants in the
boycott confiscated goods at loading docks in inland cities. In Tsinan,
capital of Shantung province in North China, boycott leaders ordered
carters not to deliver any incoming BAT goods and arrested those who

violated the order.[25] And at the lowest levels in the marketing hierarchy, activists wearing armbands inscribed "Examiners of British and Japanese Goods" checked the stocks of local jobbers, shopkeepers, and hawkers and punished anyone caught carrying contraband. People found guilty of breaking the May Thirtieth boycott by selling BAT goods were known to have been beaten, fined, imprisoned, and in one case, dragged through the streets in chains.[26]

Rather than meet this formidable campaign head on, BAT tried to salvage sales by asserting that some of its brands should be spared from the anti-British May Thirtieth boycott because they had been made in China or had come from the United States rather than Britain. BAT's advertising concentrated on defending the American origins of one cigarette above all others, Ruby Queen, a low-priced brand which the Chinese distributor, Yung-t'ai-ho, had made BAT's largest selling cigarette in China and the second largest selling cigarette in the world by the time of the May Thirtieth movement. Ruby Queen had been immediately identified as contraband by participants in the May Thirtieth movement because its Chinese name, *Ta-ying*, could be construed to mean Great English, a phrase that immediately caught the eye of a Chinese boycotting British goods. As a result, according to a leading BAT dealer, the rate of monthly sales of Ruby Queen in Shanghai fell from the usual 50 million to a mere 5 million within six weeks after the May Thirtieth incident. Alarmed at the drop in sales, BAT immediately cut back production of this brand at its factories in Virginia by about one half.[27]

In hopes of restoring the popularity of Ruby Queen in China, BAT directors responded to the boycott by changing the Chinese name of this brand and by carrying out a massive advertising campaign during the first weeks of the May Thirtieth movement in which the old Ruby Queen posters and signboards all over China were replaced with new ones emphasizing that it was made in America, not England, and publicizing its innocuous new Chinese brand name, Red Pack (*Hung-hsi pao*). To substantiate their advertising claims, BAT directors persuaded Edwin Cunningham, the American Consul-General in Shanghai, to supply a letter in early July 1925 certifying that Ruby Queen was manufactured in the United States, and upon receipt of his letter they published it with banner headlines in Chinese newspapers. According to the accompanying gloss, supplied in Chinese by BAT, this official document proved that Ruby Queen was American rather than English and provided the basis for legal action against anyone who dared to interfere with the transport or sale of it. Cunningham was shocked to see his letter incorporated into BAT advertisements, and his fellow American consular officials in China lamented that BAT had misused the letter because they felt that "this form of American official support of the British-American Tobacco Co.,

Ltd., [was] perhaps somewhat undignified and unbecoming" for members of the diplomatic corps, but the company had widely circulated the letter before American officials could do anything about it.[28]

Maneuvers such as these by BAT did not prevent the May Thirtieth boycott from driving down the company's sales in every region of China proper. In the words of one troubled BAT executive, the company received "peculiar attention." In fact, British and American consular officials reported that BAT was singled out and boycotted more intensively than any other foreign firm.[29] Throughout China BAT's sales came crashing down. In North China the boycott cut BAT sales in half at Tsingtao and Tsinan and by 90 percent in Peking. In South China, according to a British official, sales of the company's cigarettes at Canton "completely stopped." In the Upper Yangtze region, another British official observed that BAT cigarettes had "vanished" from Chungking. In the Middle Yangtze region BAT's sales at Changsha dropped by 85 million, valued at $350,000, between July and September in 1925, compared to sales there during the same period in the previous year. And in the Lower Yangtze region, the boycott was most effective of all; the company's sales decreased by 95 percent in Shanghai and Chekiang province during the six weeks following May 30 and by almost as much in the provinces of Kiangsu (outside Shanghai) and Anhwei. The boycott began to wane in the autumn of 1925, but it was revived sporadically in all these regions during the winter of 1925-26 and never ceased to make BAT its chief target.[30]

As the market for BAT's goods collapsed, the market for cigarettes made by Chinese-owned firms boomed. New Chinese companies were hastily founded to produce substitutes for BAT's boycotted brands—raising the number of Chinese cigarette companies in Shanghai from 16 to 52 in 1925—but most of these remained small and none benefited from the boycott so richly as Nanyang. In regions where Nanyang's brands were well established, British consular officials estimated that its sales shot up at rates ranging from 60 to 300 percent during the May Thirtieth movement, and in at least one region that Nanyang had not previously penetrated, the Upper Yangtze, it took advantage of this opportunity to introduce its cigarettes for the first time. The company's total sales for the month of June 1925 were valued at between ¥4 and ¥5 million, more than twice the average for monthly sales during the previous year.[31] Moreover, its sales reflected only a fraction of the growth in the demand for its goods, for its manufacturing system was able to fill only one fourth of the orders that poured in from the various branches in its distributing system.

Disappointed not to be taking full advantage of the dramatic upswing in demand, the Chiens desperately tried to expand production. They recruited additional workers (including some on strike at the time against BAT), gave raises in pay, added night shifts, and kept machines in factories running around the clock.[32] On July 6, 1925, their problem of underproduction was further aggravated when foreigners on the Municipal Council of the International Settlement in Shanghai suddenly shut down the Settlement's electric power plant, cutting off the electricity on which Nanyang's factories and other industries in the area depended. Since British and Japanese manufacturing companies in the International Settlement had already been idled by the May Thirtieth strikes, this action affected only Chinese-owned factories. It was initiated by foreign industrialists who resented the success of their Chinese competitors in cigarettes, matches, textiles, and other industries during the May Thirtieth boycott.[33]

The Chiens did not, however, permit the loss of electricity in the Settlement to stop them for long. Moving swiftly, they rented a large warehouse from French merchants in the Lu-chia-ts'ai (Louza) area of Shanghai outside the International Settlement, where electricity was still available, and called upon the M.I.T.-educated manager, Ch'en Ch'i-chün, to set up an improvised factory where production could continue without delay. Almost overnight machinery was transferred to this Little Nanyang, as it was called, and 2,700 workers soon were back at their jobs, making Nanyang cigarettes. But even though the Chiens made every effort to keep their manufacturing system operating at peak levels, they were still unable to satisfy the rapidly rising demand for their goods. When their machinery fell behind or they ran out of tobacco and other raw materials, they distributed every cigarette that they had in storage— down to and including stocks of moldy ones—but the incoming orders still far exceeded the outgoing shipments.[34]

As the Chiens' sales soared, so did their spirits, for they felt that their emerging alliance with political leaders of the May Thirtieth movement and their success at exploiting the mass support for May Thirtieth strikes and boycotts changed the fundamental basis of their rivalry with BAT. Confident that this combination of political and social support guaranteed that Nanyang would grow rapidly, they announced in July 1925 a proposal for large-scale expansion of the company. According to the plan, they expected to decentralize their manufacturing system by building new factories in Hankow, Tientsin, Tsingtao, and Manchuria in addition to their existing ones in Hong Kong and Shanghai. In adopting this approach, the Chiens hoped to make their business more equivalent with BAT, which had already decentralized its manufacturing system (see Chapter 6). The Chiens had become painfully aware of the advantages of this kind of decentralization during the early 1920s when floods and warfare had cut Nanyang off from markets that were distant from its factories in Hong Kong and Shanghai while BAT, relying on its regionally based manufacturing system, was able to market its goods in most of China's major markets without interruption.[35] Though the Chiens' proposal thus represented a possible solution to a pressing problem and though it might have made Nanyang more competitive with BAT in the long run, it was never carried out because the Chiens had to reckon with competition from BAT in the short run.

When the May Thirtieth movement showed its first signs of subsiding in August 1925, the Chiens were forced to face up to the fact that the basis for their rivalry with BAT had not changed so permanently as they had at first supposed. Despite the leadership of political parties and the breadth of the mass support for the May Thirtieth movement, this boycott, like previous ones, did not have a lasting effect on consumers' preferences. Accordingly, as the Chiens anticipated the return of BAT to the market at the end of the summer of 1925, they had to reconsider their ambitious plans and had to postpone construction of any new factories except the one in Hankow, they acknowledged, "until the market for national goods once again becomes favorable" as it had been at the height of the May Thirtieth boycott.[36]

In the months that followed, as the Chiens had expected, BAT reclaimed its place in the market. All May Thirtieth strikes against BAT outside Shanghai were settled by mid-September, and within that same month BAT was "doing such a roaring business," according to one newspaper, that it was compelled to work its big Hankow plants overtime.[37] Aware of BAT's recovery, the Chiens lowered their expectations further by the end of 1925, jettisoning a plan to expand an existing plant, the improvised Little Nanyang in Shanghai. And in the spring of 1926, they retrenched still further. Claiming that BAT's use of heavy weapons—the advertising blitz, dumping of free samples, underpricing of equivalent brands—had made them casualties in a commercial war (shang-chan), they felt compelled to relinquish two pieces of property in Shanghai on which factories might have been built if the Chiens had not needed the 210,000 taels in cash they received for the sale.[38]

And yet, even if the May Thirtieth movement did not give Nanyang permanent advantages over BAT, it raised the Chiens' estimation of political leaders in the reorganized Kuomintang because it showed that militants who organized strikes and Kuomintang members who participated in the boycott were willing and able to lead mass movements that would hold big foreign businesses such as BAT in check. The Chiens' profits from the movement did not catapult it back to the halcyon days of the postwar golden age—for the year 1925, profits totaled ¥1.22 million, about 30 percent as much as the annual average profits between 1920 and 1923—but profits for 1925 were three times greater than for

1924, and according to the Chiens, the improvement from one year to the next was entirely attributable to the May Thirtieth movement.[39]

If members of the reorganized Kuomintang had mobilized support for a movement that profited Nanyang, wouldn't a government under the Kuomintang, once in power throughout China, enable the Chiens to sustain Nanyang's growth and make their company consistently more competitive with BAT? In the aftermath of the May Thirtieth movement, the Kuomintang government at Canton encouraged the Chiens to believe that this would be the case by continuing to tax Nanyang less heavily than BAT. Though Chinese-owned cigarette companies (including Nanyang) ceased to be exempt from Kuomintang taxes in late 1925 and early 1926 (as they had been during the May Thirtieth movement) and had to buy Kuomintang tax stamps for all goods (whether imported or made in China), they were required to pay only 20 percent ad valorem for the stamps while foreign-owned firms (including BAT) had to pay 40 percent.[40]

In deciding whether to join forces with the Kuomintang, the Chiens had to balance the prospects of benefiting from such preferential Kuomintang tax policies against the prospects of suffering from the policies of militant Kuomintang labor leaders. In 1924 the Chiens had battled with the militants not only during the Nanyang strike but also in the anti-Kuomintang uprising of the Hong Kong-Canton Merchant Volunteers (which was organized by a member of the Nanyang board of directors, Ch'en Lien-po) and on other occasions. But within a year after Sun Yat-sen's death in March 1925, the Chiens no longer had reason to feel threatened by these militant labor policies because the Kuomintang's major leftist leaders suddenly disappeared from the ranks of the party. Liao Chung-k'ai, for example, was murdered by an assassin in August 1925, and a number of other militants were purged from the party and arrested in March 1926 by the Kuomintang's new leader, Chiang Kai-shek.[41] If Chiang's purge of militants in Canton appealed to the Chiens, then his plans to extend his power outside South China apparently appealed to them still more, for in late 1925 and 1926 they began actively participating in preparations for a military and political campaign that was intended to defeat warlords and imperialists and unite China under a national Kuomintang government: the Northern Expedition.

The Northern Expedition: From Canton to Shanghai, July 1926-April 1927

Just prior to the Northern Expedition, which started at Canton in July 1926, Chien Ch'in-shih, a nephew of Chien Yü-chieh and a manager for Nanyang, identified himself, his family, and his company with Chiang

Kai-shek's cause. In December 1925 he accepted a post as senior advisory officer in the Kuomintang government, and during the early summer of 1926, he was named president of the Kuomintang's Alliance of Workers, Merchants, Intellectuals, and Peasants of Kwangtung Province. In these positions, his job was to appeal to businessmen in Canton and Hong Kong on behalf of the Kuomintang government, urge them to make financial contributions to the Kuomintang, and persuade them to settle strikes and boycotts so that they and workers in their factories would unite in support of the Northern Expedition.[42]

Once the Northern Expedition was under way, the Chiens' service to the Kuomintang government was well rewarded, for they profited even more from anti-foreign protests against BAT during Chiang Kai-shek's drive northward than they had during the May Thirtieth movement. As Kuomintang troops advanced, taking Hankow by the end of 1926 and Shanghai in March 1927, strikes again closed down BAT's manufacturing system. Preparing for the arrival of the Northern Expedition, BAT's workers in Shanghai held strikes in July and October 1926 and February 1927; and during the Kuomintang's takeover of Hankow in November and December 1926, the two BAT factories there were closed and were kept closed while BAT directors bargained with labor leaders and Kuomintang officials throughout the following year.[43] Meanwhile, BAT's purchasing system felt the effects of the Northern Expedition too. Upon reaching Honan province, troops attacked the company's tobacco leaf reception station near Hsuchang, burning the building, auctioning the machinery, expropriating the remaining property, and arresting the comprador who had originally bought the land on behalf of BAT's foreign management.[44] At the same time, BAT's marketing system was disrupted by the boycott which spread throughout provinces in South China and the Middle and Lower Yangtze as the Northern Expedition passed through these regions. Like previous boycotts, this one was probably strongest in Canton, where, according to the United States commercial attaché, Julean Arnold, the company was able to get only about 20 of its 6,000 "B.A.T. hucksters on the streets." Investigating the city in mid-April, he noted that "B.A.T. has recovered but 1/15 or 1/20 of its Canton trade." The boycott might not have hit BAT quite as hard in other regions, but the company felt its effects wherever there was support for the Northern Expedition.[45]

The strikes, boycotts, and other actions against BAT created commercial opportunities for its competitors, which the Chiens, in turn, fully exploited. The Chiens supported the strikes against BAT at least as vigorously as the ones during the May Thirtieth movement. This time, besides contributing to a fund for BAT workers' strike pay, the Nanyang owners also dispatched agents to distribute anti-BAT handbills around the Western company's factories. The Chiens capitalized on BAT's losses in the countryside too, replacing it as the leading purchaser of tobacco leaf in Hsuchang as soon as the Western company was forced to evacuate.[46] In the cities the Chiens also turned the boycott to Nanyang's advantage. The boycott featured such slogans as, " 'The blue and white sun flag [of the Kuomintang]' should smoke, but see that the cigarettes are Chinese made,"[47] and boycott supporters urged smokers to abandon the very BAT brands with which Nanyang competed most directly. Behind such anti-BAT propaganda, a British consular official noted, "The hand of the Nanyang Brothers Tobacco Company would seem to be evident."[48]

By now experienced at this kind of commercial manipulation, the Chiens obtained briskly rising sales for Nanyang during the fall of 1926 in Kwangtung, Kwangsi, Kiangsi, Hunan, Hupeh, and other provinces that the Northern Expedition reached. As during the May Thirtieth movement, such campaigns seemed to be well worth Nanyang's investments of capital and effort, for the benefits were reflected in the Chinese company's higher profits. In 1926 Nanyang's profits were nearly twice as high as in 1925, ¥2.3 million as compared with ¥1.22 million.[49]

As the Northern Expedition progressed, the Chiens profited not only from the anti-foreign strikes, boycotts, and other mass movements that Chiang Kai-shek's advancing armies touched off in late 1926 but also from Chiang's decision to end labor protests in April 1927 after he had taken control of Shanghai. In January and February 1927, the strikes in Shanghai had begun to turn against Chinese industrialists, causing the closure of Chinese-owned as well as Western-owned factories, including Nanyang's.[50] As the strikes had spread throughout Shanghai prior to Chiang Kai-shek's arrival in the city, Chinese businessmen had begun to panic, fearing a social revolution in which workers would seize control of factories under Kuomintang rule. A few weeks after Chiang arrived, however, he put the Chinese industrialists' fears to rest on April 12, 1927, by unleashing the "white terror," an anti-Communist crusade in which Chiang's troops and secret police suddenly descended upon strikers and brutally attacked and arrested or killed militant labor leaders—non-Communists as well as Communists. In return, businessmen in Shanghai reportedly paid Chiang Kai-shek ¥10 million.[51]

The Chiens were undoubtedly pleased with Chiang Kai-shek's repression of militant labor leaders in Shanghai, but this event was a confirmation rather than the origin of their conversion to the Kuomintang's cause, for their commitment to the Kuomintang had grown gradually in 1925 and 1926, and at least one of them had placed himself unequivocally in Chiang Kai-shek's camp by joining and serving the Kuomintang government in Canton several months before the Northern Expedition had reached Shanghai. Nonetheless, the white terror of April 1927 was im-

portant to the Chiens, for it showed that government under the Kuomintang in Shanghai would not sanction militant strikes and boycotts
against Chinese companies of the kind that Liao Chung-k'ai had promoted against Nanyang in 1924. Such a policy appeared to eliminate any
basis for doubting that the Kuomintang would establish a government
that would be ideal for Chinese businessmen. By taking stands against
foreign business during the May Thirtieth movement and Northern Expedition and by wreaking havoc with militant labor in April 1927, the
Kuomintang's leaders seemed to have aligned the party against Chinese
businessmen's enemies exactly as the Chiens had hoped. But the Chiens
soon discovered that the Kuomintang's opposition to one's enemies did
not guarantee the Kuomintang's support for oneself, for in May and June
1927 the alliance the Chiens had carefully built up with the Kuomintang
over the past two years suddenly fell apart.

The Establishment of Kuomintang Rule in Shanghai, May 1927-1930

In May 1927, just weeks after Chiang Kai-shek's government had penetrated Shanghai and cracked down on labor, it began to crack down on
business. Apparently because the Kuomintang leaders were desperate for
funds to extend their political and military power and thus unify the
country, their government began extracting large amounts of revenue
from businessmen in Shanghai.[52] The signal of a change in the Kuomintang's relations with Chinese businessmen in Shanghai, at least in retrospect, was Chiang's dismissal of T. V. Soong as Minister of Finance in
April 1927.[53] Prior to that time, Soong had tried to simplify and stand-
ardize the enormous variety of taxes being levied by provincial and local
governments without drastically raising taxes, and as part of this program, he had instituted a "consolidated tax" (t'ung-shui) on tobacco of
12.5 percent ad valorem soon after the Northern Expedition had reached
Hankow in December 1926.[54] But after Chiang removed Soong from
office, the Kuomintang's demands on big businessmen in Shanghai escalated sharply. In May 1927 the government asked Chinese bankers and
businessmen in Shanghai to subscribe to a loan of ¥30 million, and of
this amount, the largest single portion was assigned to the Chiens,
¥500,000.[55] At the same time, the government raised taxes on all industrial commodities, bringing the tax on tobacco and tobacco products to a
peak of 50 percent ad valorem in July 1927. By international standards,
this rate of taxation on cigarettes was not exorbitant, but Chinese cigarette companies—accustomed to light tax burdens and lax enforcement
by provincial and local officials—were unable to bear it.[56]

Complaining bitterly about Kuomintang cigarette taxes in the late
1920s, all Chinese-owned cigarette companies went down and most went

under. Of the 182 such firms operating in Shanghai in 1927, only about one third survived until 1930.[57] Since all but 14 of the 182 were new small companies that had sprung up in response to anti-BAT boycotts during the May Thirtieth movement and the Northern Expedition, many might possibly have failed for lack of capital and expertise regardless of the Kuomintang government's tax structure, but the Chiens agreed with the owners of the failed firms that the demise of the industry was the fault of the Kuomintang government.[58]

Despite the Chiens' earlier work within the Kuomintang and their support for the Northern Expedition, they were openly hostile toward the government for introducing higher taxes in the late 1920s. When the new excise tax of 50 percent ad valorem was announced in June 1927, for example, they publicly criticized it, complaining that it raised retail prices and lowered consumer demand so drastically that they were forced to cut production in half and would soon have to close their factories in Shanghai altogether. Within the next three years, Nanyang's production in Shanghai did in fact come to a complete halt. In January 1929 the Chiens closed the company's smaller factory there and a year later its larger one.[59] In February 1930, at the time of the latter closure, Chien Ying-fu and Lao Ching-hsiu met with reporters, and in the stillness and silence of a factory that had been bustling almost continuously since opened by Chien Chao-nan thirteen years earlier, they placed the blame for Nanyang's failure to remain competitive with BAT in Shanghai squarely on the Kuomintang government. "Taxation," they complained, "is more than five times what it used to be. We have had to compete with the better financed foreign cigarette companies. We have been unable to increase our prices because, every time we increase them, the Government, who has promised to do everything to encourage Chinese trade, has hampered us by imposing heavier taxation."[60] Staggering under the weight of this fiscal burden, Nanyang's manufacturing system in Shanghai was sinking more deeply into debt every day it operated, they said, so that it was "only fair to our shareholders that we should suspend business instead of being forced to lose more money. We can't afford to lose any more."[61]

While the Chiens publicly criticized the government's tax policies, they privately discussed the problems of making covert payoffs to people in power. In January 1929, as they shut down the first of the two Shanghai factories they eventually closed, Chien Yü-chieh, Ch'en Lien-po, and Lao Ching-hsiu considered trying to win support from political and military officials by giving them stock in Nanyang gratis, but the Nanyang directors were dubious about the idea because of the unstable political atmosphere at the time. Government was in such a state of flux, they observed, that it was like a "wayside inn (ch'uan-she). Today someone is an official, tomorrow he is out." As a result, they feared that any official whom they bribed might fall from power before he had a chance to lobby with the government on their behalf. They couldn't afford to bribe every official, and they felt that "the so-called important people invariably have big eyes" (yen-k'ung pi-ta)—a tendency to seek larger payoffs than anyone is ever prepared to make.[62] Nonetheless, between 1929 and 1931, besides paying taxes, the Chiens spent ¥240,000 on "fees to assist the government" (yuan-tsu cheng-chih fei) and "gifts" (sung-jen li-wu), and they lent another ¥130,000 to two ranking Kuomintang officials, Wang Ching-wei and Ch'en Kung-po.[63]

Besides levying higher taxes and extracting unofficial exactions, Kuomintang officials further antagonized the Chiens by supporting the demands of workers who were idled by the closures of Nanyang's factories in Shanghai. Initially, these workers seem to have expressed their demands without assistance from the government, angrily attacking Nanyang's management for giving neither sufficient notice nor sufficient compensation for throwing them out of work. During the first month of 1929, following the closure of the smaller factory on January 1, some of the 2,700 workers who had lost their jobs appealed to the Kuomintang Municipal Council to force the Chiens to resume production; others distributed leaflets urging shopkeepers to boycott Nanyang's goods; and the majority continued to report for work, with the male workers (who lived in the factory dormitory) sounding the starting whistle and admitting the female workers to the factory at seven o'clock every morning as in the past. The Chiens did not respond to these demonstrations until the workers managed to get inside the Chiens' offices and cover doors, walls, and windows with handbills protesting the lockout. At that point, Chien Ying-fu called the police, who dispersed the workers, injuring several in the process.[64] In closing the other factory, the Chiens tried to avoid this kind of an incident by accommodating the demands of the workers' representatives at the outset, and they succeeded in signing an agreement whereby each of the 4,500 workers formerly employed in the factory was to be paid two months' wages plus a ¥5 bonus. But the crowd of more than a thousand unemployed Nanyang workers that gathered in the streets around the factory vehemently repudiated the agreement that its negotiating team had approved. When a Nanyang staff member emerged to post a copy of the agreement, the workers closed in on him, and according to a newspaper reporter who was present at the time, they shouted, " 'Darng! Darng!' (Hit him! Hit him!). Tearing the poster to a thousand pieces, they assaulted the unfortunate fellow, who was forced to flee for his life. He managed to get inside, but not before he had been badly beaten."[65] The workers then stormed the iron gate at the factory's entrance and were in the process of pushing it down when police arrived and drove them back.

Following each of these incidents in 1929 and 1930, the Kuomintang joined the Chiens and representatives of the workers at the bargaining table for several months of negotiations. These negotiations stirred political controversies reminiscent of the Nanyang strike of 1924, for leftist journalists again, as in 1924, took up the Nanyang workers' cause and accused conservatives in the Kuomintang of conspiring with the Chiens against the interests of labor.[66] But in 1929 and 1930, unlike 1924, the workers, assisted by the Bureau of Social Affairs (*She-hui chü*) of the Kuomintang Municipal Council, forced Nanyang's management to retreat from its original bargaining position and make concessions, obtaining, in both cases, severance pay for workers that was two or three times higher than the Chiens had initially offered. In the end, each worker from the smaller factory received five months' wages or an extra month's wages for every year of employment, and each worker from the larger factory received six months' wages.[67] Moreover, at the time of the settlement in September 1930 the district branch of the Kuomintang acted on other workers' demands by ordering Nanyang to resume production in the larger factory—an order the Chiens reluctantly obeyed, putting a token labor force of 300 back to work in a plant that had previously employed 4,700.[68] In these negotiations, though Kuomintang labor leaders may have gained less for workers than their leftist critics wanted, they extracted more from the Chiens than Liao Chung-k'ai had been able to do a few years earlier. No longer could the Chiens defy orders from the Kuomintang government in Shanghai as they had once defied Liao's orders from Canton.

In retrospect, the establishment of Kuomintang rule in Shanghai had bitterly ironic consequences for the Chiens. In 1930, just three years after welcoming the Kuomintang to Shanghai, they felt that the new government had driven them out of the city. Retreating to Hong Kong, they took extraordinary measures to cover their enormous losses which totaled ¥2.25 million in 1928 and ¥3.2 million in 1929. To pay some of their debts, for example, they liquidated their holdings in "Captain Jack" Gravely's China-America Tobacco Company, selling all $300,000 of their stock in the firm in 1929. In the same year they tried to keep the Shanghai office functioning by transferring ¥800,000 from other branches to Shanghai (¥200,000 from Singapore and ¥600,000 from Hong Kong). When these sums proved insufficient, they finally resorted to borrowing ¥2 million from the British-owned Hong Kong and Shanghai Banking Corporation. At first they had difficulty persuading the bank to lend them such a large amount, for, as the bank's directors pointed out, the China Merchants' Steam Navigation Company—once China's leading shipping line—had recently collapsed and defaulted on a comparable loan. Only after the Chiens agreed to pay 7 percent interest and to mortgage their land, buildings, and tobacco was the loan approved. Though these debts weighed heavily upon the Chiens (and continued to weigh heavily upon them for several years), they prevented their China trade from lapsing by continuing to operate their factories in Hong Kong and by distributing Hong Kong-made cigarettes in China. Eventually they resumed production in Shanghai, but they never again employed as many regular (nonpiece-rate) workers in their main Shanghai factory as they had before 1928, nor did they ever again earn as much in annual profits as they had in the early 1920s.[69]

BAT meanwhile was supposedly subject to the same new higher Kuomintang taxes as Chinese companies were, but unlike Chinese companies, it refused to pay these taxes or even deal with the Kuomintang government while T. V. Soong was out of office between May and December 1927. BAT's directors in Shanghai saw no point in negotiating because, according to Archibald Rose, the Kuomintang's policies showed that the new government leaders wished not to tax but to wipe out BAT. "Taxation has passed out of all control," Rose declared in June 1927, "and is now merely a successful device for destroying our business." The only hope for BAT, Rose went on, writing to the British minister in China whom he had known as a colleague for more than twenty years while they had both served in the British diplomatic corps before Rose joined BAT, was Western military intervention in China. "You know that I am no fire-eater and that I have no belief that trade can be kept going by the indiscriminate 'shooting-up' of Chinese. But we have reached a point where we have to face a plain question—do we intend to maintain the right to trade in China? . . . The Naval and Military Authorities . . . could control the situation very quickly in their own way, and probably without firing a gun."[70] Foreign troops had been brought into China during the earlier phases of the Northern Expedition, and the British had recently stationed a contingent of 16,000 men in Shanghai, but despite pleas from other Western businessmen as well as Rose, there was no foreign military intervention at this time.[71]

When Western military intervention was not forthcoming, BAT directors started their own war of resistance against the Nanking government by refusing to buy the new tax stamps at the rate of 50 percent ad valorem, which the Kuomintang required companies to place on all cigarettes sold in China (whether imported or domestically made) beginning in July 1927. For the next few months the company flouted the tax (while Chinese companies grudgingly paid it) and continued manufacturing in Shanghai until BAT's 9,300 workers went on strike there in September 1927.[72] The strike was initiated by tens of thousands of workers in several industries with the aim of raising wages and improving conditions, not as a protest against BAT's tax evasion. But once the strike was underway, two Kuomintang members who were leaders of the union at BAT persuaded the strikers from their factory to adopt an additional demand, which was expressed in the slogan "Until the foreigners pay tobacco taxes, we'll do no work."[73] The Kuomintang thus used the strike to try to coerce BAT into paying taxes, and it kept BAT under this kind of pressure as the strike continued for 109 days from September 30, 1927, until January 16, 1928.

This strike in Shanghai marked the culmination of a series of strikes in BAT's factories in Hankow, Tientsin, and Tsingtao as well as Shanghai which virtually immobilized BAT's manufacturing system in 1926 and 1927, but BAT's directors still refused to acquiesce to the Kuomintang's demands.[74] Instead, they counted on the financial strength of BAT in China and the backing of their parent company to sustain them while they waited for the Kuomintang to lower its demands.

In the meantime, BAT resorted to illegal as well as legal distributing techniques to circumvent governmental opposition. One of the Western company's illegal techniques was to ship goods from its Shanghai factories to parts of China not controlled by the Kuomintang via the British colony of Hong Kong or the Japanese colony of Taiwan disguised as exports to the latter places. Although using such circuitous routes added to BAT's shipping costs, the additional expense was offset by savings on taxes the company would have had to pay the Kuomintang government if it had revealed that the cigarettes were for domestic consumption in China. Among the legal changes in its distributing system, the company temporarily stopped sending new American representatives to China and permanently appointed Chinese to fill many posts in the company's marketing system previously held by Westerners.[75] On the basis of illegal maneuvers and internal adjustments such as these, BAT continued to supply its distributors while its directors kept the Kuomintang at bay and waited for the party's leaders to make a financial proposition that the Westerners considered acceptable.

In January 1928, shortly after T. V. Soong resumed his position as Minister of Finance in the Nanking government, he finally produced the kind of proposal that BAT directors wanted. American-educated and a Harvard graduate, Soong personally conducted the negotiations with BAT's American lawyers and concluded a private agreement which achieved the goals of everyone concerned—except BAT's workers. BAT's management was relieved because the agreement ended the strike and lowered the cost of Kuomintang tax stamps that the company had to buy and place on all of its cigarettes from 50 to 27.5 percent ad valorem. The Kuomintang government was satisfied because the agreement finally brought revenue needed to finance the Northern Expedition (which continued on to Peking in 1928) not only from BAT's regular tax payments, but also from a payment of between $1 million and $5 million which the company made in advance. Soong himself was pleased because the agreement consolidated taxes on tobacco products and set the precedent for a program of consolidating taxes on other commodities which he proceeded to develop over the next few years. As for the workers, though their rice allowance and medical coverage were improved, their original strike demands for higher wages were not met. Instead of securing the material benefits the workers had sought from BAT, the Kuomintang gave them symbolic compensation by conducting a ceremony in which T. V. Soong's deputy, Ch'en Jui-tsung, presented a silver shield, a tablet of honor, and several scrolls to the BAT workers' union and silver medals to 600 of the union's members for "maintaining order" in a "spirit of patriotism" during the strike.[76]

Once BAT signed this agreement and received assurances that the Nanking government would not levy arbitrary taxes on it or inspire popular protests against it, the company attacked the China market as vigorously as it had after signing similar agreements with the Ch'ing government in 1904 and the Peking government in 1921. Unhindered by the strikes and taxes that had slowed its operation between 1925 and 1927, BAT set out in 1928 and 1929, according to American government officials in China at the time, "to recapture the market and re-establish their brands at any cost." Cutting its prices and allowing its Chinese dealers six months' worth of credit with guarantees that they would not lose by overordering, the company spared no expense in restoring its full circulation. The company's determination to defeat its Chinese competition was expressed in a notice distributed by BAT's management in Shanghai to all Chinese BAT agents at the end of the decade. Written in Chinese, it ordered them to gather information on the Chinese competition (by asking a list of questions that were included in the notice), to convince dealers throughout China to sign exclusive dealing agreements with BAT (by offering them bonuses, higher profit margins, and other inducements), and to make every effort to drive Chinese rivals out of the market.[77]

Rebounding from the defeats during the Northern Expedition of 1926 and 1927, BAT not only recovered but surpassed the peak levels of distribution it had reached in the past. In 1928 and 1929 it marketed cigarettes at the rate of 80 billion per year, the largest volume in its history and more than twice the rate of distribution prior to the curtailment of its operations in 1926 and 1927.[78] This aggressive and expensive BAT selling campaign contributed directly to its Chinese rivals' losses in the late 1920s according to officials in the United States Department of Commerce who were in China at the time. It was, they said, "manifestly impossible for native factories to hold up under this competition."[79]

Ironically, the agreement between BAT and Soong, which opened the way for the Western company's conquest of the market, appeared, on the face of it, to offer Chinese companies opportunities to become more rather than less competitive with their Western rival. By lowering the cost of tax stamps for Chinese-owned cigarette companies to 22.5 percent ad valorem, the Kuomintang government seemed to be setting this tax 5 percent below the one on BAT—at least in theory. But the Chinese cigarette manufacturers complained that in practice it discriminated against them because it exempted BAT's foreign imports from all other Kuomintang taxes (for example, import duties on raw materials, surtaxes, and luxury taxes), which added at least another 10 percent ad valorem to the Chinese companies' taxes and brought their total tax rate to 32.5 percent ad valorem, 5 percent higher than that levied against BAT.[80]

Whether equitable or not, the reduction in taxes on cigarettes in January 1928 was too little and too late to prevent a collapse of BAT's Chinese competition. To maintain Nanyang's rivalry with BAT in the late 1920s, an inequitable tax policy strongly favoring Chinese-owned industry was needed because the new tax policies of 1927 had a more devastating and lasting effect on Chinese companies than on BAT. In 1928 the Kuomintang government arranged tariff autonomy with the Western powers and thus secured the legal right to raise tariffs on Western goods for the first time in seventy years, but Chinese leaders in Nanking chose not to set high tariffs on Western cigarettes—perhaps because, as Cheng Yu-kwei has observed, "the primary purpose of the Chinese tariff [under the Kuomintang government] was the increasing of revenue collections, whereas protection of domestic industries was only its secondary aim."[81]

Lacking tariff protection, owners of Chinese firms were vulnerable to BAT competition. Lacking BAT's foreign diplomatic protection, they had not dared to produce any cigarettes without paying taxes. And lacking BAT's financial backing, they had not been able to absorb the costs of maintaining idle factories for months at a time without depleting their capital reserves. Crippled by the taxes of 1927, the Chinese cigarette manufacturers were in no position to jump back into full production, marketing, and purchasing as BAT did when taxes were lowered in January 1928. Instead, between 1927 and 1930, most small firms failed, Nanyang lost more than ¥5.7 million, BAT continued to expand, and an era of Sino-Western rivalry in the cigarette industry came to an end, for no Chinese-owned company ever again competed with BAT as successfully as Nanyang had before 1927.

Postscript: From New Rivalries to the End of Rivalry, 1931-1952

In the 1930s Japan's military and commercial invasion of China created opportunities for the Japanese Imperial Tobacco Monopoly (Tōa) and decisively changed the competitive structure of the cigarette industry in

China. Prior to 1931, Tōa had shown an interest in the Chinese cigarette market but had not given BAT or Nanyang sustained competition. Soon after its founding in 1904 (see Chapter 2), Tōa had begun to sell and manufacture cigarettes on the mainland, largely confining itself to Korea and South Manchuria and had been much more successful in Korea than in South Manchuria. Building factories at Seoul in 1908, Pyongyang in 1911, and Kwangju in 1912, it had received strong backing from the Japanese government in Korea—especially after Japan's annexation of Korea in 1905. Tōa had initially encountered resistance in Korea from BAT, but the Japanese monopoly had defeated its Western rival by 1914 when BAT —complaining about discriminatory new Japanese taxes in Korea—had closed its factory at Chemulpo (Inchon) and withdrawn from the Korean market. But in Manchuria and other parts of China, where Japan had not yet acquired political dominion, Tōa had experienced little success before 1931. Outside Manchuria Tōa had not given either BAT or Nanyang serious competition, and even in Manchuria it had been less competitive than either BAT or Nanyang. More than once Tōa had begun to push its sales upward in Chinese markets, but each time its gains had been interrupted by anti-Japanese boycotts, which had occurred in 1908, 1909, 1915, 1919-21, 1923, 1925-26, 1927, 1928-29, and 1931-32—far more often than Chinese boycotts against any other country.[82] Thus, even though Tōa had marketed and manufactured cigarettes in China for more than two decades, it had not challenged the BAT-Nanyang duopoly prior to 1931.

After Japan invaded Northeast China in 1931 and created the puppet state of Manchukuo in 1932, Tōa began to compensate for these years of frustration by making its first significant inroads into Chinese cigarette markets. Building large new factories in Dairen and Liaoning, Tōa virtually eliminated Chinese competitors from the Northeast in the 1930s and became BAT's chief rival there. As the Japanese army advanced into North China and the Lower Yangtze regions in the latter years of the decade, Tōa followed, adding more factories in Peking, Tientsin, Tsingtao, and Shanghai, expropriating formerly Chinese-owned cigarette companies and equipment along the way, and leaving the Manchurian market to a newly founded Japanese tobacco company called Manshū. To finance these operations, Tōa was capitalized at 10 million *yen* and Manshū at 5 million *yen*. In addition, Tōa, which had ranked third behind BAT and Nanyang as a purchaser of Chinese-grown bright tobacco in the late 1910s and 1920s, became the leading tobacco purchaser in Manchukuo and Japanese-occupied Shantung in the late 1930s and early 1940s.[83] Benefiting in these ways from the military conquest, Tōa displaced Nanyang as BAT's leading rival in China.

Forced to retreat in the 1930s, the Chiens kept Nanyang alive but felt compelled to surrender control of its management by 1937. In the early 1930s they reopened Nanyang's factories in Shanghai and produced cigarettes there at a higher rate than ever before—averaging 5.36 billion per year between 1932 and 1936—but they were never able to earn profits as high as they once had. After the large losses of the late 1920s, Nanyang's profits averaged ¥.7 million between 1930 and 1936—a small fraction of the ¥4 million in annual profits it had made during the early 1920s.[84] Plagued by high taxes, family dissension, the depreciation of silver in world markets (which raised the cost of importing tobacco from the United States),[85] and competition from the Japanese as well as BAT, the Chiens tried in 1933 to persuade the Kuomintang government to invest in Nanyang and bring it under "national management" (*kuo-ying*); and when this arrangement failed to materialize, they sought protection from official and unofficial exactions by appointing to the Nanyang board of directors in 1934 two men who helped to determine the Kuomintang government's economic policies, Tu Yueh-sheng and T. L. Soong (Sung Tzu-liang), T. V. Soong's brother.[86] Two years later, in 1936, as Nanyang's profits dwindled to ¥.3 million and Kuomintang taxes took 38.7 percent of the company's total income, Chien Yü-chieh reluctantly approached T. V. Soong for financial assistance, and in 1937 the two men concluded a deal whereby Soong became owner of about 27 percent of Nanyang's stock and chairman of the company's board of directors—all in exchange for about ¥1 million. The combined holdings of the members of the Chien family were still 8 percent larger than Soong's share, but he gained power of attorney and became Nanyang's chief decision maker between 1937 and 1945.[87] For the first time since Nanyang's founding in 1905, the Chiens thus lost control over the company.

Under Soong's leadership during the Sino-Japanese War (1937-1945), output at each of Nanyang's factories sharply declined. As the Japanese invaded China, Nanyang's owners resorted to a variety of strategies to maintain production, but where Nanyang was able to continue to produce at all, it did so on a very small scale. In Shanghai, after Nanyang's main factory was destroyed by Japanese bombing during the August 13 incident of 1937, the company's management subcontracted (*tai chüan ch'ang*) with several small independent Chinese cigarette factories, which made cigarettes for Nanyang using tobacco and wrappers supplied by Nanyang,[88] but the company's annual production in Shanghai fell from an average of 5.36 billion cigarettes in the mid-1930s to 1.5 billion between 1937 and 1941 and to a mere 0.26 billion between 1942 and 1945. Hoping to compensate for this reduction in the Yangtze River regions, Nanyang increased the output of its factory in Hankow—a small plant opened in 1934—which produced 1.38 billion cigarettes during the year beginning October 1937 but which had to be closed when the Japanese took Hankow in October 1938.[89] As the Japanese army advanced, some of Nanyang's owners retreated farther up the Yangtze River and built a new factory in Chungking, the city in Szechwan province where Chiang Kai-shek had taken refuge and had established a wartime capital for the Kuomintang government. Capitalized at ¥1 million, Nanyang's Chungking factory employed 6,000 unskilled workers and 100 technicians and used machinery brought from Hankow and Hong Kong and tobacco grown locally in Szechwan and Yunnan, but it added little to Nanyang's productive capacity.[90] Between the factory's opening in March 1939 and the end of the war in August 1945, it produced 0.013 billion cigarettes annually—a minute fraction of Nanyang's total output during these years. Retreating southward as well as westward in the late 1930s, Nanyang also stepped up production at its factory in Batavia (Djakarta) in Indonesia, but this plant was scarcely more productive than the one in Chungking. Operating sporadically throughout the 1930s, its peak capacity was about 0.075 billion cigarettes per year near the end of the decade. Even in Hong Kong, previously the Chiens' mainstay and a British colony not taken by the Japanese until December 1941, Nanyang's annual production declined from an average of 1.2 billion in the late 1930s to an average of 0.83 billion in 1940 and 1941. In 1940 T. V. Soong acknowledged the general diminution of Nanyang's trade and announced a policy of deliberately curtailing the company's production and sales until prices rose higher.[91]

Thus, Japanese invaders delivered Nanyang's coup de grace. Despite the Chinese owners' maneuvers, they were not able to keep their company very productive—let along competitive—in China's major markets during the Sino-Japanese War. The Japanese by no means started Nanyang on its downward course, for the decline of the company was traceable to previous developments such as mismanagement of Nanyang since the death of Chien Chao-nan, new taxes imposed by the Kuomintang government, and heavy competitive pressure from BAT. But the Japanese forced Nanyang to contract to the point where it ceased to be a big business.

While Nanyang languished, BAT grew, expanding its operations on all fronts in the 1930s. As a manufacturer, BAT built new cigarette factories, giving it a total of ten in China and one in Hong Kong by 1937. As a distributor, it penetrated still more deeply into Chinese markets, operating warehouses in 169 of China's two thousand counties (*hsien*) under the auspices of Cheng Po-chao's Yung-t'ai-ho Company and in another 378 counties under all its other distributors combined.[92] One indication of the huge scope of BAT's distributing system, according to a Chinese journalist, was that its commercial campaigns touched more peasants in China's countryside than the Kuomintang government's political propaganda did. Writing in 1934, after seven years of Kuomintang indoctrination according to the Three Principles of the People (*San min chu-i*), he re-marked that "many rural Chinese villages still don't know who in the world Sun Yat-sen is, but very few places have not known Ruby Queen (*Ta-ying*) cigarettes."[93] As a purchaser of Chinese-grown bright tobacco, BAT operated six curing plants in Shantung, Anhwei, and Honan prov-inces where a total of about 300,000 Chinese families comprising two million peasants cultivated the crop in the mid-1930s.[94]

BAT was encouraged to make these investments by the Kuomintang government's policies on tobacco taxes. Between 1928 and 1937 taxes became increasingly regressive—lower on high-quality cigarettes (of which BAT was the leading producer) and higher on low-quality ciga-rettes (on which small Chinese companies tended to concentrate). More-over, BAT received tax rebates as large as 20 percent in return for paying tobacco taxes in advance (which Chinese firms could generally not afford to do).[95] Steadily building up its investments during the Nanking decade, by the time of the Japanese invasion of 1937 BAT had registered thirty-three various enterprises in China and Hong Kong capitalized at ¥288.4 million ($84.8 million) and had holdings in China whose total book value was ¥461.8 million ($168.3 million).[96]

In the late 1930s and early 1940s, BAT adjusted deftly to the Japanese invasion of China and continued to flourish despite wartime conditions. In 1936 it moved its headquarters from Shanghai to Hong Kong and at the same time set up under Manchukuo law two new companies in the Northeast which ran cigarette factories at Mukden, Liaoyang, and Har-bin, and in 1939 it opened a new one in Ying-k'ou to replace the one at Liaoyang. On the eve of the Pacific War, these factories produced nearly 10 billion cigarettes per year. In Shanghai and other Chinese cities, while the Japanese confiscated Chinese-owned factories and equipment, BAT retained its property and continued to do business in occupied as well as unoccupied China. Between 1931 and 1941 BAT held two thirds of the Chinese cigarette market (including the Northeast), and between 1934 and 1940 it earned $178.6 million in profits—a return during this seven-year period that was $18.8 million larger than the company's total book value in China as estimated in 1935. Not until 1941, the year of Pearl Harbor, did BAT finally withdraw its Western representatives and sus-pend its operations in China for the duration of the Pacific War.[97]

After the war, Sino-foreign rivalries in the cigarette industry finally came to an end. Following Japan's defeat in 1945, Tōa withdrew, retreat-ing from China along with the Japanese troops even as it had earlier ad-vanced into China with them. BAT and Nanyang returned to the market and for the first time in their history formally collaborated—sharing raw materials, manufacturing jointly, and dividing markets between them—during the Chinese civil war between 1945 and 1949. But both had diffi-culty reestablishing their businesses, and both ceased to operate in China as private enterprises soon after the founding of the People's Republic in 1949.[98] On February 1, 1951, the Chiens and representatives of the new government signed an agreement making Nanyang one of the first private companies to be reorganized under "joint supervision of public and private management" (kung-ssu ho-ying) in China, and in 1952 BAT's factories and other property were taken over by the government. Officially BAT's property was not confiscated but was transferred from the Western company to the government by agreement. According to this agreement, BAT owners and the Chinese government deemed the value of BAT's assets to be equivalent to its back taxes and debts, so its Western owners received no additional compensation. A British BAT manager who was in China at the time recently recalled that BAT felt coerced into relinquishing its property and that it wrote off a loss of about £50 million on the transaction, but he admitted that the company consented to the agreement and that its business had been declining for some time because of rampant inflation, changes in currency, and general instability in China.

And so, BAT left China half a century after the company had been founded—a half century in which its total profits there had amounted to at least $380.8 million and probably more. The parent company was undoubtedly weakened by the loss of its China trade, but as a large multinational corporation it absorbed the loss without impairing its overall operations. Since then, it has grown steadily and is now (in the late 1970s) the single largest manufacturer of cigarettes in the West.[99]

Unlike BAT's Western managers, the Chiens were encouraged to stay in China, and they became active participants in the country's economic, social, and political life under the new government. Chien Yü-chieh was acclaimed as "an exemplary member of the national bourgeoisie," an officially approved group of businessmen (as opposed to the officially censured comprador bourgeoisie), and he served in official posts at the Chinese People's Political Consultative Conference, on the financial-economic committee of the Government Administrative Council, in the Central-South Military and Administrative Commission, on the provincial government council of Kwangtung, and as deputy to the National People's Congress representing Kwangtung.[100] As a septuagenarian living the last years of his life in the People's Republic, he undoubtedly found these honors and appointments gratifying, but perhaps his greatest satisfaction came from the outcome of Nanyang's long rivalry with BAT. "In the past," he reflected in 1956, one year before his death, "I felt that the greatest joy in life would come from seeing foreign cigarettes completely eliminated from Chinese markets. And today this goal toward which I have strived for many years has finally been reached."[101]

8.

Conclusion: Imperialism, Nationalism, and Entrepreneurship

THIS HISTORY of the cigarette industry shows not only that BAT and Nanyang were rivals but that they were both highly successful big businesses. As rivals, they did not succeed simply by dividing the market between them, the foreigner selling only expensive cigarettes and the Chinese only cheap ones or the foreigner selling only in one region and the Chinese only in another. Instead, both companies made and distributed cigarettes at all price levels and competed directly in all of China's major markets. From the standpoint of economic theory, the intensity of their rivalry is not surprising in light of the fact that they were battling for a rapidly growing market, for, as the economist Richard Caves has observed, companies have generally been willing to use price cutting and other competitive tactics in an expanding market because "even if cutting the price or raising the quality of the product sacrifices profits this year, the returns from having a bigger share of next year's bigger market may more than compensate for this year's profit reduction."[1] But why did the market expand so rapidly for both companies, bringing them a period of almost uninterrupted rises in sales and profits before, during, and after World War I?

The usual explanations for rising demand in terms of increases in population and per capita income or decreases in a substitute product do not fully explain the growth of the market for cigarettes in early twentieth century China. If economists are correct in their recent assessments of the early twentieth century, then changes in population and per capita income may partially account for the rise in cigarette consumption, for, according to Dwight Perkins' estimates, China's population rose from 385 million in 1893 to 500 million in 1933 and "there was no pronounced downward trend in per capita GDP [Gross Domestic Product] in the first half of the twentieth century, and . . . there may have been a slight increase in the decades prior to the Japanese attack [of 1937]."[2] But these increases are not large enough to explain a rise in China's annual consumption of cigarettes from a negligible number in the 1890s to about 100 billion in the early 1930s.[3] Nor is China's long history of tobacco growing and pipe smoking sufficient to explain the popularity of cigarettes in China. As noted in Chapter 2, the cigarette companies benefited from this history by taking advantage of Chinese familiarity with smoking, but available evidence on hand-made tobacco products suggests that BAT and Nanyang did not succeed merely by persuading China's pipe smokers to switch to cigarettes, for Chinese consumption of hand-made tobacco products survived competition with machine-made cigarettes in the early twentieth century. If the Chinese did not simply substitute machine-made cigarettes for hand-made tobacco products, then the question remains: why did Chinese smoke so many cigarettes (in addition to as many or more than the usual amount of hand-made tobacco products) during the first third of the twentieth century?

The findings in this book suggest that a more complete explanation for the expansion of the cigarette industry may be found in the business practices of BAT and Nanyang. To place these findings in broader perspective, the end of this inquiry will be to analyze the business practices of the two companies in terms of three concepts: imperialist exploitation, economic nationalism, and entrepreneurial innovation. Clearly, whether these terms apply depends on how they are defined. Accordingly, I will compare first, the behavior of BAT with commonly accepted definitions of imperialist exploitation; second, the behavior of Nanyang with commonly accepted definitions of economic nationalism; and finally, the behavior of both companies with commonly accepted definitions of entrepreneurial innovation.

Imperialist Exploitation

According to one definition of imperialist exploitation which has been used by mercantilists in the West since the seventeenth century and by Chinese historians and social commentators at least since the late nineteenth century, it is the process whereby foreign capitalists drain (*lou-chih*) wealth from the economies of poor countries for use in the imperialists' home countries.[4] BAT had the opportunity to transfer capital from China to the West in this way, for its business generated large amounts of sales revenue in China (for example, a net profit of $3.75 million annually by 1916), but available records do not show what fraction of BAT's net profits went to foreigners in the West, what fraction was paid to foreigners in China, what fraction was paid to Chinese in China, and what fraction was reinvested in China. According to Wang Hsi, a

Chinese historian who has assessed BAT's performance on the basis of records in China, the company plowed a "comparatively high" share of these profits back into BAT's operations in China during the first decade in the market (1902-1912) but maintained a very low reinvestment ratio of 7.2 percent between 1913 and 1941. Assuming that Wang has defined reinvestment ratio (which he renders *li-jun tsai-t'ou-tzu lü*) as Western economists generally do—the ratio of a business' undistributed profit to its total profit—then BAT's reinvestment ratio fell far below the median figure for other foreign firms that have been studied by Hou Chi-ming. So, assuming that BAT distributed no larger share of its profits in China than other foreign firms did, its reinvestment policies appear to have made it more imperialistic in this sense of the term than other firms were.[5]

Another definition of imperialist exploitation has been advanced by David Landes to serve as a basis for evaluating foreign management's relations with local labor. According to Landes, to show imperialist exploitation of labor, it is not enough to demonstrate that a foreign firm paid local workers wages that were low in relation to wages in other countries or in other industries within the local economy. He insists that "this kind of imprecision simply will not do" and asserts that the only significant definition of imperialist exploitation is "the employment of labor at wages lower than would obtain in a free bargaining situation; or in the appropriation of goods at prices lower than in a free market."[6]

So defined, imperialist exploitation of BAT's factory workers is impossible to document. Wang Hsi has shown that the income of most BAT workers was so low that they were barely able to buy enough rice to survive day by day in the 1910s and 1920s. And John Gittings has noted that, because of wide wage differentials, "a minority of trusted [BAT] workers was comparatively well-paid while the majority was not much better off" than other Chinese cigarette workers in the 1920s.[7] But there is no evidence that BAT workers received less than the "going wage" for comparable work in China's other cigarette factories at any time in the company's history. In fact, as Gittings' comment implies, BAT workers were slightly better paid than workers in Chinese-owned cigarette factories. Moreover, there is no evidence that BAT exploited Chinese compradors, distributors, and other BAT commercial agents, for BAT's wages, commissions, and mercantile credit proved more than lucrative enough to persuade them to cooperate with the company.

There is evidence, however, that Chinese peasants may have suffered from BAT's imperialist exploitation in Landes' sense of the term. As shown in Chapter 6, only during the late 1910s, while BAT was willing to subsidize production and offer peasants special inducements to plant bright tobacco, was this crop more profitable for peasants than the crops that it displaced. In the 1920s and 1930s, after BAT ceased to give peasants subsidies, the company appears to have benefited from exploitation, for peasants continued to plant bright tobacco not because they found it more profitable than other crops but either because they were unable to obtain other seeds and loans at all or because they were unable to extricate themselves from lending arrangements that they had made with Chinese BAT agents and local landlords or because they were willing to discount their family members' labor costs. Peasants undoubtedly calculated profit and loss very carefully, but they continued to grow bright tobacco even though they knew it was less profitable than other crops (unless labor costs are ignored).

A third definition of imperialist exploitation has been used to analyze the consequences of foreigners procuring raw materials in agrarian economies. According to this usage of the term, imperialists have distorted the development of agrarian economies by introducing extractive industries (plantations, mines, oil wells) and exporting raw materials for processing in Western factories; or, as a variation on this pattern, raw materials have sometimes been processed in foreign-built factories within agrarian countries, after which the finished product is then exported for sale abroad. In either case, according to Ernest Mandel's formulation, the agrarian countries' economies have tended to become monocultures which abandon various food crops in favor of the one cash crop demanded by the imperialists, and as a result, the agrarian economies are made dependent on the viscissitudes of foreign markets, for "the capital exported to the under-developed countries specializes in production *for the world market*."[8]

This definition is applicable to BAT to the extent that it (perhaps more than any other foreign company in China) induced peasants to plant imported seed for the purpose of growing a commercial crop, and it led more and more peasants in certain localities (Wei-hsien in Shantung, Hsuchang in Honan, and Fengyang in Anhwei) to abandon food crops in favor of this cash crop in the 1920s and 1930s. But BAT did not significantly alter the overall structure of China's agricultural markets, for bright tobacco at its most widespread in the first third of the twentieth century occupied no more than one million *mou* which was less than 9 percent of China's total tobacco land and a mere 0.007 percent of China's total cultivated acreage at the time.[9] Within the comparatively small area where bright tobacco was introduced, BAT did not set up plantations, transform the landscape, or "modernize" tobacco growing and marketing. Instead, it capitalized on existing Chinese agricultural expertise, which had been developed while growing a different type of tobacco, and it procured bright tobacco leaf through Chinese middlemen in the existing market structure. Moreover, at no time in its history did BAT extract tobacco leaf from agrarian China for use in the industrialized West. In fact, from 1905 until the 1940s, it did the reverse, exporting substantial amounts of tobacco from the United States for use in its factories in China. And its Chinese-made cigarettes, like its Chinese-grown tobacco, were distributed and consumed within China. Like many American multinational corporations in the early twentieth century, BAT thus showed more concern for exporting goods, capital, and technology abroad than for importing raw materials into its home country.[10]

A fourth definition of imperialist exploitation has been used to characterize foreign businesses' relations with governments. According to this definition, foreign businesses use the support of their own governments against weaker local governments to protect or further foreign businesses' material interests. Stephen Endicott has implied his acceptance of this definition in his study of British foreign relations with China in the 1930s where he has stated that "the pioneers of British enterprise in the Far East [among whom Endicott includes managers of BAT, Asiatic Petroleum Corporation, Imperial Chemical Industries, and other firms] had always been able to count upon effective military-diplomatic support [from the British government] to back their claims."[11]

In its dealings with Chinese governments, BAT engaged in this kind of exploitation in that it consistently benefited from Western treaties with China which maintained a low ceiling on tariffs. In addition, sometimes BAT received American or (more often) British diplomatic support for its campaigns to reduce Chinese tariffs or taxes to levels still lower than the limit set in the treaties. But BAT was not always able to count on this kind of support from its home governments. In promoting its cherished and potentially lucrative scheme for a cigarette monopoly in 1915, to cite one example, and in calling for military intervention during the Northern Expedition in 1927, to cite another, BAT was not able to win support from Western governments and was forced to abandon its proposals.

A fifth definition of imperialist exploitation has been used by Chinese since 1895 to evaluate the impact of a foreign company on its local rivals. According to this definition, a foreign industrial trust maintains a monopoly of the market that exploits customers as well as competitors in the local economy. Customers suffer because the foreign monopolist charges high prices and earns excess profits, and local producers suffer because the foreign monopolist has better access to capital markets and uses superior technology and aggressive competitive tactics to crush local handicraft industries and bar the entry of local industrial rivals into the market.[12]

Of all these definitions of imperialist exploitation, this one most aptly characterizes BAT's behavior in China. Though BAT had little effect on Chinese handicraft industries, it made every effort to secure a monopoly of China's cigarette market. Driving its new Chinese rivals to the wall after the boycott of 1905 and 1906, it blocked the development of any significant competitor within China's industrial sector prior to 1915. Using its monopoly control over the market, it set prices that yielded high profits—18.75 percent on net sales in 1916, for example, a profit rate higher than that of most foreign manufacturing firms in China, and 6.65 percent higher than James Duke's American Tobacco Company earned on the average in the United States at the time.[13] After 1915, when Nanyang penetrated South China, the Lower and Middle Yangtze, and North China, BAT tried to bar its entrance into these regional markets, and once the Chinese company had overcome barriers to entry, BAT tried to reduce Nanyang to submission by using a variety of competitive tactics its Western owners had previously used in the United States, England, and elsewhere—tactics that ranged from coercive campaigns, such as price wars, to appeals for cooperation, such as negotiations for a merger. As the competition in this industry heightened, Chinese customers pre sumably benefited from price cutting (though BAT's price wars seem to have been brief and localized—just long enough and extensive enough to stop and roll back Nanyang's penetration of urban centers in major regional markets wherever possible). But if BAT's price cutting benefited smokers, it and other BAT campaigns placed Chinese rivals under immense competitive pressure.

Thus, BAT's behavior in China meets the specifications of some of these definitions of imperialist exploitation and does not meet the specifications of others. And yet, after acknowledging that BAT's behavior did not conform perfectly to all these definitions, I think it is important to note that the company's presence in China had adverse consequences for Chinese industrialists that were economic as well as political and psychological.[14] Some of BAT's Chinese critics might have underestimated its financial investments in China or exaggerated its unfavorable economic effects on China because they resented its political privileges and its Western managers' arrogance, but they did not exaggerate when they accused BAT of crushing or absorbing potential competitors in China's industrial sector. Whatever redeeming features BAT may have had, its use of coercive tactics to destroy or buy out Chinese cigarette companies or block their entrance into the market clearly had adverse economic effects on the development of Chinese-owned industry. (For that matter, BAT's use of similar tactics against non-Chinese competitors in China, such as the American-owned Liggett and Myers Tobacco Company, the Tobacco Products Corporation, and F. E. Soter Company, also thwarted these firms' efforts to penetrate the market, but BAT's opposition to foreign-owned rivals does not fit any commonly accepted definition of imperialist exploitation.)[15]

In light of BAT's behavior as an imperialist—especially by the last of these definitions—how did one Chinese company, Nanyang, break into the market, survive, grow, and develop into a big business? One broad answer worth testing is economic nationalism, for, even as definitions of imperialist exploitation are at least partially applicable to the record of BAT, so too are definitions of economic nationalism potentially useful in analyzing the record of Nanyang.

Economic Nationalism

According to one definition of economic nationalism, as it is used by historians in both China and the West, it refers to businessmen who permit no foreign capital to be invested in their firms.[16] This definition applies to Nanyang in that its founders were born in China, its initial capital was raised among investors who were racially Chinese, and its record at attracting Chinese stockholders after going public in 1919 was very impressive. It eventually had approximately 7,000 Chinese stockholders—more than any other corporation in China during the first half of the twentieth century.[17]

At the same time, Nanyang's history was by no means devoid of financial relations with foreigners. Its founder and president, Chien Chao-nan, raised capital outside China and, after becoming a naturalized Japanese citizen, was vulnerable to the charge that his company was under (his own) Japanese ownership—a charge that embarrassed Nanyang and hurt its business during the anti-Japanese boycotts of 1915 and 1919. Furthermore, the Chiens helped finance Nanyang by borrowing from two British-owned banks, the Chartered Bank of India, Australia, and China in the late 1910s and the Hong Kong and Shanghai Banking Corporation in the late 1920s.[18] Thus, although Nanyang ultimately earned a justified reputation for its success at mobilizing capital among Chinese investors within China and Hong Kong, it nonetheless received financial support from Overseas Chinese in Japan and Southeast Asia (including some with foreign citizenship) and from foreign-owned banks.

A second definition of economic nationalism has been used to analyze the rationale behind demonstrators' decisions to participate in strikes, boycotts, and other popular protests against foreign imperialism. According to this definition, economic nationalism was a motivating force which caused Chinese workers to go on strike against Chinese-owned firms less often than against foreign firms and caused Chinese businessmen, merchants, and consumers to join in boycotts against foreign goods in almost every year between 1905 and the early 1930s. In James Sheridan's words, this kind of economic nationalism was what caused people to resolve "the conflict between nationalist and class interests" in favor of the former.[19]

The commitment of workers in the cigarette industry to this form of economic nationalism was reflected in their willingness to go on strike more often against BAT than Nanyang even though wages were sometimes higher and never lower at the Western company. In Shanghai, for example, the site of both companies' headquarters and largest factories, Chinese workers went on strike thirty-one times against BAT between 1918 and 1930—more often than against any other company in China except two Japanese-owned corporations, Naigai Wata and the Japan-China Spinning and Weaving Company (*Nikka bōseki kabushiki kaisha*) —and struck only five times against Nanyang.[20] Though particular labor protests against Nanyang—especially the ones in 1924, 1926-27, and 1929-30—may have contributed directly to the company's decline, even these lasted for briefer periods, had less ambitious objectives, and achieved more modest benefits for workers than several strikes in BAT's factories during the 1920s.

If Nanyang's workers often permitted this kind of economic nationalism to override their material interests, Nanyang's management was spared such an agonizing choice. By contrast with workers, merchants, and consumers, the Chiens were able to participate in anti-foreign strikes, boycotts, and the national goods movement without facing any conflict between nationalist and class interests because their nationalist and class interests were mutually reinforcing; that is, the more this kind of economic nationalism impelled Chinese workers to go on strike against BAT or persuaded Chinese merchants and consumers to boycott BAT's goods, the better it was for Nanyang's business. Accordingly, the Chiens actively promoted the national goods movement and other anti-foreign protests not because they felt any need to put the interests of the nation ahead of their own interests but because the interests of the nation and the interests of their own business seemed to them to coincide.

A third definition of economic nationalism—one popular in China throughout the twentieth century—describes a manufacturer that relies on raw materials from his home country. This definition bears almost no relation to Nanyang's policy on purchasing between 1905 and 1915, for the Chiens imported nearly all their equipment, cigarette paper, packaging, dyes, spices, and other materials from Japan and much if not all their tobacco from the United States during these years. At the time of the anti-Japanese boycotts in 1915 and 1919, however, Chinese activists who called for this kind of economic nationalism demanded that the Chiens place fewer orders abroad and more in China. Perhaps for this reason (but more likely for business reasons), Nanyang's owners began in the early 1920s to invest in their own supply houses in China. They founded a large printing establishment in China, for example, and according to Chien Chao-nan's son, Chien Jih-lin, Chien Chao-nan had plans to build a paper mill, a can factory, and other supporting facilities in China as well, but after the Nanyang founder's death in 1923, these plans were never carried out. The Chiens further became economic nationalists in this sense by purchasing tobacco grown in China and by importing American bright tobacco seed to be cultivated in China. But Nanyang's agricultural system was never as successful as that of BAT, and the Chinese-owned company, unlike the Western-owned one, continued to import the bulk of its tobacco from the United States throughout the 1920s.[21]

A fourth definition of economic nationalism has been used to analyze local businesses' relations with their countries' governments. According to this definition, economic nationalists are those who ally with their national government and benefit from its "retaliation" against foreign competition.[22] The Chiens sought to form this kind of alliance with Chinese political leaders and governments, but their three bids for official support did not, on balance, work out to their advantage. In the first of these attempts, during 1916 and 1917, they negotiated about a jointly managed enterprise with the Peking government but ceased to pursue this possibility when the government collapsed and successive governments in the late 1910s and early 1920s at Peking also appeared to be unstable. Making a second effort to ally with political leaders, the Chiens tried in the early 1920s to curry favor with the Kuomintang, which was then establishing a power base at Canton, and in 1925 and 1926 at least one member of the family joined the party and raised funds to finance the Northern Expedition. But this liaison was short-lived, for Nanyang benefited from Kuomintang protection only during brief anti-foreign strikes and boycotts in mid-1925, late 1926, and early 1927, and these gains were heavily outweighed by the losses Nanyang suffered as a result of taxes and other exactions the Kuomintang government imposed in Shanghai throughout the late 1920s and early 1930s. In their third major effort to establish a favorable relationship with officials in government, the Chiens relinquished control over the company to T. V. Soong in 1937, but during the wartime period that followed, Nanyang's sales and profits decreased rather than increased under his executive leadership. Thus, throughout the Republican period, the Chiens were generally unsuccessful in their attempts to secure protection from Chinese governments and officials. Overall, they derived few benefits from this kind of economic nationalism.

A fifth definition of economic nationalism has been used by Mao Tsetung and a number of historians in China to distinguish between patriotic and unpatriotic Chinese businesses and businessmen in the history of China's economic development. According to this definition, economic nationalism was exemplified by national capitalists (*min-tsu tzu-pen chia*) with their national capitalist enterprises (*min-tsu tzu-pen ch'i-yeh*), who merit praise for remaining independent of foreign capitalists, for competing with foreign rivals, and for resisting imperialist exploitation. They are distinguished from comprador capitalists (*mai-pan tzu-pen chia*), who, by contrast, are condemned for lacking a proper sense of national loyalty, for collaborating with foreigners, for becoming dependent on foreign capital, and for facilitating imperialist exploitation.[23]

Though accepted by many historians in China, this distinction has been sharply criticized by historians in the West. Hao Yen-p'ing, for example, has faulted it because it seems not to allow for the possibility that a person who began his career as a comprador working for a foreign firm might eventually have left that job to open his own business in direct competition with his former foreign employer—a pattern that was common, according to Hao, in nineteenth century China.[24] A still more searching criticism of the distinction between national capitalists and comprador capitalists has been made by Marie-Claire Bergère. "The distinction," she has concluded,

> is superficially obvious, but surely rather artificial. In early twentieth-century China with its semicolonial economy dominated by the presence of imperialist powers, there could be no such thing as independent national enterprise. One way or another, whether from the aspect of finance, supply, equipment, or distribution, all Chinese businesses of any size operated within a context of foreign domination . . . In these circumstances, "national capital" is meaningless as an economic term. It can be said that in all relations established with foreigners, the Chinese bourgeoisie was in a position of total economic dependence . . .
>
> The real contradiction lies not between a "national" and a "compradore" bourgeoisie but between the economic dependence of the entire bourgeoisie and its unanimous nationalist aims.[25]

These criticisms have brought to light the dangers of positing an ideal type, the national capitalist with his national capitalist enterprise, which ignores the historical fact that the success of virtually all Chinese industrialists was rooted in or dependent upon their relations with foreigners. And yet, the definition of economic nationalism used by historians in China is perhaps more useful in analyzing the attitudes of Chinese industrialists than Hao and Bergère have supposed.

The virtue of the distinction between national and comprador capitalist is that it calls attention to an urgent dilemma that Chinese who were engaged in Sino-foreign rivalries frequently faced: to merge or not to merge with one's foreign competitor. In the cigarette industry, the dilemma was particularly acute. As emphasized throughout this book,

Chinese cigarette companies were under immense pressure to sell out to BAT because of BAT's coercive tactics in the battle for markets and because of its offer to provide a profitable and financially secure future for those who consented to merge with it. At least a dozen Chinese firms and probably many more responded to this combination of punishment and promises from BAT by selling out to it,[26] and if Nanyang's owners were representative of other Chinese industrialists, then even those who did not sell seriously considered taking this step. In light of these cases, it is clear that the question of whether to retain a measure of financial independence (that is, preserve one's identity as a "national capitalist") or relinquish control to foreign competitors (that is, become a "comprador capitalist") was a pressing concern for Chinese in the cigarette industry and probably for Chinese in other industries as well.[27] For this reason, historians in China have more justification for proposing the distinction between national and comprador capitalists than historians in the West have suggested. These two terms, to be sure, have deficiencies as tools for historical analysis because they have accumulated layers of meaning that are not historically derived, but the distinction that they represent is nonetheless historically useful insofar as it identifies two horns of a dilemma that faced many Chinese businessmen in Sino-foreign commercial rivalries.

This discussion of economic nationalism suggests that some definitions of it fit Nanyang better than others even as some definitions of imperialist exploitation fit BAT better than others. Comparing each set of definitions with the behavior of each of the two companies provides a historical test for imperialist exploitation and economic nationalism that shows their possibilities and limitations as concepts to be used in the analysis of Sino-foreign rivalry in the cigarette industry. The results of the test suggest that the two concepts are partially applicable to this case, but is it possible, on the basis of these concepts, to explain fully the rapid growth and extraordinary commercial success of BAT and Nanyang? Certainly the notion of imperialist exploitation partially explains BAT's success by emphasizing that it had many advantages over its Chinese rivals. And the notion of economic nationalism partially explains Nanyang's success by emphasizing that it had advantages over its foreign rival. But if our analysis is confined to the concepts of imperialist exploitation and economic nationalism, some important questions about the history of the cigarette industry remain unanswered. If foreign firms generally possessed advantages similar to those of BAT, why was BAT more successful than most other foreign firms? And if Chinese firms generally possessed advantages similar to those of Nanyang, why was Nanyang more successful than most other Chinese firms? Why, in short, did these two companies become two of China's biggest businesses? Though a partial answer to these questions may be that BAT benefited more from imperialist exploitation than other foreign companies and that Nanyang benefited more from economic nationalism than other Chinese companies, a more complete answer must take into account other possible explanations. Of these other explanations, an important one lies in the two companies' entrepreneurship.

Entrepreneurial Innovation

Like imperialist exploitation and economic nationalism, entrepreneurship has been defined in a variety of ways, but the concept of entrepreneurship most commonly used by scholars in general and by China specialists in particular is the one formulated by Joseph Schumpeter. Schumpeter postulated that "the defining characteristic [of entrepreneurship] is simply the doing of new things or the doing of things that are already being done in a new way (innovation)."[28] He regarded entrepreneurial innovation as a "creative response" to economic opportunities, and he offered it as an ideal or model for judging particular actions. This ideal has been used by scholars for analyzing the economic history of many countries—including China[29]—but it is often interpreted very broadly with reference to nothing more specific than the general definition just quoted. It is possible, however, to apply Schumpeter's concept of entrepreneurship more precisely, for he specified five consequences of individual actions that constitute evidence of entrepreneurial innovation: (1) the introduction of a new good or a new quality of a good; (2) the opening of a new market for goods; (3) the conquest of a new source of supply of raw materials; (4) the introduction of a new method of production; (5) the organization of an industry along new lines, whether creating or breaking up a monopoly.[30] These criteria are helpful as a basis not only for analyzing what kinds of entrepreneurship BAT and Nanyang possessed but also for determining who the entrepreneurs in each company were.

Applying Schumpeter's criteria to BAT highlights the sharp distinction the company maintained between Western and Chinese entrepreneurs. The second and third of Schumpeter's five entrepreneurial functions were performed for BAT primarily by Chinese; the other three were generally handled by Westerners. Though Western journalists and governmental officials supposed that BAT depended on Western "pioneers" to open new markets for cigarettes and create new sources of tobacco, in fact BAT relied primarily on its Chinese compradors and agents to carry out these two tasks. Westerners like the American James Thomas were responsible for the entrepreneurial task (1) above, the introduction of cigarettes to certain treaty ports and the transportation of bright tobacco seed to certain provinces, but compared to the company's Chinese compradors and agents, they played minor roles in penetrating new markets for cigarettes and persuading peasants to grow bright tobacco. In this respect, BAT distinguished itself from other foreign-owned firms in China. Others (including BAT) necessarily employed Chinese compradors, but few if any benefited so richly from Chinese entrepreneurship in the marketing of goods and procurement of raw materials as BAT did.

And yet, though taking advantage of Chinese entrepreneurship in cigarette marketing and tobacco procurement, BAT preferred to rely on Westerners for entrepreneurial functions (4) and (5), and fortunately for BAT, its Western management made "creative responses" to opportunities for innovation in manufacturing and industrial organization. As manufacturers, BAT's Western management installed in China the latest models of cigarette making machinery. And as industrial organizers, they introduced two major innovations. Prior to 1923, led by the Americans James Duke and James Thomas, they integrated mass production and mass distribution, achieving national distribution in China and thereby making BAT one of the first businesses ever to distribute a trademarked product throughout the entire country. Then later, after BAT came under predominantly English management in 1923, its executive leadership introduced a second organizational innovation: a decentralized system with regional offices at the cores of each of China's major marketing areas, each equipped to perform its own manufacturing, marketing, and administrative functions independently of the others.

This kind of entrepreneurial creativity in China was possible because of another innovative policy maintained by the top management of BAT's parent company in the West. Unlike the top management of many other multinational corporations manufacturing in poor countries at the time, BAT's top management permitted all goods BAT produced locally to be distributed locally.[31] Without this policy, BAT's Western management could not have attempted to integrate production and distribution or to decentralize operations in China as it did.

The distinction between the entrepreneurial functions performed by Westerners and the ones performed by Chinese in BAT is important, for it suggests that Chinese at BAT gained far less experience as industrial managers than as commercial managers. It would be wrong to draw this distinction too sharply, for Chinese BAT employees certainly had some exposure to machinery in BAT's factories, and, as a BAT executive recently recalled, the Chinese had sufficient technical competence to operate BAT's factories after 1949 without once consulting the four British managers of BAT who were detained in Shanghai between 1949 and 1952.[32] The value of the distinction is that it properly calls attention to BAT's general tendency to keep industrial management in Western rather than Chinese hands, and it may help explain why no Chinese from the company ever established a reputation as an industrial manager within BAT or outside it. At least one leading BAT agent, Cheng Po-chao, expressed an interest in building a new cigarette factory, but he and other Chinese BAT agents lacked experience as industrial managers, and neither they nor salaried Chinese compradors (who would have had better opportunities to become familiar with BAT's technology and industrial organization) nor any other Chinese ever broke away from BAT to open Chinese-owned cigarette companies in competition with the Western one.

Whereas BAT's entrepreneurship came partly from Westerners and partly from Chinese, almost all of Nanyang's entrepreneurial skill came from Chinese. There were exceptions to this generalization, for the Chiens relied on Japanese for technical advice before entering the China market in 1915, and they recruited an American, "Captain Jack" Gravely, to act as their purchasing agent in the tobacco market of the United States between 1918 and 1929. But otherwise virtually all of their entrepreneurial leadership was provided by Chinese.

One man, Chien Chao-nan, was Nanyang's foremost entrepreneur, and in many ways he embodied Schumpeter's entrepreneurial ideal. He was unquestionably the most innovative Chinese entrepreneur in China's cigarette industry and one of the most innovative entrepreneurs in any of China's industries. In the course of his career he was responsible for innovations that fit all five of Schumpeter's categories of entrepreneurship. First, he introduced a new quality of good by identifying Nanyang cigarettes with patriotism and thus persuading customers that his company's product possessed a patriotic quality which BAT cigarettes lacked. Second, he devised new methods of production largely on the basis of his experience in Japan and the United States—risking capital on imported, up-to-date machinery, accommodating workers to avoid strikes, hiring foreign-educated Chinese as factory managers, and even adopting a new system of industrial organization that had recently been conceived by the American advocate of "scientific management," Frederick Winslow Taylor. Third, he opened new markets for Nanyang's cigarettes, traveling widely, setting up branch offices, and recruiting Chinese agents who, in turn, introduced his product into regions previously dominated by BAT. Fourth, he found new sources of raw materials for Nanyang not only in China but also in the United States. And fifth, he succeeded in changing the competitive structure of the industry by challenging BAT's monopoly of the market, developing Nanyang into a big business, and making it the only serious commercial rival BAT ever faced in China.

These findings on the entrepreneurship of Chinese within BAT and

Nanyang add to the growing body of evidence documenting the existence of entrepreneurship in the history of China's industrialization, but this interpretation of entrepreneurship in the cigarette industry is not, paradoxically, irreconcilable with the conclusions of Albert Feuerwerker and Marion Levy, who have sought to explain the absence of entrepreneurship in China.[33] In fact, the works of each of these scholars offers a basis for helping explain the role of entrepreneurship in the cigarette industry. Feuerwerker's explanation for the lack of entrepreneurship among late Ch'ing official-industrialists, for example, is helpful in analyzing Chien Chao-nan's career, for it highlights the contrast between him and his late Ch'ing predecessors. Unlike these official-industrialists, Chien Chao-nan was free from the institutional restraints that stifled their entrepreneurial potential, for he received neither a classical Confucian education nor a post in the imperial bureaucracy and was, therefore, able to devote himself to the practice of business without being distracted from it by the concerns of the official elite. Similarly, Levy's analysis of nepotism in the Chinese family is relevant to Nanyang's history because it helps account for some of the mismanagement of the company after Chien Chao-nan's death—especially the malfeasance of Chien Ying-fu. Chien Ying-fu's presence in the company well illustrates the power of nepotism, for he showed no signs of entrepreneurial talent and hastened the company's decline by stealing from it, and yet he still was able to hold a high managerial position simply by virtue of his family ties.

Though entrepreneurship was crucial to the growth and success of BAT and Nanyang, its presence in the cigarette industry does not resolve the issue of how widespread it was in late nineteenth and early twentieth century China. The most one may do on the basis of this case is suggest certain inferences about Chinese entrepreneurship: first, that Chinese employees working for foreign firms may have gained exposure to commercial techniques (such as advertising) that offered them opportunities for entrepreneurial innovations as salesmen without necessarily being exposed to industrial technology, which would have offered them opportunities for entrepreneurial innovation as manufacturers; second, that entrepreneurship was more likely to develop among Chinese living outside the imperial elite or in post-imperial China than among official-industrialists during the late Ch'ing; and third, that entrepreneurship might either have been encouraged by the strength of family ties (through which capital was raised and trustworthy business organizations such as Nanyang were formed) or it might have been limited by the strength of family ties (which placed unentrepreneurial managers in high positions). These are tentative hypotheses based largely on the study of one industry, but they touch upon an issue of paramount importance. Surely, as research in the young field of Chinese business history progresses, one of its major aims should be to evaluate further the significance of entrepreneurship in China's response to opportunities for industrialization.

The Significance of Sino-Foreign Rivalry

Interpreting Sino-foreign commerical rivalry in terms of imperialist exploitation, economic nationalism, and entrepreneurial innovation brings to light some of the characteristics of BAT and Nanyang that enabled them to grow and prosper as big businesses. But it leaves unanswered a final question to be considered in this book. What was the significance of these businesses in Chinese history?

Historians and social scientists have tended to judge the significance of Sino-foreign commercial rivalries according to the effects that foreign and Chinese companies had on each other's growth. Advocates of the "oppression" argument like Cheng Yu-kwei and Jean Chesneaux have regarded it as significant that foreign firms had advantages that enabled them to "oppress" their Chinese competitors. And, in reply, critics of the "oppression" argument such as Hou Chi-ming and Robert Dernberger have regarded it as significant that Chinese firms had advantages of their own that offset foreigners' advantages.

If the significance of Sino-foreign commercial rivalries is to be assessed within the context of this debate, then the rivalry in the history of the cigarette industry is significant both for advocates and for critics of the "oppression" argument. On the foreign side of the rivalry, BAT possessed every advantage that has been attributed to foreign firms—privileges under the unequal treaties, access to foreign capital abroad, factories with superior technology, and foreign managers with entrepreneurial skill. And still more significantly for advocates of the "oppression" argument, BAT deliberately used coercive business practices—especially price wars—to crush or slow the growth of its Chinese rivals.

On the other hand, on the Chinese side of the rivalry, Nanyang possessed all of the advantages over BAT that critics of the "oppression" argument have claimed Chinese firms had over foreign ones—knowledge of its owners' home region (South China), rising sales during boycotts (with the exception of the May Fourth boycott), and nationalistic loyalty from workers (who held fewer strikes than BAT workers even though they received equal or sometimes lower wages). And still more significantly for critics of the "oppression" argument, Nanyang fought back and grew into a big business despite competitive pressure from BAT.

So if "oppression" (which has often been imprecisely defined in this debate)[34] means using aggressive business tactics to promote and develop one's firm at the expense of one's rival, then BAT unquestionably oppressed Nanyang, but to a lesser extent, Nanyang also oppressed BAT.

Critics of the "oppression" argument have proposed an alternative to it as a basis for judging the true significance of Sino-foreign commercial rivalry, namely, the "imitation" effect, and the history of the cigarette industry also provides examples that might be cited to support or debunk this argument. According to the "imitation" argument, the presence of foreign industrial enterprises in relatively unindustrialized countries has caused local entrepreneurs to imitate the foreigners by opening and developing their own factories.[35] The cigarette industry followed this pattern to the extent that foreign-owned firms manufactured cigarettes in China before Chinese-owned firms were founded, set an example for Chinese entrepreneurs, and demonstrated the usefulness of business practices that Chinese (notably Chien Chao-nan) subsequently adopted.

The history of the cigarette industry, however, does not unambiguously confirm the "imitation" argument any more than it confirms the "oppression" argument. The presence of foreign firms in China undeniably induced the Chiens to open industrial enterprises, but the one foreign-owned firm that dominated the market, BAT, successfully prevented its Chinese employees from opening their own companies in competition with it. Moreover, the one Chinese-owned firm that became a big business, Nanyang, was started without assistance from BAT or former BAT employees. Unable to copy BAT's technology in China, its founders went abroad for technological training and expertise, first to Japan and later to the United States. Not until the early 1920s, after Nanyang was well underway, did the Chiens manage to hire Chinese who had gained experience in marketing and finance in BAT's headquarters at Shanghai, Wu T'ing-sheng and Ch'en Ping-ch'ien, and even these two former BAT compradors brought no technological expertise with them to Nanyang. Perhaps most significant of all for an evaluation of the "imitation" effect, whereas the advocates of this argument have contended that Chinese imitators grew as fast as their foreign rivals in China, Nanyang and other Chinese-owned cigarette firms did not match BAT's rate of growth in the long run.[36]

Thus, this case does not confirm the "oppression" argument to the exclusion of the "imitation" argument or vice versa. Instead, it provides illustrations of both concepts that suggest the two are not mutually exclusive. So one may say that BAT both stimulated its Chinese competitors to "imitate" it (for example, by its introduction of cigarette factories and mass advertising) *and* "oppressed" them (for example, through price wars and smear campaigns).

However advocates of the "oppression" or "imitation" arguments may choose to use the findings presented here, the history of this one industry underscores the need for more case studies of Sino-foreign commercial rivalries. Not until these are done will it be possible to determine whether Chinese-owned industrial firms generally grew as fast as foreign ones.

Judging from this case, such a hypothesis is not likely to be substantiated. Moreover, as Albert Feuerwerker has perceptively remarked, "From the available evidence, it might just as reasonably be argued that in the absence of foreign competition Chinese firms might have grown even faster and carried the whole modern sector of the economy along with them."[37]

If this Sino-foreign rivalry illustrates both "oppression" and "imitation," it is perhaps still more significant as an illustration of a process that historians of China have not associated with businesses: Sinification. The two big businesses in this industry—both the foreign-owned BAT and the Chinese-owned Nanyang—grew steadily more dependent on Chinese and less dependent on foreigners throughout the history of their rivalry. Whereas Westerners like James Thomas introduced cigarettes into China and at first imported all cigarettes from the West, BAT soon came to rely on Chinese to distribute and advertise its product, to manufacture between 50 and 70 percent of the cigarettes it sold in China, and to produce most of the tobacco it used to make cigarettes in China for Chinese consumption. And whereas Nanyang initially relied on Chien Chao-nan's Japanese citizenship and overseas connections to open a business in Hong Kong and Southeast Asia, it severed most of its Japanese ties (including Chien Chao-nan's Japanese citizenship) during its first five years in the China market (1915-1920) and became heavily dependent on Chinese in all phases of its operation in China. Only in their purchasing of American tobacco in the United States did both companies consistently rely on non-Chinese representatives and agents.

Besides becoming more Sinified, both companies tried to appear even more Sinified than they actually were. In this Sino-foreign rivalry, each firm repeatedly charged that it was more "Chinese" and that its competitor was more "foreign"; and each used publicity saying that it was contributing more to China's welfare and that its competitor was more concerned with siphoning off profits and expatriating them abroad. Various criteria for judging nationality were invoked at the time, and depending on which criteria one uses, the question of which firm contributed more to the material welfare of people in China may be debated even now.[38] But whichever company was truly more "Chinese" or truly contributed more generously to the welfare of the Chinese people, both recognized the need to disguise their foreign ties and appear as "Chinese" as possible.

BAT and Nanyang made responses to the need for Sinification that proved to be extraordinarily effective. Each company adapted to the circumstances that faced businesses in China by recruiting the appropriate Chinese to give its operation social respectability, guide its product through the existing market structure, adjust its advertising to suit the political atmosphere and cultural milieu, and perform other tasks that made a business acceptable to Chinese society at all levels—regional and local as well as national. Unable to ignore or avoid dramatic events in early twentieth century China such as the collapse of the central government and the rise of social movements, BAT and Nanyang relied on their Chinese staffs to cope with these events and, if possible, capitalize on them. Thus, these two big businesses did not concentrate exclusively on each other or compete in a vacuum (as the "oppression" and "imitation" arguments might lead us to suppose) but also became involved in the social, economic, and political context of their time. They were, in short, businesses in history.[39]

The success of these two big businesses at adapting to Chinese conditions and Sinifying their operations is perhaps the single most persuasive explanation for the growth of the cigarette market in early twentieth century China; and perhaps the best evidence of their adaptability and Sinification may be found in their advertising. Designed and distributed by Chinese in both companies, it showed not only the companies' abilities to use existing commercial structures but also their awareness of popular traditions, regional differences, political developments, and other features of Chinese life. This advertising seems to have been the key to BAT's and Nanyang's commercial success, and, pressed on consumers in intensive campaigns, it attracted enough smokers to make the market for cigarettes in China during the early twentieth century almost as large as the one in the United States.[40]

In light of BAT's and Nanyang's record at building up the market for cigarettes in China, no one should be surprised to discover that Chinese continue to smoke cigarettes on a large scale today. The product that BAT and Nanyang promoted in China during the first half of the twentieth century is now routinely served to guests (along with tea) as part of Chinese etiquette and is consumed by people throughout Chinese society. According to the most recent figures, 725 billion cigarettes were produced in China in 1977, 60 billion more than in the United States.[41] So a habit that started and spread in China during an era of Sino-foreign commerical rivalry is now pervasive there.

In relying largely upon state-owned enterprises to supply this market, the government of the People's Republic seems to have demonstrated that it is possible for a strong socialist state to eliminate privately owned enterprises and thus to avoid international commercial rivalries. And yet, it is still not clear what roles big businesses will play in China's future.

In late 1978 and early 1979, big businesses began to pursue more promising prospects in China than at any time since the founding of the People's Republic, and, after once appearing to be permanently cut off from markets in China, Nanyang and BAT are both presently selling cigarettes there. Nanyang, still based in Hong Kong and managed by the Chien family, resumed its trade with China sooner than BAT. It has exported cigarettes from Hong Kong to China since the early 1950s and now supplies about 10 percent of the country's market. BAT, now the world's leading tobacco company in sales, assets, and number of employees, began in March 1979—after a hiatus of almost thirty years—once again to sell its cigarettes in China. It is important to note that Nanyang and BAT are now operating under far stricter regulation in China than they did before 1949; for example, according to BAT's contract with the government of the People's Republic, the distribution of its cigarettes is specifically restricted to places visited by tourists. But the very presence of BAT's cigarettes in China—a possibility that seemed inconceivable until recently—further illustrates the extraordinary resilience that the company has shown throughout its history in China.[42]

BAT's recent reentrance into China's markets raises for Chinese once again the question with which this book began: is it possible to compete with big and expanding multinational corporations? And, whatever answers to this question are reached in China, the question will continue to confront businesses and governments in other countries throughout today's world. Asking this question of the history of China's cigarette industry, as I have done here, has not produced a definitive answer to it. But insofar as a generalization may be formulated on the basis of this Sino-foreign rivalry, the answer to the question in this case seems inescapable: in the short run, it is possible to compete with such a corporation, but in the long run it is difficult to remain competitive, for, as the economist Stephen Hymer has remarked, big businesses like BAT "do not die like ordinary trees; they are like California redwoods."[43]

Appendix: Statistical Tables

TABLE 1. Cigarettes imported into China, 1904-1930 (quantity in thousands, value in Haikwan taels)

Year	United States		Great Britain		Hong Kong	
	Quantity	Value	Quantity	Value	Quantity	Value
1904	—	1,042,886	—	402,596	—	254,425
1905	—	2,203,636	—	955,923	—	225,941
1906	—	3,245,177	—	1,686,692	—	460,412
1907	—	1,348,650	—	1,169,876	—	364,805
1908	578,480	991,337	967,113	1,794,067	212,303	404,086
1909	677,403	1,171,160	1,336,696	2,448,455	234,403	453,463
1910	532,112	968,216	2,122,566	4,049,501	217,616	441,463
1911	244,274	424,451	2,493,681	4,827,403	186,233	609,324
1912	207,164	364,954	2,819,317	5,534,130	188,987	616,095
1913	381,685	683,876	4,248,729	8,643,415	221,214	653,284
1914	162,500	322,550	4,480,957	9,806,899	368,934	932,319
1915	378,868	782,579	3,771,016	8,667,284	522,107	1,347,084
1916	1,613,106	6,454,671	3,082,344	12,959,201	902,129	3,078,966
1917	3,784,218	15,142,875	872,343	4,146,067	874,912	2,805,168
1918	4,563,129	11,413,053	234,914	1,491,190	951,350	3,042,334
1919	4,239,735	11,006,437	153,433	1,021,148	707,005	2,774,243
1920	4,629,471	11,721,647	463,211	2,693,077	650,579	2,116,048
1921	6,285,714	17,600,207	916,306	4,000,612	423,361	1,279,145
1922	8,003,139	21,431,980	485,345	3,282,934	575,019	1,535,779
1923	8,197,159	20,752,039	518,825	3,703,789	400,637	1,306,021
1924	7,994,740	20,389,613	543,142	3,829,388	625,219	2,017,746
1925	4,951,721	11,967,156	654,675	3,262,004	636,162	1,960,782
1926	6,613,291	15,386,014	533,362	3,984,213	80,689	337,570
1927	3,915,843	9,234,877	309,567	2,014,859	267,371	971,521
1928	6,785,572	15,929,526	913,054	4,757,626	506,877	1,606,749
1929	4,472,247	10,050,613	2,796,892	8,873,253	458,467	1,185,658
1930	1,733,636	6,599,293	3,177,282	16,319,487	1,147,134	2,542,192

Sources: China, Imperial Maritime Customs, *Returns of Trade and Trade Reports* (Shanghai, issued annually) (1908), pt. 3, vol. 1, p. 136; (1912), pt. 3, vol. 1, p. 162; (1915), pt. 3, vol. 1, p. 273; (1918), pt. 3, vol. 1, p. 268; (1921), pt. 3, vol. 1, p. 247; (1924), pt. 3, vol. 1, p. 258; (1926), pt. 3, vol. 1, p. 364; (1929), pt. 3, vol. 1, p. 345; (1931), pt. 2, vol. 1, p. 353.

a. Apparently not all cigarettes exported from the United States to China reached China. Note the discrepancy between American imports into China in this table and exports from the United States to China in table 2.

pan (including Taiwan)		Korea		Canada		Total (including others)	
Quantity	Value	Quantity	Value	Quantity	Value	Quantity	Value
—	1,100,869	—	9,789	—	14,914	—	2,909,073
—	871,530	—	25,110	—	288	—	4,427,171
—	395,307	—	1,260	—	22,500	—	6,019,129
—	653,079	—	33,862	—	22,950	—	3,777,512
410,732	685,531	20,010	34,250	—	—	2,762,724	4,963,890
439,554	593,988	3,725	6,499	—	—	3,147,529	5,636,935
325,134	604,438	20,050	36,536	—	—	3,782,114	7,061,137
447,355	799,046	24,689	44,500	54,265	94,752	3,905,476	7,731,557
232,685	436,383	46,104	82,279	321,100	569,625	4,400,991	8,773,638
265,332	501,409	53,747	100,352	447,550	785,601	6,262,745	12,668,861
207,022	413,095	127,798	274,568	156,875	295,730	6,155,994	13,517,060
124,900	278,422	106,879	236,659	138,350	287,419	5,284,098	12,234,535
353,789	1,415,484	297,946	838,348	174,239	696,956	6,686,228	26,102,630
778,263	3,113,463	381,798	1,261,505	1,191,848	4,771,283	7,974,072	31,467,278
,386,348	3,460,093	651,854	1,053,252	1,455,306	3,650,549	9,305,898	24,250,039
380,004	992,100	557,225	905,617	1,819,590	4,605,727	7,895,056	21,442,328
95,645	88,041	268,863	368,845	1,975,916	5,008,273	8,187,863	22,353,749
12,852	23,287	87,189	279,483	722,750	2,003,256	8,509,938	25,410,905
52,508	128,963	47,024	73,445	687,315	2,050,544	9,982,754	28,749,530
43,820	118,461	80,692	210,642	907,065	2,277,866	10,179,369	28,493,514
74,821	192,502	98,628	216,035	493,610	1,237,302	9,883,810	28,056,435
67,697	182,125	98,475	242,882	68,583	199,294	6,546,749	17,936,282
10,403	40,406	161,726	401,464	291,829	511,235	7,772,207	20,903,232
34,447	64,254	109,171	262,974	93,450	148,825	4,806,114	12,870,142
64,430	89,189	139,605	336,662	981,563	2,217,384	9,571,742	25,219,749
90,879	79,196	36,775	81,696	149,650	337,951	8,158,664	20,829,209
78,581	197,311	9,213	34,517	—	—	6,243,602	25,889,170

TABLE 2. Cigarettes exported from the United States to China, 1914-1930[a]

Year	Total cigarettes exported from the U.S. (thousands)	Cigarettes exported from the U.S. to China (thousands)	Percentage to China
1914	2,407,226	861,387[b]	35.8[b]
1915	2,076,178	1,083,518	52.2
1916	4,258,664	2,551,772	59.9
1917	7,019,723	4,949,137	70.5
1918	12,145,539	6,792,428 (+122,455)[c]	55.9 (+1.0)[d]
1919	16,211,769	6,191,765 (+43,710)	38.2 (+.3)
1920	15,833,870	8,506,600 (+9,070)	53.7 (+.06)
1921	8,543,676	6,443,727 (+112,595)	75.4 (+1.3)
1922	11,470,179	8,551,338 (+617,960)	74.5 (+5.4)
1923	12,252,528	9,227,030 (+878,065)	75.3 (+7.2)
1924	10,495,883	7,449,448 (+609,797)	71.4 (+5.8)
1925	8,145,639	5,381,761 (+7,455)	66.1 ('+.1)
1926	9,539,335	6,873,305 (+23,575)	72.1 (+2.5)
1927	7,093,039	4,418,498	62.3
1928	11,706,110	8,669,591	74.1
1929	8,455,851	4,854,586	57.4
1930	4,927,223	1,369,667	27.8

Sources: U.S. Department of Commerce, Bureau of the Census, Stocks of Tobacco Leaf, Bulletins 139 (1919), 30; 146 (1921), 40; 149 (1922), 32; 151 (1923), 30-31; 155 (1924), 36-37; 157 (1925), 28-29; 159 (1926), 30-31; 163 (1928), 32-33; 165 (1929), 30-31; U.S. Bureau of Foreign and Domestic Commerce, Monthly Summary of Foreign Commerce (December 1926), p. 17; (December 1930), p. 16; (December 1931), p. 15. Cf. Pan, p. 241; and Cox, p. 71.

a. Apparently not all cigarettes exported from the United States for China reached China. Note the discrepancy between American exports to China in this table and American imports into China in Table 1.

b. Any entry in this column that has no figure in parentheses next to it includes exports from the United States to Hong Kong.

c. Figures within parentheses in this column indicate the amounts of cigarettes exported from the United States to Hong Kong.

d. Figures within parentheses in this column indicate the percentages of cigarettes exported from the United States that went to Hong Kong.

TABLE 3. The British-American Tobacco Company's sales in China and profits worldwide, 1902-1928 (quantity in billions of cigarettes and value in U.S. dollars)

Year	Sales in China		Net profits	
	Quantity	Value	In China	World wide[a]
1902	1.25[b]	—	—	677,600
1903	1.54	—	—	—
1904	—	—	—	—
1905	—	—	—	—
1906	—	—	—	—
1907	—	—	—	—
1908	—	—	—	—
1909	—	—	—	—
1910	—	—	—	—
1911	—	—	—	—
1912	9.70	—	—	13,870,000
1913	—	—	—	—
1914	—	—	—	—
1915	—	—	—	—
1916	12.00	20,750,000[b]	3,750,000[b]	—
1917	—	—	—	—
1918	—	—	—	—
1919	22.50[c]	—	—	18,127,000
1920	—	—	7,610,000[d]	23,420,000
1921	—	—	—	20,705,000
1922	31.00[e]	—	—	21,124,000
1923	—	—	—	20,677,000
1924	—	57,900,000[f]	—	21,509,000
1925	—	—	—	24,852,000
1926	—	—	—	30,112,000
1927	—	—	—	30,881,000
1928	80.00[g]	—	—	—

Total profits in China, 1902-1948: $380,785,120[h]

a. U.S. Department of Commerce, *Tobacco Markets and Conditions Abroad*, 237 (January 21, 1930); *Yin-hang yueh-k'an*, 7.2 (May 3, 1927), 10, cited by Yen Chung-p'ing, *Chung-kuo chin-tai ching-chi shih t'ung-chi tzu-liao hsuan-chi* (Shanghai, 1961), p. 169; speech by Hugh Cunliffe-Owen, chairman of the board of BAT, published in *The Times*, a newsletter distributed to BAT stockholders (January 17, 1928), p. 3, Brown Collection.

b. FO 371/3189, no. 126013, Rickards of BAT to Grey, July 17, 1918.

c. This figure is the sum of all cigarettes imported into China from the United States and Great Britain (see table 1) and my estimate that BAT produced 16.5 billion cigarettes in 1919. The figure of 16.5 billion was reached by assuming that

(continued)

TABLE 3 cont.

the working year was 300 days long and extrapolating from a report by William Morris of BAT to James Thomas in 1919 in which Morris indicated that BAT was producing 243,562,000 cigarettes per week in Shanghai, 10-12 million per day in Hankow, and 75 million per month in Mukden. See Morris to Thomas, October 28, 1919, Thomas Papers. Also in 1919, Julean Arnold, the U.S. attaché in China, made a lower estimate, concluding that all factories in China produced a total of 10 billion cigarettes during that year. See his "China's Commercial and Economic Progress and Prospects," *The Trans-Pacific*, 1.2 (October 1919), 14.

d. Chien Chao-nan to Chou Shou-chen et al., February 6, 1921, *Nanyang*, p. 425.

e. The estimate that the addition of the Tientsin plant (built between 1920 and 1922) raised the company's total production by 6 billion cigarettes per year is based upon figures in China, Inspectorate General of Customs, *Decennial Reports, 1922-1931* (Shanghai, 1933), p. 361. This source states that BAT's Tientsin plant produced 30 billion cigarettes per year in the 1920s, but Thomas Wiens has pointed out that the 60 machines available in that factory (according to the same report) were probably capable of producing only about 6 billion per year. Cf. Wiens, pp. 11-12.

f. FO 371/10940, no. 5505, Cunliffe-Owen of BAT to the Foreign Office, November 13, 1925.

g. Thomas to Hornbeck, box 412, August 9, 1928, Hornbeck Papers.

h. Wang Hsi, p. 94. This figure is probably too low. It represents an average of only $8.1 million per year, or, if the highly lucrative years between 1934 and 1940 are excluded, only $5 million per year. In light of the reliable figures of $3.75 million for profits in 1916 (before the company's postwar expansion) and $7.61 million in 1921, it seems likely that BAT's total profits between 1902 and 1948 were higher.

TABLE 4. Distribution of cigarettes in China's treaty ports, 1908, 1916, and 1930 (quantity in thousands, value in Haikwan taels)

Customs district	1908		1916		1930	
	Quantity	Value	Quantity	Value	Quantity	Value
Ai-hui	—	—	5,799	25,431	—	—
Harbin	—	—	—	—	50,349	89,799
Hunchun	—	—	7,445	19,008	—	—
San-hsing	—	—	3,946	9,573	—	—
Ling-ching-ts'un	—	—	9,774	39,321	9,180	34,476
Antung	30,757	52,902	144,153	225,937	8,571	60,313
Ta-t'ung-k'ou	—	—	200	200	—	—
Dairen	286,141	648,965	670,739	2,804,996	1,556,296	6,491,910
Manchouli	478,121	812,346	177,854	444,637	—	—
Sui-fen-ho	38,178	96,433	—	—	—	—
Ying-k'ou	86,172	158,257	26,863	109,315	35,350	153,969
Chinwangtao	108,441	196,206	117,923	479,903	6,577	25,480
Tientsin	258,381	463,501	1,140,088	4,667,269	392,459	1,907,143
Lungkow	—	—	3,967	15,973	2,400	5,581
Chefoo	36,423	67,587	34,885	143,287	4,260	22,864
Weihaiwei	—	—	—	—	2,440	10,405
Kiaochow	43,325	89,960	321,683	1,302,854	34,380	270,361
Chungking	1,146	2,995	17,082	76,896	48,255	249,866
Ichang	702	1,200	23,346	108,219	40,588	176,572
Shasi	253	437	22,286	99,389	15,146	55,420
Changsha	—	19,820	54,506	233,564	12,501	47,472
Yo-chou	—	300	15,354	63,525	1,960	7,064
Hankow	92,225	188,304	520,641	2,162,235	301,601	1,450,064
Kiukiang	6,164	12,560	64,322	267,239	139,095	374,588
Wuhu	31,121	54,427	173,842	690,324	475,347	1,526,499
Nanking	52,282	93,424	299,761	1,232,504	56,458	376,418
Chinkiang	44,127	76,879	252,948	1,024,335	77,095	387,191
Shanghai	491,994	873,154	1,481,385	6,110,205	1,895,245	8,906,779
Soochow	68,290	108,813	148,190	609,110	65,850	222,265
Hangchow	94,353	161,537	438,102	1,775,373	29,510	71,575
Ningpo	175,221	299,629	474,372	1,903,996	52,520	146,442
Wenchow	2,350	4,019	12,902	53,673	34,320	129,896
San-tu-ao	—	—	45	159	6,470	12,941
Foochow	2,966	5,919	14,811	50,015	54,489	157,246
Amoy	19,865	35,614	46,179	147,805	98,495	393,812
Swatow	8,461	31,444	42,600	133,169	463,126	729,414
Canton	118,905	216,189	289,157	854,686	394,372	923,173
Kowloon	1,323	2,646	3,564	18,533	1	2
Kowloon Railway Traffic	—	—	255	1,326	34	289

(continued)

TABLE 4 cont.

Customs district	1908		1916		1930	
	Quantity	Value	Quantity	Value	Quantity	Value
Lappa	75	212	361	758	95	256
Kongmoon	3,372	5,767	23,131	73,389	6,260	39,485
Samshui	1,858	3,186	23,346	65,698	2,130	8,457
Wuchow	9,002	15,844	19,802	56,041	42,507	107,941
Nanning	2,756	4,713	10,210	38,267	—	—
Ch'iung-chou	883	1,983	5,667	18,934	57,186	125,860
Pakhoi	2,744	4,927	2,851	8,126	23,488	47,676
Szemao	10	64	—	—	20	75
Mengtze	4,380	14,859	90,600	160,288	6,473	20,199
Tengyueh	3,499	7,908	147	160	14	205
Lungchow	—	—	10	28	2	24

Source: China, Inspectorate General of Customs, *Returns of Trade and Trade Reports* (Shanghai, 1912), pt. 3, vol. 1, p. 162; ibid. (Shanghai, 1919), pt. 3, vol. 1, pp. 259-260; ibid. (Shanghai, 1931), pt. 2, vol. 1, pp. 353-354.

TABLE 5. Chinese-owned cigarette companies and their machines and workers in Shanghai, 1902-1936

Year	Companies	Machines	Workers
1902	1[a]	8	240
1903	1	8	240
1904	1	8	240
1905	8 (2)[b]	16	480
1906	9 (1)	16	480
1907	6 (1)	8	240
1908	1 (1)	1	—
1909	1 (1)	1	—
1910	1 (1)	1	—
1911	1 (1)	1	—
1912	3 (2)	2	30
1913	3 (2)	2	30
1914	3 (2)	2	30
1915	5 (4)	5	120
1916	5 (7)	6	150
1917	7 (8)	18 (119)[c]	628
1918	8 (9)	22	748
1919	9 (9)	84	4,832
1920	14 (9)	105	5,568
1921	13 (9)	104	5,512
1922	12 (9)	97	5,232
1923	13 (10)	107	5,552
1924	16 (14)	113	5,721
1925	52 (51)	176 (207)[d]	8,615
1926	64 (105)	318	14,215
1927	67 (182)	344	15,781
1928	101 (94)	414	17,913
1929	100 (79)	416	17,427
1930	94 (65)	543	19,683
1931	79 (64)	540	—
1932	75 (60)	535	—
1933	58	519	17,483
1934	53	495	17,875
1935	49	482	—
1936	44	474	16,078

a. All figures except those in parentheses are from *Nanyang*, pp. 254-255.

b. "Chinese Cigarette Factories in Shanghai," *Chinese Economic Journal*, 9.6 (December 1932), 426; Lieu, *The Growth and Industrialization of Shanghai*, pp. 44-45; Shang-hai-shih she-hui chü, *Shang-hai chih chi-chih kung-yeh*, pp. 250-254.

c. Ch'en Chen, col. 4, vol. 1, p. 447. According to this source, Nanyang alone had 119 machines in Shanghai during 1917.

d. NA 693.1112/104, Cunningham to State, July 20, 1925. According to this source, 140 of these 207 machines belonged to Nanyang.

TABLE 6. Nanyang Brothers Tobacco Company's sales and profits, 1912-1948 (quantity in thousands, value and profits in *yuan* for 1912-1936 and in *fapi* for 1937-1948)

Year	Hong Kong factories			Shanghai factories			Other factories[a]	Total		
	Quantity	Value	Profits	Quantity	Value	Profits	Quantity	Quantity	Value	Profits
1912	237,900	433,031	52,000	—	—	—	—	237,900	433,031	52,000
1913	347,900	—	117,000	—	—	—	—	347,900	—	117,000
1914	526,700	—	175,000	—	—	—	—	526,700	—	175,000
1915	930,450	2,323,467	324,000	—	—	—	—	930,450	—	324,000
1916	1,310,850	3,494,463	1,152,000	—	—	—	—	1,310,850	3,494,463	1,152,000
1917	1,691,250	4,812,959	1,091,000	—	—	400,000	—		—	1,491,000
1918	1,850,900	5,622,644	1,194,000	—	8,377,356	800,000	—		14,000,000	1,994,000
1919	1,948,000	6,318,109	1,047,000	—	—	—	—		—	—
1920	—	15,497,000	2,339,000	—	9,516,000	2,519,000	—	—	25,013,000	4,858,000
1921	—	15,837,000	1,857,000	—	12,175,000	2,185,000	—	—	28,012,000	4,042,000
1922	—	16,166,000	2,271,000	—	12,069,000	1,814,000	—	—	28,235,000	4,085,000
1923	—	16,249,000	2,390,000	—	15,670,000	705,000	—	—	31,919,000	3,095,000
1924	—	13,187,000	1,187,000	—	12,024,000	-707,000	—	—	25,211,000	480,000
1925	—	16,256,000	648,000	5,343,050	20,200,000	572,000	—	—	36,456,000	1,220,000
1926	—	11,669,000	1,104,000	6,050,655	17,052,000	1,197,000	—	—	28,721,000	2,301,000
1927	—	7,203,000	49,000	4,250,275	20,525,000	238,000	—	—	27,728,000	287,000
1928	—	6,510,000	-286,000	—	11,033,000	-1,961,000	—	—	17,543,000	-2,247,000
1929	—	5,133,000	-750,000	—	8,314,000	-2,452,000	—	—	13,447,000	-3,202,000
1930	—	7,693,000	552,000	—	6,457,000	-858,000	—	—	14,150,000	-306,000
1931	—	9,328,000	538,000	—	14,452,000	224,000	—	—	23,780,000	762,000

Year										
1932	—	10,161,000	565,000	4,936,535	18,372,000	490,000	—	—	28,533,000	1,055,000
1933	—	10,699,000	313,000	5,823,680	19,555,000	1,047,000	—	—	30,254,000	1,360,000
1934	—	9,423,000	570,000	5,653,205	19,227,000	635,000	—	—	28,650,000	1,205,000
1935	—	4,539,000	498,000	5,494,375	22,033,000	103,000	—	—	26,572,000	601,000
1936	—	4,345,000	171,000	4,888,140	23,483,000	131,000	—	—	27,828,000	302,000
1937	1,075,450	5,130,000	—	3,153,950	19,417,000	844	388,745	4,618,145	24,547,000	1,032,000
1938	1,408,200	6,824,000	—	1,371,520	17,960,000	2,307	1,058,317	2,603,669	24,783,000	2,726,000
1939	1,171,300	6,965,000	—	1,428,682	15,763,000	1,532	123,360	2,723,342	22,728,000	2,315,000
1940	832,400	5,618,000	—	1,023,683	24,949,000	4,186	227,700	2,083,783	30,567,000	5,090,000
1941	826,200	—	—	776,963	—	—	166,680	1,769,843	9,551,000	9,551,000
1942	—	—	—	183,300	—	—	141,840	—	—	19,520,000
1943	—	—	—	296,700	—	—	164,700	—	—	19,802,000
1944	—	—	—	361,150	—	—	76,920	—	—	65,839,000
1945	—	—	—	212,300	—	—	52,860	—	—	27,258,000
1946	167,650	—	—	—	—	—	—	—	—	—
1947	435,275	—	—	1,163,134	—	—	824,010	2,422,419	—	—
1948	738,290	—	—	1,058,630	—	—	786,370	4,380,210	—	—

Source: Nanyang, pp. 25, 37-38, 52, 114, 156, 171, 220-221, 275-276, 467, 521, 523, 528, 531-532, 534, 550, 562, 597.
a. "Other" includes Chungking (March 1939-1945 and 1947-1948); Hankow (October 1937-September 1938 and 1947-1948); Djakarta (1938); and Canton (1937-1938).

TABLE 7. Paid-up capital of the British-American Tobacco Company and Nanyang Brothers Tobacco Company, 1902-1928 (U.S. dollars)

	BAT		
	Headquarters (London)	China	Nanyang
1902	29,040,000	—	—
1903	—	—	—
1904	—	—	—
1905	—	—	49,000
1906	25,369,000	2,490,000	—
1907	—	—	—
1908	—	—	—
1909	—	—	55,000
1910	—	—	—
1911	—	—	—
1912	58,465,000	—	42,000
1913	65,480,000	—	75,000
1914	44,630,000	—	118,000
1915	44,630,000	—	189,000
1916	44,630,000	—	515,000
1917	44,630,000	—	1,240,000
1918	44,630,000	—	2,290,000
1919	62,409,000	231,480,000	3,353,000
1920	98,412,000	—	3,493,000
1921	98,475,000	—	3,316,000
1922	98,612,000	113,760,000	4,507,000
1923	94,110,000	—	4,527,000
1924	90,866,000	—	4,574,000
1925	99,337,000	99,385,000	4,916,000
1926	135,936,000	—	5,328,000
1927	136,106,000	—	5,562,000
1928	136,449,000	—	6,300,000

Sources: Bureau of Corporations, *Report*, pt. 1, pp. 16 and 306; British-American Tobacco Company, Limited, *Report of the Directors*, published annually, Brown Collection; Ch'en Chen, col. 2, p. 94; FO 371/3189, no. 126013, Rickards of BAT to Grey, July 17, 1918; FO 371/9205, no. F127, Rickards of BAT to the Colonial Office, December 21, 1922; FO 371/11688, no. F1593, Rickards of BAT to the Colonial Office, November 4, 1925; *Nanyang*, pp. 2, 4, 38, and 145.

TABLE 8. Imported and domestic tobacco leaf in China, 1908-1937 (thousands of catties)

Year	Imported, flue-cured bright tobacco	Domestically grown, flue-cured bright tobacco	All kinds of domestically grown tobacco
1908	8,426.2	—	—
1909	6,204.9	—	—
1910	10,246.7	—	—
1911	9,769.8	—	—
1912	14,293.1	—	—
1913	16,158.6	—	—
1914	11,835.6	—	
1915	7,674.6	—	1,590,000
1916	14,713.2	2,640	(1914-1918)
1917	15,392.7	14,300	
1918	18,109.1	25,300	
1919	15,982.4	50,050	—
1920	22,732.3	59,400	—
1921	22,128.1	26,950	—
1922	25,403.3	21,450	—
1923	30,925.4	46,200	—
1924	67,757.8	77,000	—
1925	55,168.5	57,200	—
1926	75,508.3	25,630	—
1927	63,300.3	19,800	—
1928	106,985.1	36,300	—
1929	91,094.0	42,020	—
1930	92,593.8	91,850	—
1931	124,207.0	122,100	
1932	58,769.4	114,400	
1933	40,523.7	163,900	
1934	48,991.8	154,000	1,830,000
1935	13,431.1	179,300	(1931-1937)
1936	18,849.0	198,000	
1937	27,960.2	231,000	

Sources: Chen Han Seng, Appendices 3 and 4, pp. 93 and 95; Perkins, *Agricultural Development*, p. 284, table D.25.

TABLE 9. Consumption of cigarettes in the United States and China, 1900-1928 (billions)

Year	United States	China
1900	2.5	0.30
1901	—	—
1902	2.5	1.25
1903	—	—
1904	—	—
1905	3.6	—
1906	—	—
1907	—	—
1908	—	—
1909	—	—
1910	8.5	7.5
1911	—	—
1912	11.0	9.7
1913	—	—
1914	—	—
1915	20.0	—
1916	21.0	13.0
1917	—	—
1918	—	—
1919	—	—
1920	45.0	25.0
1921	—	—
1922	—	—
1923	—	—
1924	60.0	40.0
1925	80.0	—
1926	—	—
1927	—	—
1928	100.0	87.0

Sources: Tennant, pp. 4 and 16; "Cigarette and Tobacco Trade in China," *Chinese Economic Bulletin*, 225 (June 13, 1925), 338; H. M. Wolf, "The Tobacco Industry in China," *Chinese Economic Journal*, 14.1 (1934), 91; tables 2 and 5 above.

A Note on the Sources

IN THIS BOOK, I have relied on several compilations of primary sources on Chinese industries (Ch'en Chen, cols. 1, 2, and 4, and Wang Ching-yü); on Chinese handicraft industries (P'eng Tse-i); and on Chinese agriculture (Li Wen-chih and Chang Yu-i). In the notes I have not given full citations of the original sources contained within the compilations except at a few points where I felt that the nature of the original sources was particularly important to my interpretation. To suggest the range of the various sources I have used from the compilations, let me identify the principal ones here: business records of BAT, which are now in the archives of the Economic Research Institute of the Chinese Academy of Sciences (*Chung-kuo k'o-hsueh yuan ching-chi yen-chiu so*) in Shanghai and were left behind when the company withdrew from China; a BAT internal Chinese language publication, *The BAT Monthly, republished in a twenty-first anniversary edition (Ying-Mei yen-ts'ao kung-ssu yueh-pao, erh-shih-i chou-nien chi-nien k'an;* 1923); a Japanese study of BAT in China by the Asian Development Board (Nihon Koain; 1941); Chinese, British, and American diplomatic correspondence; Maritime Customs reports; local gazetteers (*ti-fang chih*); scholarly journals; journals published by Chinese chambers of commerce; Chinese, Japanese, and Western periodicals and newspapers; Chinese folk songs; and interviews that Chinese researchers conducted in the 1950s and 1960s with Chinese who had formerly worked for BAT.

Only in my references to one Chinese compilation, *Nanyang*, have I given the full citation of each original source. An exception was made for this volume to show the intimate nature of many of the documents cited in it—especially the private correspondence and accounting records.

Abbreviations

Chan-tou	Chung-kuo kung-ch'an tang (Chinese Communist Party), *Chan-tou ti wu-shih nien—Shang-hai chüan-yen i ch'ang kung-jen tou-cheng shih-hua* (Fighting for fifty years—a history of the workers' struggle based on interviews in Shanghai Cigarette Factory No. 1; Shanghai, 1960).
Chang Yu-i	Chang Yu-i, comp., *Chung-kuo chin-tai nung-yeh shih tzu-liao, ti-erh-chi, 1912-1927* (Historical materials on modern Chinese agriculture, second collection, 1912-1927; Peking, 1957)
Ch'en Chen, col. 1	Ch'en Chen, comp., *Chung-kuo chin-tai kung-yeh shih tzu-liao, ti-i-chi* (Historical materials on modern Chinese industry, first collection; Peking, 1957), 2 vols.
Ch'en Chen, col. 2	_____, comp., *Chung-kuo chin-tai kung-yeh shih tzu-liao, ti-erh-chi* (Historical materials on modern Chinese industry, second collection; Peking, 1958), 2 vols.
Ch'en Chen, col. 4	_____, comp., *Chung-kuo chin-tai kung-yeh shih tzu-liao, ti-ssu-chi* (Historical materials on modern Chinese industry, fourth collection; Peking, 1961), 2 vols.
FO	Great Britain, Records of the British Foreign Office, Public Records Office, London
HKRS	Hong Kong Record Series, Public Records Office, Hong Kong
Li Wen-chih	Li Wen-chih, comp., *Chung-kuo chin-tai nung-yeh shih tzu-liao, ti-i-chi, 1840-1911* (Historical materials on modern Chinese agriculture, first collection, 1840-1911; Peking, 1957)

NA   National Archives of the United States, Washington, D.C.

Nanyang   Chung-kuo k'o-hsueh yuan Shang-hai ching-chi yen-chiu so
                   Shang-hai she-hui k'o-hsueh yuan ching-chi yen-chiu so
                   (The Shanghai economic research institute of the Chinese
                   academy of sciences and the economic research institute of
                   the Shanghai academy of social sciences), comps., *Nan-yang
                   hsiung-ti yen-ts'ao kung-ssu shih-liao* (Historical materials
                   on the Nanyang Brothers Tobacco Company; Shanghai,
                   1958)

NCH   *The North China Herald and Supreme Court and Consular
                   Gazette*

P'eng Tse-i   P'eng Tse-i, comp., *Chung-kuo chin-tai shou-kung-yeh shih
                   tzu-liao, 1840-1949* (Historical materials on modern Chinese
                   handicraft industries, 1840-1949; Peking, 1957), 4 vols.

Wang Ching-yü   Wang Ching-yü, comp., *Chung-kuo chin-tai kung-yeh shih
                   tzu-liao, ti-erh-chi, 1895-1914* (Historical materials on mod-
                   dern Chinese industry, second collection, 1895-1914; Pe-
                   king, 1957), 2 vols.

Wu-hu   Kuang-tung kuo-huo chiu-cheng hui (Society for the protec-
                   tion of national goods in Kwangtung province), ed., *Wu-hu
                   exposé of the swindle by the Nanyang Brothers Tobacco
                   Company; Canton, 1919).

Notes

1. Introduction: Sino-Foreign Commercial Rivalries

1. These estimates are cited in Richard J. Barnet and Ronald E. Müller, *Global Reach* (New York, 1974), p. 76.

2. See, for example, two splendidly researched historical studies by Mira Wilkins, *The Emergence of Multinational Enterprise* (Cambridge, Mass., 1970); and *The Maturing of Multinational Enterprise* (Cambridge, Mass., 1974).

3. See Kang Chao, "The Chinese American Textile Trade, 1830-1930," paper presented at the Conference on the History of American-East Asian Economic Relations, Mt. Kisco, New York, June 1976; Kwang-ching Liu, "British-Chinese Steamship Rivalry in China, 1873-85," in C. D. Cowan, ed., *The Economic Development of China and Japan* (New York, 1964), pp. 49-78; Yen-p'ing Hao, *The Comprador in Nineteenth Century China* (Cambridge, Mass., 1970); Edward LeFevour, *Western Enterprise in Late Ch'ing China* (Cambridge, Mass., 1968); Rhoads Murphey, *The Treaty Ports and China's Modernization* (Ann Arbor, 1970).

4. Murphey, *Treaty Ports*, p. 26.

5. Chi-ming Hou, *Foreign Investment and Economic Development in China, 1840-1937* (Cambridge, Mass., 1965), pp. 7-8, 147, 98, and 14; Albert Feuerwerker, "China's Nineteenth Century Industrialization," in Cowan, ed., *The Economic Development*, pp. 80-81. On the basis of these statistics, Hou has reached the conclusion (contrary to mine) that foreign investment in China "was not large." (Hou, *Foreign Investment*, p. 97). My view has been informed by Cheryl Payer's perceptive reading of Hou's statistics and by Robert Dernberger's revision of Hou's interpretation. See Payer, "Harvard on China II: Logic, Evidence, and Ideology," *Bulletin of Concerned Asian Scholars*, 6.2 (April-August 1974), 66; and Dernberger, "The Role of the Foreigner in China's Economic Development, 1840-1949," in Dwight H. Perkins, ed., *China's Modern Economy in Historical Perspective* (Stanford, 1975), pp. 28-30.

6. Albert Feuerwerker, *China's Early Industrialization* (Cambridge, Mass., 1958), p. 242.

7. Hou, *Foreign Investment*, pp. 138 and 139.

8. For general surveys which describe Chinese and foreign rivals in these and other industries, see D. K. Lieu (Liu Ta-chün), *China's Industries and Finance* (Peking, 1927); Lieu, *The Growth and Industrialization of Shanghai* (Shanghai, 1936); G. C. Allen and Audrey G. Donnithorne, *Western Enterprise in Far Eastern Economic Development* (New York, 1954); Hou, *Foreign Investment*; Wellington K. K. Chan, *Merchants, Mandarins and Modern Chinese Enterprise in Late Ch'ing China* (Cambridge, Mass., 1977); Chang Yen-shen, *Jih-pen li-yung so-wei "ho-pan shih-yeh" ch'in-Hua ti li-shih* (Peking, 1958); Ch'ien I-shih, *Chin-tai Chung-kuo ching-chi shih* (Chungking, 1939); Kuo-huo shih-yeh ch'u-pan she, comp., *Chung-kuo kuo-huo kung-ch'ang shih-lueh* (Shanghai, 1935); Akira Nagano, *Development of Capitalism in China* (Tokyo, 1931); Wang Ching-yü; Ch'en Chen, cols. 1, 2, and 4. On the cotton textile industry, see Yen Chung-p'ing, *Chung-kuo mien-fang-chih shih-kao* (Peking, 1955), chs. 5-6 and appendix 1; Kang Chao, *The Development of Cotton Textile Production in China* (Cambridge, Mass., 1977), chs. 5-6, especially pp. 120-129; and Fang Hsien-ting (H. D. Fong), *Chung-kuo chih mien-fang-chih yeh* (Shanghai, 1934). On the textile machinery industry, see Thomas G. Rawski, "Producer Goods and Industrialization in Twentieth Century China" (manuscript, May 1977), pp. 8-12; and two studies cited by Rawski: George Sweet Gibb, *The Saco-Lowell Shops* (Cambridge, Mass., 1950), pp. 478-482 and 671; and Thomas R. Navin, *The Whitin Machine Works since 1831* (Cambridge, Mass., 1950), pp. 324-333. On the match industry, see *Matchi kōgyō hōkokusho* (Nanking, 1940), a translation of *Lin-ts'un kung-yeh pao-kao shu*; Z. T. Kyi, "Match-making Industry," *Chinese Economic Journal*, 4.4 (April 1929), 305-311; and Kyi, "Match Industry in China," ibid., 10.3 (March 1932), 197-211. On the soap industry, see Charles Wilson, *The History of Unilever* (London, 1954), I, 140, 192, and 225; and II, 354 and 364. On the chemical industry, see W. J. Reader, *Imperial Chemical Industries*, I, *The Forerunners, 1870-1926* (London, 1970), 224-227 and 333-347, and II, *The First Quarter Century* (London, 1975), 331; Carroll Lunt, *Some Builders of Treatyport China* (Los Angeles, 1965), pp. 138-143; the biographical sketch of Hou Te-pang in Howard L. Boorman and Richard C. Howard, eds., *Biographical Dictionary of Republican China* (New York, 1968), II, 84-86. On the cement industry, see Edward Friedman, "The Failure to Modernize China's Most Modern City, Canton, 1911-1914," paper presented at the Conference on Urban Society and Political Development in Modern China, St. Croix, Virgin Islands, December 1968-January 1969; and Albert Feuerwerker, "Industrial Enterprise in Twentieth Century China" in Feuerwerker, Rhoads Murphey and Mary C. Wright, eds., *Approaches to Modern Chinese History* (Berkeley, 1967), pp. 304-341. On the iron and steel industry, see Feuerwerker, "China's Nineteenth Century Industrialization," in Cowan, ed., *The Economic Development*, p. 93.

9. Yu-kwei Cheng, *Foreign Trade and Industrial Development of China* (Washington, D.C., 1956), ch. 3; Jean Chesneaux, "The Chinese Labour Force in the First Part of the Twentieth Century," in Cowan, ed., *The Economic Development*, p. 125.

10. Hou, *Foreign Investment*, p. 154; Dernberger, "The Role of the Foreigner," in Perkins, ed., *China's Modern Economy*, p. 45.

11. See Wang Hsi, "Ts'ung Ying-Mei yen kung-ssu k'an ti-kuo chu-i ti ching-

chi ch'in-lueh" (The British-American Tobacco Company as a case study in imperialist economic exploitation), *Li-shih yen-chiu*, 4 (August 1976), 77-95; and Huang Ch'eng-ching, "Ts'ung Nan-yang hsiung-ti yen-ts'ao min-tsu tzu-ch'an chieh-chi ti hsing-k'o (Insights into the character of the national capitalist class based on the Nanyang Brothers Tobacco Company)," *Hsueh-shu yueh-k'an*, 10.22 (October 10, 1958), 34-35; Pien Chieh, " 'Nan-yang hsiung-ti yen-ts'ao kung-ssu shih-liao' p'ing-chieh," *Hsueh-shu yueh-k'an*, 6.30 (June 10, 1959), 62-63; *Chan-tou*, pp. 60-61.

12. Feuerwerker, *China's Early Industrialization*, pp. 22-26 and 58-59.

13. Marion J. Levy, Jr., "Contrasting Factors in the Modernization of China and Japan," *Economic Development and Cultural Change*, 2.3 (October 1953), 170, 186, and 196; Levy, "The Social Background of Modern Business Development in China," in Levy and Kuo-heng Shih, *The Rise of the Modern Chinese Business Class* (New York, 1949), p. 12.

14. Hao, pp. 215, 102 and 146-153.

15. Thomas G. Rawski, "The Growth of Producer Industries, 1900-1971" in Perkins, ed., *China's Modern Economy*, p. 208.

16. Dwight H. Perkins, "Introduction: The Persistence of the Past," in ibid., p. 5.

17. Joseph A. Schumpeter, *The Theory of Economic Development*, trans. Redvers Opie (Cambridge, Mass., 1959), p. 66, cited by Albro Martin, "Towards a More Precise Definition of Entrepreneurship," paper presented to the Columbia University Seminar in Economic History, March 4, 1976, p. 3.

18. Jean Chesneaux, *The Chinese Labor Movement, 1919-1927*, trans. H. M. Wright (Stanford, 1968), p. *vii*; Hui-min Lo, "Some Notes on Archives on Modern China," in Donald D. Leslie, Colin Mackerras, and Wang Gungwu, eds., *Essays on the Sources for Chinese History* (Canberra, 1973), p. 212. The problem of gaining access to business records is by no means unique to the study of China. For a discussion of efforts to gain access to the records of American businesses, see James Lawton, ed., *Shoptalk: Papers on Historical Business and Commercial Records in New England* (Boston, 1975).

19. On the movement in which documentation was gathered for histories of families, factories, villages, and communes, see Shih Ch'eng-chih, "Shih-lun 'ssu-shih' yü 'wen-ko,' " *Ming-pao yueh-k'an*, 72 (December 1971), 5-17, and 73 (January 1972), 37-43. An English translation of these articles appears in *Chinese Sociology and Anthropology*, 4.3 (Spring 1972), 175-233.

20. In 1968, Albert Feuerwerker observed that this compilation provided historians with a kind of data on a Chinese enterprise that was not available elsewhere. As far as I know, his statement still holds true. See his "China's Modern Economic History in Communist Chinese Historiography," in Feuerwerker, ed., *History in Communist China* (Cambridge, Mass., 1968), p. 243.

2. Penetrating the China Market

1. Paul A. Varg, "The Myth of the China Market," ch. 3 in his *The Making of a Myth: The United States and China, 1897-1912* (East Lansing, 1968); Marilyn Blatt Young, *The Rhetoric of Empire: American China Policy, 1895-*

1901 (Cambridge, Mass., 1968); Tyler Dennett, *Americans in East Asia* (New York, 1922), ch. 30; George F. Kennan, *American Diplomacy, 1900-1950* (Chicago, 1951), chs. 2 and 3.

2. Richard P. Dobson, *China Cycle* (London, 1946), p. 18.

3. James A. Thomas, *A Pioneer Tobacco Merchant in the Orient* (Durham, N.C., 1928), p. 42.

4. See Appendix, tables 3, 2, and 9.

5. *Tobacco*, January 7, 1915, p. 4. See Appendix, table 3.

6. On the cigarette industry in China before 1902, see China, Inspectorate of Customs, *Decennial Reports, 1892-1901* (Shanghai, 1906), pp. 12 and 45; Wang Ching-yü, pp. 206-207, 209-211.

7. The full text of the agreement between the American Tobacco Company and the Imperial Tobacco Company has been reproduced in Bureau of Corporations, *Report of the Commissioner of Corporations on the Tobacco Industry* (Washington, D.C., 1909), pt. 1, pp. 440-447. For details on the negotiations, see B. W. E. Alford, *W. D. & H. O. Wills and the Development of the U.K. Tobacco Industry, 1786-1965* (London, 1973), pp. 268-269; Maurice Corina, *Trust in Tobacco: The Anglo-American Struggle for Power* (New York, 1975), pp. 101-103 and 129; and Wilkins, *Maturing*, pp. 146-151.

8. BAT stockholders' newsletter called *The Times*, January 17, 1928, pp. 2-3, in the William H. Brown Collection, Sterling Memorial Library, Yale University, New Haven, Conn.; John K. Winkler, *Tobacco Tycoon: The Story of James B. Duke* (New York, 1942), p. 147. Duke is quoted by Patrick G. Porter, "Origins of the American Tobacco Company," *Business History Review*, 43.1 (Spring 1969), 76n70.

9. Patrick Fitzgerald, *Industrial Combination in England* (London, 1927), pp. 145-146; British American Tobacco Company, *Report of the Directors* for 1923, Brown Collection; Wilkins, *Maturing*, p. 152; Winkler, pp. 147-148. See also Mira Wilkins, "The Impact of American Multinational Enterprise on American-Chinese Economic Relations, 1786-1949," paper presented at the Conference on the History of American-East Economic Relations, Mt. Kisco, New York, June 1976. Duke delegated the position of chairman to W. R. Harris from 1902 to 1912 and held it himself from 1912 until 1923.

10. Alfred D. Chandler, Jr., *The Visible Hand* (Cambridge, Mass., 1977), p. 382.

11. Thomas, *Pioneer*, p. 49.

12. Ibid., p. 50.

13. *B.A.T. Bulletin*, new series, 17.73 (May 1926), 25-31; ibid., 19.97 (May 1928), 18; Wang Ching-yü, pp. 207-208 and 212-213; Bureau of Corporations, *Report*, pt. 1, p. 443.

14. Thomas, *Pioneer*, p. 5.

15. Robert Easton, *Guns, Gold & Caravans* (Santa Barbara, 1978), p. 106.

16. James Lafayette Hutchison, *China Hand* (Boston, 1936), p. 268.

17. Thomas, *Pioneer*, p. 7.

18. James A. Thomas, "Selling and Civilization," *Asia*, 23.12 (December 1923), 896.

19. Thomas, *Pioneer*, p. 24; Corina, *Trust in Tobacco*, p. 101; Easton,

Guns, p. 106; Carl Crow, *Foreign Devils in the Flowery Kingdom* (New York, 1940), p. 60.

20. Report by Alexander Hosie, *British Diplomatic and Consular Reports on Trade: Foreign Trade in China, 1906,* 3943 (London, 1907), 46.

21. Bureau of Corporations, *Report,* pt. 1, pp. 16 and 306; FO 371/3189, Rickards of the BAT to Grey, July 17, 1918. See also Appendix, table 7.

22. Wang Hsi, p. 93; Hou, *Foreign Investment,* p. 102.

23. C. F. Remer estimates that foreign investment in manufacturing in China in 1914 was $110.6 million. See his *Foreign Investments in China* (New York, 1933), p. 70.

24. Ch'en Chen, col. 2, pp. 93-94; Wu Ch'eng-lo, *Chin-shih Chung-kuo shih-yeh t'ung-chih* (Shanghai, 1933), II, 74-75; Wang Ching-yü, pp. 215, 217-218, and 1183-92; report by Wang Shih-jen, July 1915, *Nanyang,* pp. 40 and 67; report by Alexander Hosie, *British Diplomatic and Consular Reports, 1906: The Foreign Trade of China,* 3943 (1907), 46; China, Inspectorate of Maritime Customs, *Decennial Reports, 1902-1911* (Shanghai, 1913), p. 358; Thomas, "Selling and Civilization," p. 949; Thomas, *Pioneer,* p. 50. (I have revised Wang Ching-yü's figures on BAT employment in light of the evidence cited in this note.)

25. It seems safe to assume that all cigarettes imported into China from the United States and Great Britain in 1916 (about 4.7 billion) belonged to BAT; this constituted slightly more than one third of the company's total sales for the year. If it is assumed that all imports from Hong Kong also belonged to BAT (when in fact probably only some unknown fraction of these belonged to it), the company's total imports amounted to about 5.6 billion, slightly less than one half its total sales for the year. See Appendix, table 1.

26. On the significance of Duke's innovations, see Chandler, *The Visible Hand,* pp. 382-391.

27. William Ashley Anderson, *The Atrocious Crime (of Being a Young Man)* (Philadelphia, 1973), p. 1. See also Hutchison, ch. 1.

28. Thomas, *Pioneer,* pp. 85-86.

29. Ibid., pp. 103-106; Hutchison, p. 5. For examples of the loneliness and sense of cultural isolation from which young American BAT representatives sometimes suffered, see Hutchison, pp. 18-19, 43-45, 49-50, 106-107, 114-115, 189-191, 243-244; *B.A.T. Bulletin,* 2.35 (December 11, 1915), 133 and passim; and, for the most poignant characterization, W. Somerset Maugham, *On a Chinese Screen* (New York, 1922), ch. 23.

30. Hutchison, pp. 199 and 230. One of BAT's Chinese competitors visited BAT branches north of Peking and testified to two Western BAT representatives' fluency in the local dialect. (Chien Chao-nan to Chien Yü-chieh, September 13, 1916, *Nanyang,* pp. 59-60.) The language proficiency of many English BAT representatives was also evident in their assignment during World War I. Of the 80 who returned from China to be in the military, many applied to the British government to serve as bilingual interpreters with the Chinese Labor Corps in Europe, passed the required language examinations, and were assigned to the Chinese Labor Corps for the duration of the war. See *B.A.T. Bulletin,* 4-8 (1917-1919; paginated consecutively), 789, 1325, 1391, 1579, 1652-1653, 1699, 1725, 1756, 1782, 1821, 1887, 1951, 1981-1982, 2005, 2027, 2062, 2111, 2203, 2310,

23‐26. On the role of interpreters in the Chinese Labor Corps, see Judith Blick, "The Chinese Labor Corps in World War I," *Harvard Papers on China*, 9 (1955), 117 and 123.

31. Lee Parker and Ruth Dorval Jones, *China and the Golden Weed* (Ahoskie, N.C., 1976), p. 11.

32. FO 228/2154, Jordan to the Ministry of Foreign Affairs, May 31, 1912.

33. Wang Ching-yü, p. 229; *B.A.T. Bulletin*, 16.69 (January 1926), 268-270; Thomas, *Pioneer*, p. 237; Thomas, "Selling and Civilization," pp. 948-949.

34. BAT relied, for example, on the *S.S. Cigarette* to carry goods on the Upper Yangtze. See FO 228/1946, Hewlett to Jordan, July 7, 1915. On the origins of Western shipping privileges, see Westel W. Willoughby, *Foreign Rights and Interests in China* (Baltimore, 1920), pp. 161-163.

35. For an excellent analysis of these regional systems, see G. William Skinner, "Regional Urbanization in Nineteenth Century China," Skinner, ed., *The City in Late Imperial China* (Stanford, 1977), pp. 211-249.

36. On North China, see Wang Ching-yü, pp. 215-216; H. T. Montague Bell and H. G. Woodhead, *China Yearbook* (London, 1912), p. 47; FO 371/2332, no. 156279, report by Walker; T'ung Hsu Photography Shop to Nanyang, 1914, *Nanyang*, p. 59. On the Northwest, see Chien Chao-nan to Chien Yü-chieh, September 13, 1916, *Nanyang*, pp. 59-60; Wang Shih-jen to Nanyang, April 1916, *Nanyang*, p. 64; Anderson, pp. 99-112; Hutchison, chs. 13-16; FO 371/1608, no. 51903, Alston to Grey, November 1, 1913. On the Yangtze regions, see FO 228/1659, Tours to Jordan, February 15, 1907; FO 228/1630, Werner to Jordan, November 1, 1906; FO 228/1663, Smith to Jordan, March 27, 1907; FO 228/1695, Phillips to Jordan, October 12, 1908; FO 228/1628, Giles to Jordan, November 10, 1906; Wang Ching-yü, p. 227. On the Southeast, see Wang Ching-yü, p. 229. On South China, see FO 371/1349, no. 47131, Sly to FO, November 5, 1912; Wang Ching-yü, pp. 229 and 234; Arnold Wright, ed., *Twentieth Century Impressions of Hong Kong, Shanghai, and Other Treaty Ports of China* (London, 1908), III, 795-796; U.S. Department of Commerce, Bureau of Foreign and Domestic Trade, *Tobacco Trade of the World*, Special Report no. 68 (Washington, D.C., 1915), p. 31. On the Southwest, see Thomas, *Pioneer*, p. 327; Wang Ching-yü, pp. 229-230; FO 371/640, no. 26840, Cunliffe-Owen of BAT to Grey, July 14, 1909; no. 33459, Stanley to Grey, September 11, 1909; no. 36339, Carnegie to Grey, September 30, 1909; FO 371/864, no. 39392, O'Brien-Butler to Jordan, September 25, 1910; Edwin J. Dingle, *Across China on Foot* (New York, 1911), pp. 100-101. On the Northeast, see FO 371/1349, nos. 43817 and 51150, Jeffress of BAT to FO, October 17 and November 29, 1912; FO to Jeffress, October 20, 1912; no. 47131, Sly to FO, November 5, 1912; FO 371/1688, Campbell to BAT, June 27, 1905 and August 31, 1905; FO 371/180, no. 244, Cunliffe-Owen of BAT to FO, January 1, 1906; no. 6685, Fulford to Satow, December 11, 1905; nos. 12656 and 13164, MacDonald to Grey, April 13 and March 15, 1906; no. 13338, Hood of BAT to Grey, April 18, 1906; no. 15220, Satow to Grey, March 19, 1909; Wang Ching-yü, pp. 225-227; Dana G. Munro, "American Commercial Interests in Manchuria," *The Annals of the American Academy of Political and Social Science*, 39.128 (January 1912), 157n8; T'ung Hsu to Nanyang, 1914, *Nanyang*, p. 59. On Japanese and Russian competition in Manchuria, see Wang

Ching-yü, pp. 234-235. For statistical evidence on BAT's distribution, see Appendix, table 4.

37. Quoted by Porter, pp. 64-65.

38. Thomas, *Pioneer*, pp. 10-23 and 36.

39. Ch'en Chen, col. 2, p. 125.

40. FO 371/180, nos. 2370 and 6685, Fulford to Satow, December 11, 1905; Livingston in *The Tobacco Leaf*, November 14, 1905, enclosed in FO 17/1690, Hood to Landsdowne, December 4, 1905. The Japanese administration tore down this number of posters in Ying-k'ou. The number left up was not recorded. For additional examples of BAT advertising in Manchuria, see Wang Ching-yü, pp. 222-227; and Liu Ai-jen to Chien Chao-nan, 1914, *Nanyang*, p. 68.

41. *NCH*, June 14, 1907, p. 669; and January 27, 1911, p. 198. For additional examples of BAT advertising in North China, see *NCH*, January 17, 1908, p. 124; and November 7, 1908, p. 313; J. W. Sanger, *Advertising Methods in Japan, China, and the Philippines*, Department of Commerce Special Agent Series No. 209 (Washington, D.C., 1921), photograph opposite p. 74; Thomas, *Pioneer*, pp. 161-162.

42. Arnold Wright, III, 795-796.

43. Report by H. H. Fox, acting British consul in Wuchow, September 7, 1904, republished in *Journal of the American Asiatic Association*, 4.11 (December 1904), 345.

44. FO 371/864, no. 24094, Playfair to Jordan, June 8, 1909; Dingle, p. 100. On BAT advertising in Tengyueh, Yunnan, see Wang Ching-yü, pp. 228 and 230. On Kueiyang, Kweichou, see *NCH*, June 10, 1911, p. 679; and June 13, 1914, pp. 822-823.

45. FO 371/864, no. 8889, O'Brien-Butler to FO, February 4, 1919; Wang Ching-yü, p. 230.

46. On Shanghai, Hangchow, Nanking, Anking, Hankow, Changsha, Shasi, Chungking, and Chengtu, see FO 228/1660, Fox to Jordan, June 20, 1907; FO 228/1631, Smith to Satow, January 19, 1906; FO 228/1663, Smith to Jordan, March 27, 1907; FO 228/1695, Phillips to Jordan, January 15, 1908; and Hewlett to Jordan, October 24, 1908; Wang Ching-yü, pp. 216 and 227-228; *B.A.T. Bulletin*, 14.37 (May 1923), 835; *NCH*, June 17, 1906, p. 377; September 5, 1908, p. 581; December 12, 1908, p. 641; October 7, 1910, p. 18; April 1, 1911, p. 31.

47. Arnold to Thomas, December 19, 1915, James A. Thomas Papers, Manuscript Department, William R. Perkins Library, Duke University, Durham, N.C.

48. U.S. Department of Commerce, Bureau of the Census, *Stocks of Tobacco Leaf: 1918*, Bulletin no. 139 (Washington, D.C., 1919), p. 30; Wang Ching-yü, pp. 211, 216, and 219; Ch'en Chen, col. 2, pp. 104 and 122; FO 228/1632, Fraser to Satow, January 9, 1906; *British Diplomatic and Consular Reports on Trade, no. 4366, Report on the Trade of Shanghai 1908* (London, 1909), p. 7; Chang Yu-i, p. 152; and R. H. Gregory's records of his travel and research in China, June-August 1906, in the Richard Henry Gregory Papers, Manuscript Department, William R. Perkins Library, Duke University, Durham, N.C.

49. Ch'en Chen, col. 2, pp. 140-141; *British Diplomatic and Consular Reports, China, Report on the Trade of Hankow, 1913* (London, 1914), pp. 5-6;

Gregory to Jeffress, November 12 and December 22, 1915, and Thomas to See-man, August 11, 1928, Thomas Papers; FO 228/1953, Pratt (in Tsinan) to Jor-dan, April 5, 1915; Charles E. Gage, U.S. Department of Agriculture *Yearbook for 1926*; A. Sy-hung Lee, "The Romance of Modern Chinese Industry" *China Review* (November 1923), p. 160. On efforts to transplant bright tobacco on five continents, see Nannie May Tilley, *The Bright Tobacco Industry, 1860-1929* (Chapel Hill, 1948), pp. 385-386. On subsequent production in China, see Appendix, table 8.

50. See Chandler, *The Visible Hand*, pp. 382-391.

51. Ernest O. Hauser, *Shanghai: City for Sale* (New York, 1940), p. 100.

52. Maritime Customs, *Decennial Reports 1902-1911*, p. 358.

53. "Tobacco Growing in Shantung," *The Weekly Review*, 22 (September 23, 1922), 98.

54. Julean Arnold et al., *China: A Commercial and Industrial Handbook*, U.S. Department of Commerce, Bureau of Foreign and Domestic Commerce, Trade Promotion Series no. 38 (Washington, D.C., 1926), p. 75.

55. Hauser, p. 101.

56. Thomas, *Pioneer*, pp. 41-42.

57. Porter, pp. 62 and 68-70.

58. Wang Hsi, p. 89; Chesneaux, *The Chinese Labor Movement*, pp. 59, 72, 95, and 427; Harold Isaacs, *Five Years of Kuomintang Reaction* (Shanghai, 1932), pp. 59-68.

59. Thomas to Russell, September 18, 1915, Thomas Papers.

60. Thomas to Straight, August 2, 1915, Thomas Papers.

61. Thomas, *Pioneer*, p. 40.

62. Chen Han Seng, pp. 6, 26, and 28; FO 371/3180, no. 2564, Kennett of BAT to Jordan, October 22, 1917; Chang Yu-i, p. 502; Thomas, *Pioneer*, pp. 66-69; Ch'en Chen, col. 2, pp. 141-142; Huang I-feng, "Kuan-yü chiu Chung-kuo chieh-chi ti yen-chiu," *Li-shih yen-chiu*, 87.3 (June 15, 1964), 103 and 107; Chen Han Seng, p. 11.

63. Ping-ti Ho, *Studies on the Population of China, 1368-1953* (Cambridge, Mass., 1959), p. 203; several relevant excerpts from local gazetteers have been republished in Li Wen-chih, pp. 440-442; the study was by Hsu Shu-lan, "Chung yen-yeh fa" (Techniques for the cultivation of tobacco) in *Nung-hsueh pao*, no. 14 (1898), republished in Li Wen-chih, pp. 609-611.

64. Gregory to Thomas, April 3, 1922; Gregory to Jeffress, November 12, 1915, Thomas Papers.

65. Chen Han Seng, p. 8; Ho, *Studies on Population*, p. 203; *Journal of the China Branch of the Royal Asiatic Society*, 23 (1889), 97; FO 228/1589, report by Rose, July 5, 1905; FO 228/1595, Fraser to Satow, January 4, 1905; FO 228/1632, Fraser to Satow, January 9 and April 7, 1906; and Chang Yu-i, p. 160. Lacking figures on farm prices between 1913 and 1915, I have inferred that prices were lower in China on the basis of figures that show this was the case in a slightly later period, 1919-1936. See Chen Han Seng, p. 94; and Pettitt to Thomas, March 14, 1921, Thomas Papers.

66. L. Carrington Goodrich, "Early Prohibitions of Tobacco Smoking in China and Manchuria," *Journal of the American Oriental Society*, 58 (1938),

648-657; Jonathan D. Spence, "Opium Smoking in Ch'ing China," in Frederic Wakeman, Jr., and Carolyn Grant, eds., *Conflict and Control in Late Imperial China* (Berkeley, 1975), pp. 146-154 and 161-167; Li Wen-chih, pp. 140-141.

67. I have found no direct evidence that Duke, Thomas, or others invested in China because they perceived the Chinese as opium smokers who would substitute cigarettes for drugs, but circumstantial evidence suggests that they might well have seen a connection between the two commodities. In the nineteenth century, prior to BAT's investment in China, American officials in China expressed the hope that American tobacco might replace opium. And later, eulogizing Duke at his death in 1925, a BAT executive claimed that Duke had deliberately combined "business with humanity by weaning the Chinese . . . from opium by teaching them to smoke North Carolina cigarettes." This quotation is from *B.A.T. Bulletin*, 15.58 (February 1925), 281; for a similar claim, see FO 371/ 3701, no. 171045, Hood of BAT to Amery, January 6, 1920. Officials' references to the possibility of American tobacco replacing opium are in Dennett, p. 185; and Wang Ching-yü, p. 227. On anti-opium campaigns in the early 1900s, see Mary C. Wright, "Introduction," in Wright, ed., *China in Revolution*, pp. 14-15; Roger V. DesForges, *Hsi-liang and the Chinese National Revolution* (New Haven, 1973), pp. 93-102; and Edward J. M. Rhoads, *China's Republican Revolution* (Cambridge, Mass., 1975), pp. 94-97, 124-125, 151, 253, 268.

68. Dingle, pp. 100-101.

69. U.S. Department of Commerce, *Tobacco Trade of the World*, no. 68, p. 31.

70. Wang Ching-yü, pp. 214-215 and 231-232.

71. Ibid., pp. 231-234.

72. *Who's Who in China* (Shanghai, 1920), I, 240-241; Thomas, *Pioneer*, p. 131; report by Thomas, February 23, 1927, Thomas Papers; Arnold Wright, III, 433. On Chekiang as a birthplace for compradors and a financial elite, see Hao, pp. 174-175; and Susan Mann Jones, "Finance in Ningpo: The 'Ch'ien Chuang,' 1750-1880," in W. E. Willmott, ed., *Economic Organization in Chinese Society* (Stanford, 1972), pp. 47-77.

73. Thomas, *Pioneer*, p. 131.

74. FO 228/2154, Savage to Jordan, October 23, 1912; Huang I-feng, "Kuan-yü chiu Chung-kuo," p. 103.

75. Arnold Wright, III, 662.

76. FO 228/1631, Smith to Satow, January 19, 1906.

77. Wang Ching-yü, pp. 223-224, 228-229, and 231-232; Hutchison, chs. 14-15 and passim.

78. Wang Ching-yü, pp. 222-223, 225-226, and 228-229; FO 228/2154, Cobbs of BAT to Fraser, July 15, 1912; K'ung Hsiang-hsi, "The Reminiscences of K'ung Hsiang-hsi (1880-   )" as told to Julie Lien-ying How (manuscript in the Special Collections Library, Butler Library, Columbia University, 1961), pp. 37-39; Boorman and Howard, II, 263-264; Easton, p. 116. On the development of guilds (including tobacco guilds) in China before 1850, see Peter J. Golas, "Early Ch'ing Guilds," in Skinner, ed., *The City in Late Imperial China*, pp. 555-580.

79. Hutchison, p. 221. Hutchison gives numerous examples from first-hand experience of his and other Westerners' lack of close supervision or control over

Chinese BAT dealers. See ibid., pp. 47, 107-108, 135, 137, 219-220, 281-283, 305-306, 320-321, 343.

80. Parker and Jones, p. 33.

81. Thomas, "Selling and Civilization," p. 948; Thomas to Wolsiffer, April 22, 1923, Thomas Papers; Canton branch of the Nanyang Brothers Tobacco Company to Nanyang's main office in Hong Kong, June 1915, Nanyang, p. 77.

82. Ch'en Chen, col. 2, vol. 1, p. 121. The amount received for "handling charges" varied. Around 1920 it was four yuan per case (of 50,000 cigarettes). See ibid., p. 122.

83. Liu Ai-jen in Manchuria to Chien Chao-nan, 1914, Nanyang, p. 68.

84. The British-American Tobacco Company, The Record in China of the British-American Tobacco Company, Limited (n.p., 1925?), p. 32.

85. Ōi Senzō, "Shina ni okeru Ei-Bei Tabako Torasuto no keiei keitai, zai-Shi gaikoku kigyō no hatten to baiben soshiki no ichikōsatsu," Tōa kenkyu shohō, 26 (February 1944), 12-13 and 24. Cf. Hao, p. 208.

86. Hao, chs. 2 and 4.

87. The British-American Tobacco Company, The Record in China, p. 33.

88. See Chapter 6.

89. Lu Yao-chen to Nanyang, January 23, 1915, Nanyang, pp. 40-43; Teh-lung Shop to Nanyang, July 1915, Nanyang, p. 58. While testifying under oath in a court of law, Thomas Cobbs, a BAT representative in China, admitted that the company maintained these exclusive dealing arrangements. See NCH, September 27, 1907, p. 765.

90. Arnold, China: A Commercial and Industrial Handbook, no. 38, p. 75; FO 371/10297, n. F3405, Rose of BAT to the British minister, August 24, 1924. BAT was exceptional among foreign firms in its willingness to extend credit to its Chinese agents and to set its prices in silver dollars (rather than gold or the currency of its own country). See Crow, pp. 56-58.

91. On Chungking, see FO 228/1695, Phillips (in Chungking) to Jordan, January 15, 1908; U.S. Department of Commerce, Tobacco Trade of the World, no. 68, p. 33; Wang Ching-yü, p. 219. On Shanghai, see Wang Shih-jen to Chien Chao-nan and Chien Yü-chieh, March 1916, Nanyang, pp. 65-66. On Hangchow, see FO 228/1663, Smith (in Hangchow) to Jordan, March 27, 1907. On Peking, Tientsin, Chinwangtao, Changli and Liaoyang, see Wang Ching-yü, pp. 223-224. On Foochow, see FO 371/638, no. 24094, Playfair to Jordan, May 21, 1909. On Kunming, see Wang Ching-yü, p. 229, and FO 371/864, no. 39392, O'Brien-Butler (in Kunming) to Jordan, September 25, 1910. On Canton, see Arnold Wright, III, 796.

92. Albert Feuerwerker, The Chinese Economy, ca. 1870-1911 (Ann Arbor, 1969), p. 17; see also his The Chinese Economy, 1912-1949 (Ann Arbor, 1968), p. 13.

93. P'eng Tse-i, II, appendix 4, table 1; T. C. Liu and K. C. Yeh, The Economy of the Chinese Mainland (Princeton, 1965), pp. 143, 152, 428, 513, 535; and Wu Pao-san, "Chung-kuo kuo-min so-te 1933 hsiu cheng," She-hui k'o-hsueh tsa-chih, 9.2 (December 1947), 140. For nonstatistical impressions that cigarettes were competing with hand-made tobacco products in Kwangtung, Fukien, and Chekiang provinces, see Wang Hsi, p. 88; and "Tobacco Crops of Chekiang,"

Chinese Economic Journal, 5.3 (September 1929), 806-810. Cf. Dwight H. Perkins, "Growth and Changing Structure of China's Twentieth Century Economy," in Perkins, ed., *China's Modern Economy in Historical Perspective,* p. 121.

94. This publication apparently originated in BAT's "Training School" for Chinese agents in Peking. Founded in 1921, it was called *Ying-Mei yen kung-ssu yueh-pao* (BAT monthly). Six thousand copies of this first issue were published, and, according to BAT, were circulated "from the Yellow Sea to the borders of Tibet; from Hong Kong to Siberia." See *B.A.T. Bulletin,* 12.13 (May 1921), 288. The references to Wang Ching-yü, pp. 223-229, in the following notes are to excerpts from this publication.

95. Wang Ching-yü, pp. 223-224.

96. Spence, p. 167.

97. Wang Ching-yü, p. 223.

98. Ibid., pp. 224-226 and 228. For other examples, see ibid., pp. 224 and 229.

99. Nanyang Brothers Tobacco Company agent in Manchuria to Nanyang in Hong Kong, 1914, *Nanyang,* p. 59.

100. Chien Chao-nan to Chien Yü-chieh, September 13, 1916, and October 28, 1917, *Nanyang,* pp. 59-60 and 51.

101. Hutchison, p. 102. On Hutchison's travels in rural North China in 1912 and 1913, see ibid., chs. 14-18 and pp. 320-322.

102. Rural consumption of industrial goods (including kerosene, cotton yarn, and other commodities as well as cigarettes) is a subject in need of further research. According to one isolated statistic, in the year 1935, 19 percent of Chinese farm families purchased tobacco of any kind. See Albert Feuerwerker, *The Foreign Establishment in China in the Early Twentieth Century* (Ann Arbor, 1976), p. 89.

103. Sanger, pp. 44-83; Arnold, *Commercial Handbook of China,* no. 84, II, p. 393; Thomas, *Pioneer,* pp. 160-161; U.S. Department of Commerce, *Tobacco Trade of the World,* no. 38, p. 33.

104. Sun Chia-chi, "Cigarette Cards," trans. Robert Christensen, *Echo of Things Chinese,* 6.4 (January 1977), 58-67, 71, and 74; The Cartophilic Society of Great Britain, Ltd., comp., *The British-American Tobacco Company Booklet,* pp. 59, 61, 66, 90-92, 183-189, 195-204; Arnold et al., *Commercial Handbook of China,* no. 84, II, pp. 392-393; and Arnold et al., *China: A Commercial and Industrial Handbook,* no. 38, p. 344. For examples of posters in which anti-Christian Chinese propagandists depicted mid-nineteenth century foreign missionaries as cuckolds wearing green hats, see illustrations in Paul A. Cohen, *China and Christianity* (Cambridge, Mass., 1963), pp. 140 ff.

105. Arnold et al., *China: A Commercial and Industrial Handbook,* no. 38, p. 195.

106. See Henry Doré, *Researches into Chinese Superstitions,* trans. from the French by M. Kennelly (Shanghai, 1918), V, 714-715; and Lu Hsun's amusing comparison of Chinese and Western attitudes toward bats in "On Bats," *Selected Works of Lu Hsun* (Peking, 1956), III, 281-282.

107. Sun, pp. 59 and 67; The Cartophilic Society of Great Britain, Ltd., *The British-American Tobacco Company Booklet,* pp. 195-196.

108. Hutchison, pp. 266-267.

109. *NCH*, September 27, 1907, p. 763; Crow, pp. 58-59; Edward Alsworth Ross, *The Changing Chinese* (New York, 1911), p. 86; U.S. Department of Commerce, *Tobacco Trade of the World*, no. 38, p. 33; Sanger, p. 57; Chung-hua Tobacco Company in Manchuria to Nanyang, August 1914, *Nanyang*, p. 68; report of Lu Yao-chen to Nanyang, January 23, 1915, *Nanyang*, p. 40.

110. Thomas, "Selling and Civilization," pp. 949-950.

111. Bureau of Corporations, *Report*, pt. 1, p. 335; Robert F. Durden, "Tar Heel Tobacconist in Tokyo, 1899-1904," *North Carolina Historical Review*, 53.4 (October 1976), 354. On the parallel case of Americans investing for the same reason in Europe, see Mira Wilkins, "An American Enterprise Abroad: American Radiator Company in Europe, 1895-1914," *Business History Review*, 43.3 (Autumn 1969), 342.

112. Sumiya Mikio, *Dai-Nihon teikoku no shiren*, XXII in *Nihon no rekishi* (Tokyo, 1966), 129; *Nichi-Bei bunka kōshō-shi*, II, *Tsūsho sangyo-hen* (Tokyo, 1954), pp. 428-429; Bureau of Corporations, *Report*, pt. 1, pp. 83-84, 183, 335, and 443. Cf. Robert F. Durden, *The Dukes of Durham, 1865-1929*, p. 74. The total foreign capital imported into Japan between 1897 and 1903 was only 194 million *yen*; during the subsequent decade (1904-1914) it rose to 1,857 million. See Masao Baba and Masahiro Tatemoto, "Foreign Trade and Economic Growth in Japan: 1858-1937," in Lawrence Klein and Kazuchi Ohkawa, eds., *Economic Growth: The Japanese Experience since the Meiji Era* (Homewood, Ill., 1968), p. 178.

113. Parrish to Komura, April 20, 1904, and Parrish to Murai Kichibei, April 25, 1903, Edward J. Parrish Papers, Manuscript Department, William R. Perkins Library, Duke University, Durham, N.C.; Bureau of Corporations, *Report*, pt. 2 (1911), p. 300. On Murai Brothers' advertising techniques, see Suzuki Tsutomu, ed., *Nis-Shin Nichi-Ro senso*, XIX in *Nihon rekishi shirīsu* (Tokyo, 1967), pp. 134-135 and 156-157; and Durden, *The Dukes of Durham*, p. 75.

114. For a summary of the advertising wars, see Sumiya, pp. 130-131. The largest of the Japanese tobacco companies, Chiba, earned 7,984 *yen* in 1901; 40,679 in 1902; and 72,915 in 1903. Parrish to Komura, April 20, 1904, Parrish Papers. The quotation is used by Durden in "Tar Heel Tobacconist in Tokyo," p. 358. On late Meiji attitudes toward foreign investment, see Dan Fenno Henderson, *Foreign Enterprise in Japan* (Chapel Hill, 1973), pp. 11-12. In the transaction with the government, Murai Brothers' stockholders absorbed a relatively small loss, receiving, 11,286,949 *yen* as compensation, 713,051 less than their investments (which by 1904 totaled 12 million *yen*). For details of the negotiations over compensation, see Durden, "Tar Heel Tobacconist in Tokyo," pp. 355-363. For provisions of the Tobacco Monopoly Law, see *Laws and Regulations of the Government Monopoly of Japan* (n.p., 1906).

115. As early as 1904—before Japan's annexation of Korea—Duke's representatives in East Asia were resigned to the loss of the Korean market, but they fought for a share of it between 1904 and 1914, and after finally withdrawing, demanded full compensation for their factory at Chemulpo (Inchon). See Fiske to Parrish, September 16, 1904; and Chapter 7. Though excluding American manufacturers from Japan, the Japanese continued to purchase bright tobacco from the

United States. In 1909 Japan was the world's second largest importer of this commodity (behind Great Britain), but by 1923 it imported less than 5 percent of the bright tobacco exported from the United States. See Tilley, pp. 334-335.

116. For the period 1903 to 1928, the decline of the gold value of the Haikwan tael reduced the effective rate of the tariff below 4 percent ad valorem. See Yen Chung-p'ing, ed., *Chung-kuo chin-tai ching-chi shih t'ung-chi tzu-liao hsuan-chi* (Peking, 1955), p. 61.

117. For the English text, see William F. Mayers, comp., *Treaties between the Empire of China and Foreign Powers* (Shanghai, 1906), p. 28; for the Chinese text, see Huang Yueh-po et al., comps., *Chung-wai tiao-yueh hui-pien* (Shanghai, 1935), p. 10.

118. Stanley F. Wright, *China's Struggle for Tariff Autonomy, 1843-1938* (Shanghai, 1938), pp. 361-362; Ch'eng Shu-tu et al., comps., *Yen chiu shui shih* (Nanking, 1929), II, ch. 7, pt. 1, p. 1; FO 371/861, no. 2875, Willis to Jordan, August 17, 1909; FO 371/180, no. 18884, Prince Ch'ing to Satow, December 26, 1904; Satow to Warren, December 29, 1904; Keily of BAT to Warren, January 14, 1905; Warren to Satow, January 16, 1905; Satow to Prince Ch'ing, February 8, 1905; Wai-wu pu to Satow, February 23, 1905. See also Nien Ch'eng, *Huang-pu chiang-pan hua tang-nien* (Hong Kong, 1971), p. 95.

119. Quoted by Willis to Jordan in FO 371/861, no. 2875, August 17, 1909.

120. FO 371/861, no. 2875, Jordan to the Ministry of Foreign Affairs (Wai-wu pu), September 15, 1909. On the diplomacy of Hsu and Hsi-liang, see Des-Forges; and Michael H. Hunt, *Frontier Defense and the Open Door* (New Haven, 1973).

121. FO 371/1082, no. 1593, Jordan to Grey, October 16, 1911. The Chung-hua Tobacco Company complained that total taxes on its cigarettes in Manchuria were "half again more" than taxes on BAT goods and gave this as the reason why it could not compete with BAT. Chung-hua Tobacco Company to Nanyang Brothers Tobacco Company, September 1914, *Nanyang*, p. 63.

122. Thomas, *Pioneer*, p. 48.

123. FO 371/1945, no. 9845, Hood of BAT to Grey, March 4, 1915, outlines the scheme in detail.

124. FO 371/1945, no. 21554, Stanley to FO, May 13, 1914; no. 18349, Jordan to Grey, April 11, 1914; no. 27274, Musgrave to Grey, June 16, 1914.

125. FO 371/1945, no. 56433, Jordan to Grey, July 28, 1914; BAT to Chou Tzu-chi and Liang Shih-i, June 10, 1914; FO 371/2329, no. 34426, Hood of BAT to FO, March 23, 1915; no. 37366, Jordan to Grey, March 31, 1915; *Who's Who in China*, I, 241. Negotiators for the Chinese government later privately revealed that they and BAT had discussed the possibility of BAT paying 3 million *yuan* for its share of control over tobacco tax stamps. According to American sources, the two sides also considered forming a joint corporation with 55 percent of the stock to be held by BAT and 45 percent by the Chinese government. See James C. Sanford, "Tobacco Taxation on the Eve of Tariff Autonomy," seminar paper, Harvard University, 1971, p. 22; Chien Chao-nan to Chien Yü-chieh, September 27, 1916, *Nanyang*, pp. 122-123.

126. Ch'eng Shu-tu et al., eds., *Chuan-yen t'ung-shui shih*, ch. 2, p. 7; Hou, *Foreign Investment*, pp. 104-105.

127. Mary Wright, "Introduction," in Wright, ed., *China in Revolution*, pp. 1-30; Rhoads, chs. 3-4.

128. FO 228/2155, English translation of the original Chinese document enclosed in Thomas of BAT to Pelham-Warren, August 15, 1905.

129. Ho Tso, "1905-nien fan-Mei ai-kuo yun-tung," *Chin-tai shih tzu-liao*, 1 (1956), 24 and 28; Chu Shih-chia, *Mei-kuo p'o-hai Hua-kung shih-liao* (Peking, 1959), pp. 155-156; Wang Ching-yü, pp. 737 and 1004.

130. English translations of the three posters are enclosed in FO 228/2155, Thomas of BAT to Pelham-Warren, August 2, 1905; Proclamation of the Mixed Court of the International Settlement of Shanghai, August 29, 1905.

131. FO 228/2155, Harvey of BAT to Satow, September 26, 1905.

132. A photographic reproduction of this poster appears in Wang Ching-yü, second plate following p. 732.

133. Several songs of this type were sung during the boycott and are recorded in Ting Yu, "1905-nien Kuang-tung fan-Mei yun-tung," *Chin-tai shih tzu-liao*, 5 (October 1958), 8-52.

134. On the marketing of American-made bicycles in China at the time of the boycott, see Raymond F. Crist and Harry R. Burrill, *Trade with China* (U.S. Department of Commerce and Labor, Bureau of Manufactures, Washington, D.C., 1906), pp. 89-90.

135. Che Lang (pseud.), "Tiao yin-chai," in A Ying (Ch'ien Hsien-ts'un), comp., *Fan-Mei Hua-kung chin-yüeh wen-hsüeh chi* (Peking, 1960), p. 14.

136. Cecil Clementi, *Cantonese Love Songs* (London, 1904), p. 1.

137. Cf. Arthur Waley, *The Life and Times of Po Chu-i, 772-846 A.D.* (New York, 1949), pp. 62-63.

138. Ting Yu, pp. 18, 23, and 33. On the role students and merchants played in the boycott at Canton, see Rhoads, pp. 83-90.

139. FO 228/2155, Hopkins to Pelham-Warren, July 27, 1905; report by Arnold enclosed in Rodgers to Loomis, August 17, 1905, U.S. Department of State, National Archives; Margaret Field, "The Chinese Boycott of 1905," *Papers on China*, 11 (1957), 95.

140. Shih-shan Tsai, "Reaction to Exclusion: Ch'ing Attitudes toward Overseas Chinese in the United States, 1848-1906," Ph.D. dissertation, University of Oregon, 1970, p. 307; FO 228/2155, Harvey to Satow, September 26, 1905; Field, p. 74; Howard K. Beale, *Theodore Roosevelt and the Rise of America to World Power* (Baltimore, 1956), p. 228.

141. Quoted by Beale, p. 235.

142. Quoted by Beale, p. 235.

143. Roosevelt condemned these companies for creating monopolies through unethical practices; he distinguished them from "natural" trusts, which had also created monopolies but which (in his estimation) had behaved more ethically. See John Morton Blum, *The Republican Roosevelt* (Cambridge, Mass., 1954), p. 118; and Gabriel Kolko, *The Triumph of Conservatism* (New York, 1963), pp. 122-127.

144. FO 17/1689, Jeffress of BAT to the Colonial Office, September 13, 1905; and NA, Heintzleman's consular dispatches, December 4, 1905. On American businessmen in China and their efforts to reform immigration laws in the

United States, see James J. Lorence, "Business and Reform: The American Asiatic Association and the Exclusion Laws," *Pacific Historical Review*, 39 (November 1970), 421-438. On American businessmen's promotion of progressive domestic legislation in this period, see Robert Wiebe, *Businessmen and Reform: A Study of the Progressive Movement* (Cambridge, Mass., 1962).

145. NA, Rodgers to Loomis, September 12 and 22, 1905; FO 228/2155, Fraser to Satow, December 5, 1905; FO 228/1632, Fraser to Satow, January 9, 1906; FO 228/1631, Smith to Satow, April 1, 1906; *Kuang-chou tsung shang-hui pao*, quoted in Ting Yu, p. 39; Rhoads, pp. 90-91.

146. I have based the latter two explanations on interpretations advanced by C. F. Remer, *A Study of Chinese Boycotts* (Baltimore, 1935), p. 238. The cigarette firms were founded in Peking (3), Tientsin (1), Ying-k'ou (1), Yen-t'ai (5), Shanghai (9), Hangchow (1), Hankow (2), Ichang (1), Chungking (1), Chengtu (1), T'ang-hsia (1), Hong Kong (1). See Wang Ching-yü, pp. 810-811 and 912-913; P'eng Tse-i, II, 338-340; Hsu Yen-cho, ed., *Chung-kuo kung-i yen-ko shih-lueh* (Shanghai, 1917), p. 29; Ting Yu, p. 19; FO 228/1628, Smith to Satow, July 26, 1906. In addition a few foreign tobacco companies entered Chinese markets at this time: The Japanese Imperial Tobacco Monopoly in South Manchuria; the Alliance Tobacco Company of Tientsin (established in 1904) and the Express Cigarette Company of Shanghai (established in 1906), both operated by Greeks; and the San-lin Tobacco Company of Fengtien (established in 1907), a Sino-Japanese venture. On these, see Wu Ch'eng-lo, II, 74-75.

147. Chang Ts'un-wu, *Kuang-hsu sa-i-nien Chung-Mei kung-yueh feng-ch'ao* (Taipei, 1965), pp. 245-246; Bureau of Corporations, *Report*, pt. 1, p. 306; Wang Ching-yü, p. 913; Wang Hsi, p. 86. For details on the Pei-yang Tobacco Company and a discussion of the *shang-pan* and *kuan-shang ho-pan* modes of operation, see Wellington Chan, pp. 104-105 and chs. 5-7.

148. P'eng Tse-i, II, 339-340; FO 228/1631, Smith to Satow, July 26, 1906; FO 228/1628, Giles to Jordan, November 10, 1906; FO 228/1663, Smith to Jordan, March 27, 1907.

149. Quoted by Wang Hsi, pp. 86-87. For additional evidence of the collapse of the Chinese cigarette industry in the aftermath of the boycott, see Wang Ching-yü, pp. 810 and 1004; Wu Ch'eng-lo, II, 72; *Nanyang*, p. 254.

150. For additional evidence of BAT's use of coercive tactics against Chinese companies in this period, see "Memoirs of Chien Yü-chieh," *Nanyang*, p. 3. On Duke's techniques in the United States, see Richard B. Tennant, *The American Cigarette Industry* (New Haven, 1950), pp. 28 and 41-44; and William H. Nicholls, *Price Policies in the Cigarette Industry* (Nashville, 1951), pp. 27-28. On tactics used in the "tobacco war" of 1902 in England, see Alford, pp. 264-269. On BAT's subversion of an American company interested in China, see Winkler, pp. 222-223.

151. Urban protests against BAT advertising can be documented for almost every region: Manchuria (Ying-k'ou), the Middle Yangtze (Changsha), the Southeast Coast (Foochow), and the Southwest (Chao-t'ung). See FO 371/180, no. 6685, Fulford to Satow, December 11, 1905; FO 371/180, no. 2370, Jeffress of BAT to FO, January 18, 1906; FO 371/180, no. 15212, Hosie to Satow, March 14, 1906; FO 371/476, no. 19630, Partlett to MacDonald, April 16, 1908; FO

228/1695, Hewlett to Jordan, October 24, 1908; FO 371/638, no. 19976, Hood of BAT to Grey, May 27, 1909; FO 371/638, no. 24094, Playfair to Jordan, May 21, 1909; FO 371/638, no. 36459, Jordan to Grey, August 20, 1909; Dingle, p. 100. On anti-smoking campaigns, see *NCH*, October 14, 1910, p. 106; June 17, 1911, pp. 737 and 754; August 9, 1911, p. 476; August 26, 1911, p. 539; June 29, 1912, p. 940; May 24, 1913, p. 564; June 13, 1914, p. 823. On rural protests against BAT agricultural policies in Hupeh province in 1914, see Ch'en Chen, col. 2, p. 141; and Chang Yu-i, p. 505.

152. Parrish to Fiske, October 12, 1904, Parrish Papers.

153. Ch'en Tse-shan to Chien Chao-nan, April 1915, *Nanyang*, p. 81.

3. The Rise of Commercial Rivalry

1. The Hong Kong Daily Press, comp., *The Directory and Chronicle for China, Japan, Corea, Indo-China, Straits Settlements, Malay States, Siam, Netherlands India, Borneo, The Philippines, etc. for the Year 1906* (London, 1906), pp. 827 and 907; Wang Ching-yü, p. 234; Arnold Wright, III, 796.

2. Arthur Bassett, legal adviser for BAT in China, described BAT top management's initial response to Nanyang in a letter to Thomas, February 11, 1916, Thomas Papers.

3. Cf. Durden, *The Dukes of Durham*, ch. 1.

4. Wang Ching-yü, p. 1003. Various birthdates have been given for Chien Chao-nan. *Who's Who in China* (Shanghai, 1926) sets the date at 1872. His biographer in Boorman and Howard, I, 364, gives 1875. I have used the date 1870 because Chien Chao-nan himself used it in official correspondence with the Chinese central government. Nanyang to the Ministry of Agriculture and Commerce, July 29, 1915, *Nanyang*, p. 7.

5. Chien Chao-liang et al., comps., *Yueh-tung Chien-shih ta-t'ung p'u* (n.p., 1928), ch. 11, pp. 91b-92a; *Who's Who in China* (1926), p. 83. On Cantonese motivations for moving, see Gunther Barth, *Bitter Strength* (Cambridge, Mass., 1964), pp. 1-5; G. W. Skinner, *Chinese Society in Thailand* (Ithaca, 1957), ch. 2, esp. pp. 28-35 and 67; Maurice Freedman, "Immigrants and Associations," *Comparative Studies in Society and History*, 3 (1960), 27; and Freedman, "The Growth of a Plural Society in Malaya," *Pacific Affairs*, 33 (1960), 164. In the 1870s, Li Hung-chang and Wang K'ai-t'ai, two high-ranking Ch'ing officials, each independently estimated that 10,000 Chinese had already emigrated to Japan. This estimate might have been too high. H. F. MacNair estimated that 2,580 Chinese resided there in 1895; 8,411 in 1905; 8,145 in 1911; 12,139 in 1918. See his *The Chinese Abroad* (Shanghai, 1924), pp. 35-36.

6. Huai Shu, *Chung-kuo ching-chi nei-mo* (Hong Kong, 1948), p. 104; Boorman and Howard, I, 364; Wang Ching-yü, II, 1004.

7. *Nanyang*, p. 4; *Who's Who in China* (1926), p. 84. According to the latter source, the Sheng-t'ai Navigation Company's profits "in a few years . . . increased from $3,000 to several hundred thousand."

8. Nanyang's stockholders' register, *Nanyang*, p. 2; Wang Ching-yü, p. 810.

9. Chien Chao-nan later showed his awareness of (and admiration for)

Murai Brothers in a letter to Chien Yü-chieh, July 3, 1917, *Nanyang*, p. 118.

10. Interview with Y. L. Kan (Chien Jih-lin), Hong Kong, June 28, 1977.

11. Nanyang stockholders' register, *Nanyang*, p. 2; Wang Ching-yü, p. 810; "Memoirs of Chien Yü-chieh," *Nanyang*, p. 2; Boorman and Howard, I, 364. Principal stockholders not from the Chiens' home village included Yang Chi-wu, Yuan Huan-ju, and Wang Chi-ch'eng.

12. Interviews with former employees of Nanyang's Hong Kong factory, 1957, *Nanyang*, p. 17; Huai Shu, p. 104; Wang Ching-yü, p. 1004. At the outset, Nanyang technically had no offices at all. A shop in Hong Kong owned by Chien Chao-nan's cousin, Chien K'ung-chao, called the Ming-t'ai Company, served as its headquarters. *Nanyang*, p. 27.

13. Nanyang account books, *Nanyang*, p. 27. BAT hoped to transport cigarettes across Vietnam to Yunnan in Southwest China after the completion of the French railroad in this area in 1909 but ultimately decided not to because the French transit taxes were prohibitive. FO 371/640, no. 26840, Cunliffe-Owen of BAT to Grey, July 14, 1909; no. 33459, Stanley to Grey, September 11, 1909; no. 36339, Carnegie to Grey, September 30, 1909.

14. For statistics on cigarettes exported from the United States to Southeast Asia, see U.S. Department of Commerce, Bureau of the Census, *Stocks of Tobacco Leaf, 1919*, Bulletin 143 (Washington, D.C., 1920), pp. 38-39. The Thais first advanced the idea of a national tobacco monopoly in 1920. See FO 371/4091, file 3033; FO 371/5368, file 29; and FO 371/5370, file 671. Not until 1939 did the Thai government attempt to take control of the Thai cigarette market away from Chinese manufacturers. See Kenneth P. Landon, *The Chinese in Thailand* (New York, 1941), pp. 227-229. The Dutch introduced legislation taxing tobacco products imported into Indonesia in 1932. See Lance Castles, *Religion, Politics and Economic Behavior in Java* (New Haven, 1967), p. 36.

15. "Memoirs of a Former Employee of Nanyang's Factory," *Nanyang*, p. 21; Wang Ching-yü, p. 810.

16. Rhoads, pp. 135-136; Akira Iriye, "Public Opinion and Foreign Policy," in Feuerwerker et al., eds., *Approaches to Modern Chinese History*, pp. 226-230 and 235-236; Kikuchi Takaharu, *Chūgoku minzoku undō no kihon kōzō* (Tokyo, 1966), ch. 2.

17. Quoted by Rhoads, p. 138.

18. FO 228/1694, Fox to Jordan, July 9, 1908.

19. FO 228/1694, Fox to Jordan, November 5, 1908; Rhoads, pp. 139-140; Huai Shu, p. 104; Wang Ching-yü, p. 1004.

20. "Memoirs of Chien Yü-chieh," *Nanyang*, p. 4; Kuo-huo shih-yeh ch'u-pan she, ed., *Chung-kuo kuo-huo kung-ch'ang shih-lueh*, pp. 4-5.

21. Boorman and Howard, I, 365; "Memoirs of Chien Yü-chieh," *Nanyang*, p. 4. Cf. Huai Shu, pp. 104-105.

22. "Memoirs of Chien Yü-chieh," *Nanyang*, p. 6; see Appendix, table 6; *Nanyang*, pp. 21-22 and 25.

23. Nanyang account books, *Nanyang*, p. 19. According to an article by T. C. Tsang, "A Successful Chinese Tobacco Company," in *Millard's Review*, 5 (June 15, 1918), 93, Nanyang imported "Virginia-Carolina bright leaf tobacco grown in the eastern belt of Virginia, North Carolina, and the upper part of South Carolina."

24. For numerous examples of Chinese competition with foreign-made goods in South China before 1915, see Rhoads, pp. 148-151 and 256; and Friedman, "The Failure to Modernize."

25. Prior to 1915 the Chiens observed BAT first-hand in Southeast Asia, Hong Kong, and Canton and received intelligence reports from other regions such as the one from Liu Ai-jen in Manchuria to Chien Chao-nan, 1914, *Nanyang*, p. 68.

26. U.S. Department of Commerce, Bureau of Foreign and Domestic Commerce, *Tobacco Trade of the World*, Special Consular Report no. 68 (Washington, D.C., 1915), p. 29; Shu-lun Pan, *The Trade of the United States with China* (New York, 1924), pp. 238-241. See also Appendix, table 1.

27. Julean Arnold, Weekly Report, February 22, 1919, Julean H. Arnold Papers, Hoover Institution, Stanford University, Stanford, Calif.; Thomas to Hsu En-yuen, January 22, 1924, Thomas Papers.

28. Thomas to Straight, August 2, 1915, Thomas Papers.

29. Thomas, *Pioneer*, p. 313. See also Thomas to Tewsley, October 2, 1915; Thomas to Straight, December 31, 1915, both in Thomas Papers; and Hutchison, p. 263, who reported, on the basis of first-hand observation, that BAT's cigarette business was "booming" in Shanghai in the fall of 1916 and "fairly bursting" in Hankow in the spring of 1917.

30. For a fuller discussion of the golden age thesis, see Chapter 6.

31. Nanyang's Canton branch to the Hong Kong office, June 1915, *Nanyang*, p. 65. Nanyang's investigator in Canton to the Hong Kong office, October 1915, *Nanyang*, pp. 67 and 77. FO 228/2012, Bennett (in Wuchow) to Alston, January 17, 1916. Nanyang sold the bulk of its cigarettes in the same price ranges as BAT did. About one fourth of the cigarettes produced by Nanyang at the time were high grade (selling at ¥0.70 per pack as compared with BAT's most expensive brand at ¥0.80), and the rest were evenly divided between medium and low grades. *NCH*, May 11, 1916, p. 354; Lu Yao-chen's report to Nanyang on BAT in Shanghai, January 23, 1915, *Nanyang*, p. 40; Wang Shih-jen to Nanyang, April 20, 1916, *Nanyang*, p. 47.

32. Nanyang's Canton branch to the Hong Kong office, June 16, 1915, *Nanyang*, p. 76. Nanyang's investigator Liu Shih-ch'üan to Chien Chao-nan and Chien Yü-chieh, October 8, 1915, *Nanyang*, p. 95. Shanghai Support Chinese National Goods Society to Nanyang's Hong Kong office, October 1915, *Nanyang*, p. 76.

33. This newspaper was founded a decade earlier with the avowed purpose of "arousing the spirit of citizenship and developing the feelings of patriotism" among its readers. Quoted by Rhoads, p. 83.

34. *Kuo-min pao*, June 21, 1915, enclosed in a letter from Nanyang's Canton investigators to Chien Chao-nan and Chien Yü-chieh, June 21, 1915, *Nanyang*, p. 78. Items like this one were prominently placed in newspapers partly because BAT lured Huang Chi-wen away from his job as translator and editor for Canton's popular commercial newspaper, *Ch'i-shih-erh hang*, and hired him to wine, dine, and persuade other newspaper editors to highlight stories favorable to BAT. Nanyang's investigator in Canton to Wang Shih-jen, October 1915, *Nanyang*, pp. 90-91.

35. FO 228/1941, Cecil Heinz (in Canton) to Jordan, July 10, 1915.

36. Kao Chen-pai, "Chiang t'ai-shih chih 'ku,'" *Ta-ch'eng*, 24 (November 1, 1975), 12-13. For details on Chiang, see Chapter 5.

37. The I-sheng Company in Singapore to the Hong Kong office, June 7, 1915, *Nanyang*, p. 78.

38. Nanyang investigator in Singapore to the Hong Kong office, March 31, 1915, *Nanyang*, p. 77; a petition signed by Nanyang workers in *Wu-hu*, p. 14b; Lu Shao-t'ang of Nanyang in Singapore to the Hong Kong office, May 26, 1915, *Nanyang*, p. 101; I-sheng Company in Thailand to Chien Chao-nan and Chien Ying-fu, 1915, *Nanyang*, p. 77.

39. *Nanyang*, p. 4n1. According to the newspaper *Pei-ching jih-pao*, June 1, 1919, two other Chiens, Chien Jih-hua and Chien Jih-p'eng, also took Japanese names and citizenship. Chien Jih-hua took the name Matsumoto Kyoka, and Chien Jih-p'eng took the name Matsumoto Kyoho. Newspaper clipping appears in *Nanyang*, p. 83. See also "Memoirs of Chien Yü-chieh," *Nanyang*, p. 2.

40. *Nanyang*, p. 20. After 1915 Nanyang began to have trademarks and packaging material printed in China. This one item accounted for almost half the cost of articles that Nanyang imported from Japan in 1915.

41. Chien Chao-nan and his uncle, Chien Ming-shih, the two principal stockholders in the company before 1912 (when Chien Ming-shih died), both collected debts and "arranged matters" in Japan for Nanyang. "Memoirs of Chien Yü-chieh," *Nanyang*, p. 2. They made no mention of borrowing from Japanese creditors. In August 1916 Chien Chao-nan alluded to substantial sums of money sent to and received from "Kobe." Chien Chao-nan to Chien Yü-chieh, August 29, 1916, *Nanyang*, pp. 53-54. He did not specify whether by Kobe he meant his own Sheng-t'ai Navigation Company in Kobe (as the compilers of the *Nanyang* volume indicate, editor's note, *Nanyang*, p. 54) or Japanese banks in Kobe or other Japanese creditors on whom he might have relied. Later, in June 1918, Hollington K. Tong reported that bankers in Japan had "successfully coerced the Nanyang Brothers . . . to accept from them disastrous loans . . . and have thus brought [Nanyang] under their influence." See "Japan Seeking China's Tobacco Monopoly," *Millard's Review*, 5 (June 8, 1918), 50. I have found no substantiation for this assertion.

42. Marius B. Jansen, *The Japanese and Sun Yat-sen* (Cambridge, Mass., 1954), p. 193.

43. Ch'en Tse-shan to Chien Chao-nan, April 1915, *Nanyang*, p. 81.

44. "Memoirs of Chien Yü-chieh," *Nanyang*, p. 3; Nanyang's Hong Kong office to Hsu Tsan-chou in Rangoon, August 1915, *Nanyang*, p. 75.

45. For a transcript of a legal case involving BAT and another Chinese company, the San Hsing Tobacco Company, see *NCH*, September 27, 1907, pp. 762-765; and *NCH*, October 4, 1907, pp. 44-45.

46. Nanyang's Hong Kong office to Hsu Tsan-chou in Rangoon, August 1915, *Nanyang*, p. 75; a petition signed by Nanyang workers in *Wu-hu*, p. 14a. Although the Chiens were probably unaware of him, at least one Westerner, a British official in Canton, took their side in the dispute, reporting to the British minister in Peking that in this case he saw "nothing that can be called fraudulent imitation of trademarks." FO 228/1941, Heinz to Jordan, July 10, 1915.

47. Chien Chao-nan to Wang Shih-jen, August 1915, *Nanyang*, p. 74.

48. When Nanyang was forced to withdraw the allegedly counterfeit brand,

its owners widely publicized BAT "coercion"—and then took steps to salvage as much brand loyalty as possible: first, they wrapped an additional label over the original and marketed it; then they introduced a new brand whose name and packaging resembled the banned one. See the Canton branch's advertisement concerning the switch from San Hsi to Hsi Ch'ueh brand, August 1915, *Nanyang*, p. 93.

49. For numerous examples of Western entrepreneurs using advertising to accomplish this purpose, see Joe S. Bain, Appendix D, "Product Differentiation Barriers to Entry in Individual Industries," in his *Barriers to New Competition* (Cambridge, Mass., 1956), pp. 263-317.

50. T'ung-tsu Ch'u, *Local Government in China under the Ch'ing* (Cambridge, Mass., 1962), chs. 9 and 10, especially pp. 182-183; Samuel C. Chu, *Reformer in Modern China* (New York, 1965), chs. 7 and 8.

51. Article originally published in several Canton newspapers, republished in *Wu-hu*, p. 5b; editor's comment in *Wu-hu*, p. 6a; "Memoirs of Chien Yü-chieh," *Nanyang*, p. 99; Nanyang's Hong Kong office to Hsu Tsan-chou in Rangoon, August 1915, *Nanyang*, p. 75; Chien Chao-nan to Wang Shih-jen, August 1915, *Nanyang*, p. 98. According to one eulogistic biographical sketch of the Chien brothers' mother, she was the inspiration behind the philanthropy, which included, in addition to the activities mentioned in the text, "homes for girls" (*nü-tzu chiao-yang yuan*), rice dispensaries in time of famine, and restoration of ancestral temples. Chien Chao-liang, *Yueh-tung Chien-shih ta-t'ung-p'u*, ch. 11, pp. 93a-93b.

52. Chien Chao-nan to Chien Ying-fu, August 15, 1915, *Nanyang*, pp. 93-95. Gunn and Newman are described in FO 228/1978, E. G. W. Hou to Jordan, February 9, 1916.

53. Open letter from Chiang Shu-ying to Chien Ch'in-shih in *Wu-hu*, p. 17a.

54. Kikuchi, pp. 164-165; Shang-hai shih wen-hsien wei-yuan hui, comp., *Shang-hai shih-wu so-yuan* (Shanghai, 1948), pp. 29-30, cited by Joseph T. Chen, *The May Fourth Movement in Shanghai* (Leiden, 1971), p. 93n4.

55. Nanyang's Canton branch to the Hong Kong office, June 16, 1915, *Nanyang*, p. 76; Wang Shih-jen to Chien Chao-nan, mid-July 1915, *Nanyang*, p. 33.

56. Canton Overseas Merchants General Association to the Canton Promotion of National Goods Society, August 22, 1915, *Nanyang*, p. 80; letters from Nanyang's investigators in Canton, Singapore, and Thailand to the Hong Kong office and from the Hong Kong office to the Hong Kong Overseas Chinese Merchants Association, the Canton Press Club, and the Canton Overseas Merchants Society, written in May and June 1915, *Nanyang*, pp. 77-79; Lu Chou to Ying Wu, June 21, 1915, *Nanyang*, p. 34; FO 228/2012, Bennett to Alston, January 17, 1916.

57. FO 228/1941, Heinz to Jordan, July 10, 1915.

58. Remer, *A Study of Chinese Boycotts*, pp. 47-48; Kikuchi, p. 169.

59. Report by Nanyang's investigator Liu Shih-ch'üan, October 6, 1915, *Nanyang*, pp. 31-32.

60. Ibid., p. 31; *NCH*, May 13, 1916, p. 354; Nanyang's accounting records, *Nanyang*, p. 19.

61. Nanyang's accounting records, *Nanyang*, pp. 26 and 29-30. Though

statistics are incomplete on Nanyang's trade in Southeast Asia after 1919, it seems to have continued without interruption. On competition between BAT and Nanyang in Southeast Asia during the late 1920s and early 1930s, see the correspondence between several of Nanyang's agents in Southeast Asia and the company's home office in Hong Kong in *Nanyang*, pp. 415-419 and 430-436.

62. *North China Daily News*, June 21, 1915, p. 5; report by Wang Shih-jen to Nanyang, July 1915, *Nanyang*, p. 41.

63. Chien Chao-nan to Chien Yü-chieh, August 16 and 23, 1916, *Nanyang*, pp. 62 and 92; Chien Ching-shan to Chien Yü-chieh, March 30, 1917, *Nanyang*, p. 110; Chien Chao-nan to Chien Yü-chieh, October 28, 1917, *Nanyang*, p. 51.

64. Chien Chao-nan to Nanyang, September 1916, *Nanyang*, p. 69; Wang Shih-jen to Chien Chao-nan and Chien Yü-chieh, March 1916, *Nanyang*, pp. 65-66.

65. Tientsin branch of Nanyang to the Hong Kong office, May 1916, *Nanyang*, p. 67; FO 228/2012, Bennett to Alston, January 17, 1916. The latter, a British report from Kwangsi, indicated that BAT cut prices by 50 percent in response to the competition from Nanyang.

66. Wang Shih-jen to Chien Chao-nan and Chien Yü-chieh, April 1916, *Nanyang*, p. 71. It is interesting to note that less than a year earlier (in 1914) this and some of the other competitive tactics used by BAT in China became illegal in the United States with the passage of the Clayton Act.

67. Wang Shih-jen to Nanyang, April 10, 1916, *Nanyang*, p. 46; China, Inspectorate General of Customs, *Returns of Trade, Report on Ningpo for 1917*, pt. 2, vol. III, pp. 973-974. I am grateful to Susan Mann Jones for supplying the latter reference. See also NA 693.1112/104, Cunningham to State, July 20, 1925.

68. Wang Shih-jen to Nanyang, April 20, 1916, *Nanyang*, pp. 46-47.

69. Ibid., p. 47.

70. The Chiens relied on Cantonese contacts for a variety of purposes. They operated through a Cantonese firm in Peking, the T'o kuang chi-hsiang Company, to reach officials in the capital and to inveigle for tax rebates in July 1915. (Wang Shih-jen to Nanyang, July 1915, *Nanyang*, p. 64.) Chien Chao-nan approached Cantonese officials T'ang Shao-i, Ch'en Chin-t'ao, and Wen Ying-hsing to secure support for his case at high levels in the government before Nanyang opened negotiations with the Peking government in September 1916. (Wang Shih-jen to Chien Yü-chieh, September 3, 1916, *Nanyang*, pp. 120-121.) Cantonese compradors acted as intermediaries introducing Chien Chao-nan to merchants in Shanghai while he was searching for a site for a new cigarette factory there in February 1917. (See editor's note, *Nanyang*, pp. 52-53.) It was a Cantonese, Ch'en Ping-ch'ien, who, although employed as a comprador by BAT, helped dissuade BAT's management from seizing the building which Chien Chao-nan eventually secured for this factory. (Chien Chao-nan to Chien Yü-chieh, February 22, 1917, *Nanyang*, p. 52; and see below.)

71. Chien Chao-nan to Chien Yü-chieh, August 23, 1916, *Nanyang*, p. 92; Chien Chao-nan to Chien Yü-chieh, October 28, 1917, *Nanyang*, p. 51.

72. Chien Chao-nan to Chien Yü-chieh, March 10, 1917, *Nanyang*, p. 50.

73. Chien Chao-nan to Chien Yü-chieh, August 23, 1916, *Nanyang*, p. 92; Chien Chao-nan to Chien Yü-chieh, October 28, 1917, *Nanyang*, p. 51.

74. Chung-hua Tobacco Company in Nanyang, September 1914, *Nanyang*, p. 63; Wang Shih-jen to Nanyang, July 1915, *Nanyang*, p. 64; Wang Shih-jen to Nanyang, April 1916, *Nanyang*, p. 64.

75. Chien Chao-nan to Chien Yü-chieh, October 28, 1917, *Nanyang*, p. 50.

76. Chien Chao-nan to Chien Yü-chieh, October 28, 1917, *Nanyang*, p. 51. In line with Chien Chao-nan's assessment, C. F. Remer later observed, "Southern China has usually been the region of most vigorous boycotting and it is the southerners abroad who have carried the boycott to foreign countries." See Remer, *A Study of Chinese Boycotts*, p. 243.

77. Chien Chao-nan to Chien Yü-chieh, October 1916, *Nanyang*, p. 59.

78. Chien Chao-nan to Chien Yü-chieh, October 1916, *Nanyang*, p. 66. The term "K'ung Shan" is an epithet the Chiens used among themselves to refer to BAT. It is an allusion to a poem from the T'ang period by Wang Wei. Literally it means "empty mountain," a place where no person resides, implying that it is a haunted hill where ghosts and devils congregate. See editor's note in *Nanyang*, p. 9.

79. Chien Chao-nan to Chien Yü-chieh, June 12, 1917, *Nanyang*, p. 114.

80. Ch'en Chen, col. 4, vol. I, p. 447.

81. Wang Shih-jen to Chien Chao-nan and Chien Yü-chieh, April and May, 1917, *Nanyang*, pp. 71-72; Chien Chao-nan to Chien Yü-chieh, June 29 and August 22, 1917, *Nanyang*, pp. 72-73 and 76.

82. They were established at Canton, Tientsin, Shanghai, Hankow, Chinkiang, Tsingtao, Ying-k'ou, Nanking, Hungkou, Fangtze, Peking, Pengpu, Tsinan, Swatow, Amoy, and Chia-hsing in that order. See *Nanyang*, p. 27.

83. In 1912, 60.1 percent of Nanyang's budget was devoted to tobacco purchasing; in 1914, 64 percent; in 1915, 59.3 percent; in 1917, 61.5 percent. An itemized list of Nanyang's expenditures in these years appears in *Nanyang*, p. 20.

84. Chang Yu-i, pp. 171 and 503; minutes of the meeting of the Nanyang stockholders, March 15, 1921, *Nanyang*, p. 144; Li Wen-chih, p. 412. According to a report by the Inspectorate of Customs in 1917 (quoted in Chang Yu-i), the Chiens also tried a joint venture (*ho-ku*) with Japanese partners in Tsingtao to finance experiments at tobacco growing in Shantung.

85. *Chinese Economic Monthly*, 2.6 (March 1925), 30; quotation is from Chien Chao-nan to Chien Yü-chieh, January 28, 1919, *Nanyang*, p. 422. On the Chiens' use of Chinese tobacco, see *Nanyang*, p. 19; and *Wu-hu*, p. 8a. For criticism of Nanyang on the grounds that it pretended to use more Chinese tobacco than it actually purchased, see editor's comment, *Wu-hu*, pp. 8b-9a.

86. Chien Chao-nan to Chien Yü-chieh, January 28, 1919, *Nanyang*, p. 422.

4. Motives for Merger

1. Tennant, p. 44; Alford, ch. 11; Chien Chao-nan to Chien Yü-chieh, February 22, 1917, *Nanyang*, pp. 104-105. For an analysis of the "urge to merge," see Frederic M. Scherer, *Industrial Market Structure and Economic Performance* (Chicago, 1970), pp. 103-122.

2. "Memoirs of Chien Yü-chieh," *Nanyang*, p. 103.

3. Bassett to Thomas, February 11, 1916, Thomas Papers. Bassett gradu-

ated from Washington University Law School at St. Louis in 1902 and, except for military service during World War I, was with BAT in China continuously from 1913 at least until 1933. For a biographical sketch of Bassett, see George F. Nelliest, ed., *Men of Shanghai and North China* (Shanghai, 1933), p. 19.

4. Chien Ying-fu to Chien Yü-chieh, August 18 and October 21, 1916, *Nanyang*, pp. 54-55; Chien Chao-nan to Chien Yü-chieh, October 1, 1916, *Nanyang*, p. 54; "Memoirs of Chien Yü-chieh," *Nanyang*, p. 6; Thomas to Yiu, October 7, 1925, Thomas Papers. See also Appendix, table 7.

5. Chien Chao-nan to Chien Yü-chieh, February 3, 1917, *Nanyang*, p. 55.

6. On Sheng, see Feuerwerker, *China's Early Industrialization*, ch. 3; on Chang, see Samuel Chu, ch. 3.

7. Chien Chao-nan to Chien Yü-chieh, September 20, 1916, *Nanyang*, p. 126; Chien Yü-chieh to Chien Chao-nan, copy of an undated letter from Chien Yü-chieh's private papers, *Nanyang*, pp. 107-108; Chien Ying-fu to Chien Yü-chieh, August 18, 1916, *Nanyang*, pp. 54 and 126. Chien K'ung-chao was the son of the Nanyang founders' original benefactor, Chien Ming-shih. He and his branch of the family held 47.1 percent of the stock in Nanyang (61,250 shares), the same amount Chien Chao-nan and his branch held, until 1915. See *Nanyang*, p. 5.

8. Order of the President of the Republic, decree no. 139, to Minister of Finance Ch'en Chin-t'ao and Commissioner of the Wine and Tobacco Monopoly Niu Ch'uan-shan, August 28, 1916, *Nanyang*, p. 120.

9. Chien Chao-nan to Chien Yü-chieh, September 27, 1916, *Nanyang*, p. 122. On the imperial state's cooptation of merchants, see Thomas A. Metzger, "The Organizational Capabilities of the Ch'ing State in the Field of Commerce: The Liang-huai Salt Monopoly, 1740-1840," in Willmott, ed., *Economic Organization in Chinese Society*, pp. 9-45. On the *kuan-tu shang-pan* system, see Feuerwerker, *China's Early Industrialization*, ch. 1. On the *kuan-shang ho-pan* system, see Wellington Chan, ch. 5.

10. Chien Chao-nan to Chien Yü-chieh, September 27, 1916, *Nanyang*, pp. 122-123.

11. Ibid. Niu was probably right about Wu. As late as April 3, 1915, Wu was employed by BAT and helped James Thomas pick out a wedding gift for James Duke's daughter, and as early as February 1917 Wu tried, on behalf of BAT, to engineer a merger between Nanyang and BAT. It seems likely that Wu remained affiliated with BAT in the interim. See Thomas to Angier Duke, April 3, 1915, Thomas Papers; and Chien Chao-nan to Chien Yü-chieh, February 22, 1917, *Nanyang*, p. 104.

12. Chien Chao-nan to Chien Yü-chieh, September 27 and October 17, 1916, *Nanyang*, pp. 123 and 125.

13. Chien Chao-nan to Chien Yü-chieh, October 1, 1916, *Nanyang*, pp. 126-127.

14. On Sheng's rise, see Feuerwerker, *China's Early Industrialization*, ch. 3; on Chang's rise, see Samuel Chu, ch. 3. For examples of merchants' skepticism of government, see Wellington Chan, pp. 80-84 and passim.

15. Chien Chao-nan to Chien Yü-chieh, September 27, 1916, *Nanyang*, p. 123.

16. James E. Sheridan, *Chinese Warlord: The Career of Feng Yü-hsiang* (Stanford, 1966), p. 11; O. Edmund Clubb, *Twentieth Century China* (New York, 1964), pp. 62-67; Andrew J. Nathan, *Peking Politics, 1918-1923* (Berkeley, 1976), pp. 91-92.

17. Chien K'ung-chao to Chien Yü-chieh, May 26, 1917, *Nanyang*, p. 111.

18. Chien Yü-chieh to Niu Ch'uan-shan, July 4, 1917, *Nanyang*, pp. 130-131.

19. Chien Ying-fu to Chien Yü-chieh, July 14, 1917, *Nanyang*, p. 112. On the factions that vied for supremacy in Peking during these years, see Nathan.

20. Chien Chao-nan to Chien Yü-chieh, February 22, 1917, *Nanyang*, p. 103.

21. Ibid., pp. 103-104. Thomas has a sketch of Cobbs in *Pioneer*, pp. 87-95.

22. Chien Chao-nan to the Ministry of Finance, September 1915, *Nanyang*, p. 64.

23. *Chinese Economic Monthly*, 2.6 (March 1925), 30; Chien Chao-nan to Chien Yü-chieh, January 28, 1919, *Nanyang*, p. 422.

24. Chien Chao-nan to Chien Yü-chieh, September 13, 1916, *Nanyang*, pp. 59-60.

25. Chien Chao-nan to Chien Yü-chieh, February 22, 1917, *Nanyang*, pp. 103-104; "Memoirs of Chien Yü-chieh," *Nanyang*, p. 103.

26. Chien Chao-nan to Chien Yü-chieh, February 22, 1917, *Nanyang*, pp. 103-104.

27. Corina, p. 101; Chien Chao-nan to Chien Yü-chieh, February 22, 1917, *Nanyang*, pp. 104-105.

28. Chien Chao-nan to Chien Yü-chieh, February 27, 1917, *Nanyang*, pp. 105-107.

29. This draft contract was originally drawn up in English, but no copy of the original is available. A Chinese translation of it was enclosed in Chien Chao-nan to Chien Yü-chieh, February 22, 1917, *Nanyang*, pp. 104-105.

30. Chien Chao-nan to Chien Yü-chieh, February 27, 1917, *Nanyang*, p. 107.

31. Chien Ying-fu to Chien Yü-chieh, July 14, 1917; Chien K'ung-chao to Chien Yü-chieh, July 14, 1917; Chien Yü-chieh to Chien Chao-nan, copies of two undated letters from Chien Yü-chieh's private papers, all in *Nanyang*, pp. 107-112.

32. Lu Hsi-shan to Chien Yü-chieh, March 5, 1917, *Nanyang*, p. 109.

33. Chien K'ung-chao to Chien Yü-chieh, May 26, 1916, *Nanyang*, p. 111.

34. On Cantonese kinship ties, see C. K. Yang, *A Chinese Village in Early Communist Transition* (Cambridge, Mass., 1959); Hugh Baker, *A Chinese Lineage* (Stanford, 1968); Maurice Freedman, *Chinese Lineage and Society* (London, 1966); Freedman, *Lineage Organization in Southeastern China* (London, 1957); Jack M. Potter, *Capitalism and the Chinese Peasant* (Berkeley, 1968).

35. Chien Chao-nan (quoting from Chien Yü-chieh's earlier letter) to Chien Yü-chieh, June 19, 1917, *Nanyang*, p. 117. Some of the Chiens' antagonism toward Westerners may have been attributable to their upbringing near Canton where anti-foreign feeling was evident on several occasions in the nineteenth and early twentieth centuries. See Frederic Wakeman, Jr., *Strangers at the Gate*

(Berkeley, 1966); Lloyd Eastman, "The Kwangtung Anti-foreign Disturbances during the Sino-French War," *Papers on China*, 13 (1959), 1-31; Rhoads, pp. 83-91 and passim. The Chiens might also have chafed under discriminatory laws in the British colony of Hong Kong, which restricted Chinese in matters such as where they could establish residences. According to Henry J. Lethbridge, such discrimination gave rise to "bottled up racialist hates" among Chinese businessmen in early twentieth century Hong Kong. See his "Hong Kong under Japanese Occupation: Changes in Social Structure," in I. C. Jarvie, ed., *Hong Kong: A Society in Transition* (London, 1969), pp. 88-95.

 36. Chien Yü-chieh to Chien Chao-nan, copies of two undated letters from Chien Yü-chieh's private papers, *Nanyang*, pp. 107-109.

 37. Chien Chao-nan (quoting from Chien Yü-chieh's earlier letter) to Chien Yü-chieh, July 3 and 19, 1917, *Nanyang*, pp. 117-118; Chien Yü-chieh to Chien Chao-nan, copies of two undated letters from Chien Yü-chieh's private papers, *Nanyang*, pp. 107-109.

 38. Chien Chao-nan to Chien Yü-chieh, July 3 and March 16, 1917, *Nanyang*, pp. 117-118 and 113.

 39. Chien Chao-nan to Chien Yü-chieh, June 19 and March 16, 1917, *Nanyang*, pp. 117-118.

 40. Chien Chao-nan to Chien Yü-chieh, June 19 and July 3, 1917, *Nanyang*, pp. 117-118.

 41. Chien Chao-nan to Chien Yü-chieh, July 3 and March 16, 1917, *Nanyang*, pp. 118 and 113.

 42. Chien Chao-nan to Chien Yü-chieh, June 19 and July 3, 1917, *Nanyang*, pp. 117-118.

 43. According to a Chinese advertising specialist, Liu Huo-kung, Chien Chao-nan recruited him and relied on him to introduce mass advertising techniques, especially nationalistic promotional campaigns. One of Liu's slogans that Nanyang used, for example, was the following rhyming couplet:

   Among cigarettes, the king (*Hsiang-yen chih wang*),

   For national goods, the beacon (*Kuo-huo chih kuang*).

See Liu, "Nan-yang yen-ts'ao kung-ssu Chien Chao-nan ch'uang yeh shih," *I-wen chih*, 6 (December 1965), 16-18.

 44. Chien Chao-nan to Chien Yü-chieh, March 12, 1917, *Nanyang*, pp. 112-113.

 45. Chien Chao-nan to Chien Yü-chieh, March 12, June 12, June 19, and July 3, 1917, *Nanyang*, pp. 112-116.

 46. "Memoirs of Chien Yü-chieh," *Nanyang*, p. 4; Chien Chao-nan to Chien Yü-chieh, June 12 and 19, 1917, *Nanyang*, pp. 114-115. Cf. Albert Feuerwerker's description of malpractices in the China Merchants' Steam Navigation Company in *China's Early Industrialization*, pp. 146-149.

 47. Chien Chao-nan to Chien Yü-chieh, June 29, 1917, *Nanyang*, p. 129.

 48. Chien Ying-fu to Chien Yü-chieh, July 14, 1917, *Nanyang*, pp. 111-112. Chien Ying-fu boasted that he had already discovered a Chinese merchant in Singapore willing to put up several tens of millions of Dutch guilders to finance a new factory for the Chiens. Ibid., p. 112.

 49. Chien K'ung-chao to Chien Yü-chieh, May 26, 1917, *Nanyang*, p. 111.

50. Chien Ching-shan to Chien Yü-chieh, March 30, 1917, *Nanyang*, p. 110.

51. Chien Chao-nan to Chien Yü-chieh, July 3, 1917, *Nanyang*, p. 116.

52. Chien Chao-nan to Chien Yü-chieh, July 19 and 23, and October 13, 1917; Thomas to Chien Chao-nan, October 21, 1917, all in *Nanyang*, pp. 9 and 118-119.

53. Chien Chao-nan to Chien Yü-chieh, October 18, 1917, *Nanyang*, p. 119.

54. Chien Chao-nan to Chien Yü-chieh, October 13, 1917, *Nanyang*, p. 9.

55. Chien Ch'in-shih to Chien Yü-chieh, November 6, 1917, *Nanyang*, p. 63.

56. Hao, pp. 119-201. Feuerwerker, *China's Early Industrialization*, chs. 4-6; Samuel Chu, pp. 31 and 49-52; Feuerwerker, *The Chinese Economy, 1912-1949*, p. 18. A survey of 2,435 Chinese-owned factories conducted by Liu Ta-chün in 1933 found that only 612 were organized as joint stock companies. For these and more detailed figures on the financial structure of Chinese industry, see his *Chung-kuo kung-yeh tiao-ch'a pao-kao* (Nanking, 1937), II, 33-64. The "Company Law" of 1904 may be found in *Ta Ch'ing Kuang-hsu hsin-fa-ling* (Shanghai, 1909), ch. 16, pp. 2a-11b; an English translation of it appears in Edward T. Williams, *Recent Legislation Relating to Commercial, Railway, and Mining Enterprises* (Shanghai, 1904), pp. 10-45.

57. Chien Chao-nan to Chien Yü-chieh, March 16, 1917, *Nanyang*, pp. 113-114.

58. For a perceptive comment on the family as "a positive force" in China's commercial development, see Perkins, "Introduction," in Perkins, ed., *China's Modern Economy*, pp. 13-15.

59. Chien Chao-nan to Chien Yü-chieh, June 29, 1917, *Nanyang*, p. 130. Figures on the capitalization of the publishing houses are from *Far Eastern Review*, 17.10 (October 1921), supplement, 2.

60. Kuo-heng Shih describes this episode in "The Early Development of the Modern Chinese Business Class," in Levy and Shih, *The Rise of the Modern Chinese Business Class*, pp. 55-56. For biographical sketches of Mu, see Y. C. Wang, *Chinese Intellectuals and the West, 1872-1949* (Chapel Hill, 1966), pp. 473-477; and Boorman and Howard, III, 38-40. On Chinese shareholders' suspicions that joint stock companies were bound to lead to corruption, see Lih-chung Faung, "A Chinese Corporation," in C. F. Remer, ed., *Readings in Economics for China* (Shanghai, 1922), pp. 310-311.

61. Chien Chao-nan to Chien Yü-chieh, July 19, 1917, *Nanyang*, p. 118.

62. Chien Chao-nan to Chien Yü-chieh, March 16, 1917, *Nanyang*, p. 113.

63. Lu Hsi-shan to Chien Yü-chieh, March 5, 1917; Chien K'ung-chao to Chien Yü-chieh, May 26, 1917, *Nanyang*, pp. 109-111.

64. Chien Ying-fu to Chien Yü-chieh, July 14, 1917, *Nanyang*, p. 111; Chien Chao-nan to Chien Yü-chieh, October 13, 1917, *Nanyang*, p. 9; notes from a meeting of the Chien family, March 20, 1918, Chien Yü-chieh's private papers, *Nanyang*, pp. 10-11.

65. Notes from a meeting of the Chien family, March 20, 1918, Chien Yü-Chieh's private papers, *Nanyang*, p. 10; contract dated March 28, 1918, Chien

Yü-chieh's private papers, *Nanyang,* pp. 11-12; Nanyang financial records, *Nanyang,* pp. 11 and 139. Dividends were to be paid to Chien Chao-nan within three years after he had paid for the stock. By the time that he died in 1923, Chien Chao-nan held 26.9 percent of the stock in Nanyang; the remainder of Chien Chao-nan's branch held 12.1 percent; Chien K'ung-chao's branch held 21.6 percent; and other shareholders held 4.7 percent. See Nanyang financial records, *Nanyang,* pp. 138-139.

66. Chien Yü-chieh and Chien K'ung-chao to Chien Yin-ch'u and the Nanyang board of directors, July 17, 1918; revised version of Nanyang's by-laws, Chien Yü-chieh's private papers, both in *Nanyang,* pp. 14-15. As vice president, Chien Yü-chieh received a salary of ￥8,400 and an expense account of ￥10,000 per year. *Nanyang,* p. 16.

67. Chien Yü-chieh and Chien K'ung-chao to Chien Yin-ch'u, July 17, 1918, *Nanyang,* pp. 13-14.

68. Ibid.

69. Ibid.; notes from a meeting of the Chien family, March 20, 1918, Chien Yü-chieh's private papers, *Nanyang,* pp. 10-11.

70. Chien Chao-nan to Chien Yü-chieh, January 28, 1919, *Nanyang,* p. 422.

5. Commercial Campaigns in the May Fourth Movement

1. For details on the May Fourth incident and related events, see Chow Tse-tsung's classic account, *The May Fourth Movement: Intellectual Revolution in Modern China* (Cambridge, Mass., 1960).

2. Remer, *A Study of Chinese Boycotts,* ch. 7; Chow, *The May Fourth Movement,* pp. 141 and 147-149; *Millard's Review of the Far East,* 10.6 (October 11, 1919), 236; ibid., 10.10 (November 8, 1919), 416.

3. One of these advertisements appeared in each of 35 daily issues of *Hsin-wen pao* and other major newspapers in this period. See the study prepared by the compilers of the *Nanyang* volume and included in it, pp. 95-96. It is based on only one newspaper, *Hsin-wen pao,* but it notes that the same items were also published in other newspapers. As indicated later in this chapter, I have found from other sources that newspapers in Tientsin, Peking, Canton, and Hong Kong as well as Shanghai carried the same advertisements and proclamations. The advertising in *Hsin-wen pao* alone reached a wide audience, for it had the highest circulation of any newspaper in Shanghai at the time, partly because of its low unit price. See H. P. Tseng, "China Prior to 1949," in John A. Lent, ed., *The Asian Newspapers' Reluctant Revolution* (Ames, Iowa, 1971), p. 35.

4. The circulation of Chinese newspapers in this period, as estimated by Carl Crow's advertising agency, was published by Don D. Patterson, *The Journalism of China* (Columbia, Mo., 1922), pp. 79-89. See also Arnold et al., *Commercial Handbook of China,* no. 84, vol. II, p. 388; and Sanger, p. 61. On handbills in Canton, see *Wu-hu,* pp. 14b, 17a, and 18a; in Shanghai, *Hsin-wen pao,* May 18, 1919, *Nanyang,* p. 84; in Hangchow, *Hsin-wen pao,* May 19, 1919, *Nanyang,* p. 256.

5. This was the conclusion reached by the compilers of the Nanyang volume, *Nanyang,* p. 95.

6. Thomas (quoting from a conversation with Chien Yin-ch'u) to Allen, March 27, 1920, Thomas Papers. Thomas' source for this information, Chien Yin-ch'u, was a cousin of the Chien brothers and had been a member of the Nanyang board of directors until he resigned in April 1919 and formed his own company, the National Commercial Tobacco Company, which manufactured cigarettes briefly in Shanghai between 1920 and 1922. See *Nanyang*, p. 139; and *REA's Far Eastern Manual* (Shanghai, 1923), pp. 921-922. On Nanyang's declining sales, see the summary of newspaper reports in *Wu-hu*, p. 23a. This evidence cannot, unfortunately, be corroborated statistically because quantitative data on Nanyang's production in 1919 are available only for the company's Hong Kong plant (which supplied South China where reductions in Nanyang's profits were small) and not for its Shanghai plant (where, according to nonstatistical accounts from newspapers, its losses were great). Cf. Appendix, table 6. On BAT's advertising expenditures of ¥1.8 million for 1919, see Sanger, p. 58.

7. On strikes and boycotts against BAT, see *Shen-pao*, June 7, 1919, in *Wu-ssu yun-tung tsai Shang-hai shih-liao hsuan-chi* (Shanghai, 1960), p. 311; *Hsin-wen pao*, June 9, 1919, ibid., p. 337; *Shih-pao*, June 10, 1919, ibid., p. 346; Chung-kuo k'o-hsueh-yuan li-shih yen-chiu ti-san-so, comp., *Wu-ssu ai-kuo yun-tung tzu-liao* (Peking, 1959), p. 841. On rising demand for BAT cigarettes, see *NCH*, October 18, 1919, p. 143.

8. The evidence had first appeared as a "thank you" (*ming-hsieh*) in Chinese newspapers from an employee of a British firm to members of the Chien family for helping him settle a legal dispute with Chien Chao-nan in Japan. Such public expressions of gratitude were commonly published in "personal" columns of Chinese newspapers at the time, and when this one first appeared in newspapers in Peking, Tientsin, Hankow, and Shanghai in 1918, it was received without fanfare. Not until it was republished in the charged atmosphere of May Fourth did it stir a controversy. The original "thank you" and commentary on reactions to it appear in *Wu-hu*, p. 2a.

9. Rhoads, pp. 216-217.

10. Ibid., pp. 217-222 and passim; Lin Pin, "Kuang-tung kuai-chieh Chiang Hsia-kung t'ai-shih i-shih," *I-wen chih*, 30 (March 1, 1968), 18. On Chiang K'ung-yin's later career, see also Kao, pp. 16-18.

11. Chiang K'ung-yin's public statement, originally published in various Cantonese newspapers, republished in *Wu-hu*, p. 24a.

12. Chiang Shu-ying's open letter to Chien Ch'in-shih, originally published in several Cantonese newspapers, republished in *Wu-hu*, p. 17b.

13. *Wu-hu*, pp. 24a-25b. Chiang K'ung-yin also boasted that his distributors serviced 160 or 170 tea shops and other small retailers in Canton even though such small-time business was not profitable. Although Chiang undoubtedly overstated the amount of BAT's philanthropy, his claims had some basis in fact. For example, BAT spent $200,000 on flood and fire relief in Hsuchow, Pengpu, and along the West River in the first six months of 1919 according to *NCH*, June 21, 1919, p. 769. On BAT's expansion of its agricultural and marketing systems in this period, see Chapter 6.

14. A public statement by several Cantonese social organizations, originally published in Cantonese newspapers, republished in *Wu-hu*, pp. 8a-8b. More specifically, the invited guests maintained that the Chiens held all the stock

in the company; that they procured all their tobacco from the American South or seven places in China, four of which were in South China; that they imported cigarette paper from the United States and had plans to build a cigarette paper manufacturing plant of their own in Hsuchow; that they had their printing done by the Commercial Press and Chung-hua Book Company in Shanghai and by the Tung-ya Printing Company in Canton (all Chinese firms); that they had their own tin cans and tinfoil made from tin mined in Yunnan province; that they imported spices for blending from Europe and the United States; and that all of their 4,300 employees in the Hong Kong factory were Chinese.

15. Fang-pien Hospital, public statement, originally published in Cantonese newspapers, republished in *Wu-hu*, pp. 5a-6a. According to this source, the Chiens gave ¥100,000 for flood relief in 1915 and 1918, plus incalculable regular contributions to keep dikes in good repair; ¥20,000 to help victims of warfare in 1916; over ¥300,000 for famine relief in Kwangtung province and the vicinity of Shanghai; ¥30,000 for construction of an orphanage and annual donations of ¥8,000 or ¥9,000 to help cover its operating expenses; and additional funds (of an unspecified amount) to expand the endorser's own Fang-pien Hospital in 1917. For further details on Nanyang's relief programs, see Nan-yang hsiung-ti yen-ts'ao kung-ssu, comp., *Chen tsai chi* (Canton?, 1918).

16. Li Kuan-mei et al. (three signatories), public statement, originally published in Cantonese newspapers, republished in *Wu-hu*, pp. 9a-10a.

17. Huang Hsiao-sen et al. (six signatories), public statement, originally published in Cantonese newspapers; and editor's comment, both in *Wu-hu*, pp. 6a-7b.

18. Ch'en Ch'ung-nan et al. (twenty-one signatories), public statement, originally published in Cantonese newspapers, republished in *Wu-hu*, pp. 4b-5a.

19. Li Po-t'ao, public statement, originally published in Cantonese newspapers, republished in *Wu-hu*, pp. 5a-5b; Ts'ui Tzu-ch'ü et al. (four signatories), public statement, originally published in newspapers in various Chinese cities, republished in *Wu-hu*, pp. 12a-13b. The antipathy these young Cantonese felt toward Cantonese businessmen in Japan and especially toward the latter's willingness to become Japanese citizens echoes sentiments expressed earlier by young Cantonese in Japan. The young romantic writer Su Man-shu, whose father was Cantonese and whose mother was Japanese, passionately denounced this tendency among Cantonese merchants as early as 1903. See Henry McAleavy, *Su Man-shu, 1884-1918: Sino-Japanese Genius* (London, 1960), p. 9; and Liu Wu-chi, *Su Man-shu* (New York, 1972), p. 33.

20. *Wu-hu*, pp. 13b-14b.

21. Chiang Shu-ying's open letter to Chien Ch'in-shih, republished in *Wu-hu*, pp. 15b-16a, 17a and 18b.

22. *Shen-pao*, October 15, 1936; Wu Ch'eng-lo, II, 72; minutes of the meeting of the Nanyang board of directors, August 27, 1920, *Nanyang*, p. 137. For a detailed eyewitness description of Nanyang's new manufacturing facilities in Shanghai, see *Min-kuo jih-pao*, September 12, 1919, *Nanyang*, p. 56.

23. *Hsin-wen pao*, May 14, 1919, *Nanyang*, p. 84.

24. *Hsin-wen pao*, May 19, 25, and 26, 1919, *Nanyang*, pp. 255-257. Small shopkeepers used the May Fourth boycott in a similar manner against three of the

largest department stores in China, the Sincere Company, the Sun Company, and the Chen Kwong Company. See, China, Inspectorate General of Customs, *Returns and Reports of Trade, 1919* (Shanghai, 1920), p. 1015.

25. *Hsin-wen pao*, May 17 and 22, 1919, *Nanyang*, pp. 60 and 99-100. For endorsements from thirteen of the Chiens' "fellow industrialists" (*t'ung-yeh*), see *Hsin-wen pao*, May 18, 1919, *Nanyang*, pp. 84-85. Other endorsements on behalf of Nanyang in Shanghai were given by the Association for the Support of China's National Goods (*Chung-hua kuo-huo wei-ch'ih hui*), the Federation of Overseas Cantonese Associations in Shanghai (*Shang-hai Yueh ch'iao shang-yeh lien-ho hui*), the Association to Promote the Smoking of Cigarettes that Are National Goods (*Chung-hua ch'üan hsi kuo-huo chih-yen hui*), the Kiangsu Educational Association (*Chiang-su chiao-yü hui*), the Federation of Newspaper Guilds (*Pao-chieh lien-ho hui*), the World Students Association (*Huan ch'iu hsueh-sheng hui*), the Nanhai County Inns (*Nan-hai pa-kuan*), the Ch'ang-ming School (*Ch'ang-ming hsueh-hsiao*), 158 "fellow industrialists" in the cigarette business, and 161 identifiable individuals. See *Nanyang*, p. 96.

26. *Hsin-wen pao*, May 28, 1919, *Nanyang*, pp. 86-87.

27. Ibid., p. 97.

28. Chung-hua shang-yeh hsieh-hui et al., *Ts'an-kuan Nan-yang hsiung-ti yen-ts'ao kung-ssu chi* (1926), *Nanyang*, p. 133. This edition was republished in 1926, but from its contents, it appears to have originated in the spring of 1919.

29. Introductory note appended to Nanyang's announcement that it would sell stock, *Nanyang*, p. 134. Nanyang's effort to attract stockholders through an advertising campaign in Chinese newspapers is summarized in *NCH*, August 9, 1919, p. 332.

30. Contract signed by Chien Chao-nan et al., pp. 134-137; *Nanyang*, p. 139; Wang Ching-yü, p. 963; *Ao-men hsin-sheng jih-pao*, September 28, 1938; *Ch'en Kung Ping-ch'ien fu-kao* (Macao, 1938); Thomas to Yiu, October 7, 1925, Thomas Papers. Besides Ch'en Ping-ch'ien, four new members of the board were living in Shanghai: Ch'ien Hsin-chih (the only non-Cantonese), Yang Hsiao-ch'uan, Chou Ching-ch'uan, and Chao Chou-san. The other two new board members were from Hong Kong, Chou Shou-chen (Sir Shouson Chow) and Ch'en Lien-po. *Nanyang*, pp. 433-444. All were well-known Chinese businessmen. On Ch'ien, Yang, and Ch'en, see Gaimushō jōhōbu, eds., *Gendai Chūka minkoku Manshukoku jimmei kan* (Tokyo, 1932), pp. 193, 278, and 357. On Chou Shou-chen and Ch'en Lien-po, see Lethbridge, pp. 87 and 110; Hao, pp. 194-195; and Chesneaux, *The Chinese Labor Movement*, pp. 10, 248-249, and 337.

31. On Chü, see Ho Tso, "1905-nien fan-Mei ai-kuo yun-tung," p. 30; Wang Ching-yü, pp. 965-66; Chow Tse-tsung, pp. 154-156. On Yü, see Hao, p. 124; Chow Tse-tsung, pp. 154-156; and Joseph Fewsmith, "Yü Hsia-ch'ing and the Evolution of Kuomintang-Merchant Relations, 1924-1930," paper delivered at the convention of the American Historical Association, December 31, 1973. On Ku, see Wang Ching-yü, p. 958. On Chekiang financiers in the eighteenth and nineteenth centuries, see Susan Mann Jones, "Finance in Ningpo," in Willmott, ed., *Economic Organization in Chinese Society*, pp. 47-77. For a list of 472 individuals and organizations that sponsored Nanyang, see *Nan-yang hsiung-ti yen-*

ts'ao ku-fen yu-hsien kung-ssu k'uo-chung kai-tsu chao-ku chang-ch'eng (Shang-hai, 1919), pp. 10b-16b. This appeal for subscribers was republished in *Hsin-wen pao*, August 5, 1919, and (without the list of sponsors) in *Nanyang*, pp. 97-98. Yang's actions in Shanghai on behalf of Nanyang are described in *NCH*, August 9, 1919, p. 332.

32. The contract was not signed until July 17, 1919, but Nanyang's board of directors approved it on June 28, 1919. *Nanyang*, p. 135.

33. *Hsin-chung pao* (of Shanghai), July 19, 1919; *Hsin pao* (of Canton), July 26, 1919, both in *Wu-hu*, pp. 21a-21b; Liu Shou-lin, *Hsin-hai i-hou shih-ch'i nien chih-kuan nien-piao* (Peking, 1966), pp. 559 and 564; Chou Wei-fan, public statement, published in several Shanghai newspapers, July 27, 1919, republished in *Wu-hu*, pp. 21a-21b. The term *chien-shang* was a smear word that had been used to express contempt for merchants at least since the early nineteenth century. Cf. Wakeman, *Strangers at the Gate*, pp. 48-51.

34. Articles originally published August 9 and 12, 1919, by Chinese newspapers in Peking, Canton, Shanghai, and other cities and republished in *Wu-hu*, pp. 21a and 23a-23b; "Memoirs of Chien Yü-chieh," *Nanyang*, p. 82. Huang originally entered the tobacco industry by starting his own firm, the Ta-ch'ang Tobacco Company, but as soon as it achieved a degree of success, BAT bought it out and hired Huang as an employee. See Ch'en Tzu-ch'ien and P'ing Chin-ya, "Hsiang-yen hsiao-shih," in Shang-hai jen-min ch'u-pan she, ed., *Shang-hai ching-chi shih-hua* (Shanghai, 1964), col. 1, p. 75.

35. Federation of China merchant associations and six other organizations, *Ko-kung-t'uan ts'an-kuan Nan-yang hsiung-ti yen-ts'ao kung-ssu chi* (1926), *Nanyang*, pp. 96-97; Association of the Overseas Chinese of Singapore to the cabinet of the Peking government, August 1919, *Nanyang*, p. 86.

36. A Ch'ing law to this effect was promulgated in 1712 and not annulled until 1893. See Wang Sing-wu, "The Attitude of the Ch'ing Court toward Chinese Immigration," *Chinese Culture*, 9.4 (December 1968), 75; and H. F. MacNair, "The Relation of China to Her Nationals Abroad," *The Chinese Social and Political Science Review*, 7.1 (January 1923), 23-43.

37. According to article 18, " . . . all persons who have not filed a petition for discharge or whose petition is not granted [by the Chinese government] remains Chinese for all purposes." An English translation of the full text of the law was published in Supplement to the *American Journal of International Law*, 4 (1910), 160-166. The quotation is from p. 164. Tsai Chu-tung explores the law's implications in "The Chinese Nationality Law, 1909," *The American Journal of International Law*, 4 (1910), 404-411.

38. An English translation of the full text of the Chinese nationality law of 1914 appears in Great Britain, *British and Foreign State Papers, 1921*, 114 (London, 1924), 667-671. The quotation is from p. 670.

39. Association of the Overseas Chinese of Singapore to the cabinet of the Peking government, August 1919, *Nanyang*, p. 86; Chien Ch'in-shih to Chiang Shu-ying and newspaper accounts from 1919, republished in *Wu-hu*, pp. 5a, 11b, and 20a.

40. Association of the Overseas Chinese of Singapore to T'ang Chi-yao, August 1919, *Nanyang*, p. 83; Yü-ts'ai School of Malaya to T'ang Chi-yao,

August 1919, *Nanyang*, pp. 85-86. Although this is the source given for the letter in the *Nanyang* volume, the letter's contents indicate that it was from a commercial organization in Malaya whose members were Overseas Chinese. Perhaps the letter was composed by the Overseas Chinese merchants, who hired a teacher (and good calligrapher) in the Yü-ts'ai School to copy the final draft for them. On T'ang Chi-yao, see Boorman and Howard, III, 223-225.

41. *Min-kuo jih-pao*, September 13, 1919, *Nanyang*, p. 87. The decision was officially recorded on October 27, 1919, and the full text of the endorsement from the Ministry of the Interior was enclosed in a letter from the Chinese General Chamber of Commerce of Batavia to Nanyang, November 6, 1919, *Nanyang*, p. 87.

42. The Chiens also pledged ¥200,000 in the summer of 1919 to finance the construction of a university (which was never built) and made donations to several existing educational institutions: ¥10,000 to Nankai University in Tientsin in 1920; ¥42,284 to Futan University in Shanghai in 1920 and 1921; and ¥10,000 to Chinan University in Shanghai in 1923. They made additional donations to elementary and middle schools. See Shanghai General Chamber of Commerce and five other organizations, *Chung-kuo Nan-yang hsiung-ti yen-ts'ao kung-ssu ts'an-kuan chi* (1926), *Nanyang*, p. 249; *Wu-hu*, p. 6a. The figure mistakenly recorded in the latter source was ¥20,000. Provision 6 of the contract for Nanyang's reorganization, July 17, 1919, indicates that the actual amount pledged was ¥200,000. *Nanyang*, p. 136.

43. At the time of their selection, these men were known as The Five Statesmen Going Abroad (*Wu-ta-ch'en ch'u-yang*), and after studying at leading universities in the United States and Europe in the early 1920s, they became famous for their achievements in China: Lo as a historian, educator, administrator, and popular writer; K'ang as a romantic lyric poet and military leader; Yuan as a teacher and administrator; Meng as dean of letters at Chinan University in Shanghai; Wang I-hsi as a teacher at Sun Yat-sen (*Chung-shan*) University in Canton; and at least four of the five as high-ranking Kuomintang officials. See Lo Tun-wei, *Wu-shih-nien hui-i-lu* (Taipei, 1953), pp. 24-25, cited by Wang, *Chinese Intellectuals and the West*, p. 152n25; Chow Tse-tsung, pp. 55-57, 106, 121, 164; Boorman and Howard, II, 428-431 and III, 335-336.

44. Shanghai General Chamber of Commerce et al., *Chung-kuo Nan-yang hsiung-ti yen-ts'ao kung-ssu ts'an-kuan chi*, *Nanyang*, p. 249; A. Sy-hung Lee, p. 159. Chien Chao-nan seems to have initiated the scholarship program and taken a personal interest in it. Under his direction, a total of ¥500,000 was spent on scholarships between 1919 and 1923, but after his death in 1923, his successors terminated the program. Kuo-huo shih-yeh ch'u-pan she, comp., *Chung-kuo kuo-huo*, p. 9. At least one man from outside Nanyang, Mu Hsiang-yueh, a business leader in the textiles industry, also helped finance the foreign educations of Lo Chia-lun and Tuan Hsi-p'eng. See Boorman and Howard, III, 40.

45. Ting Wen-chiang, ed., *Liang Jen-kung hsien-sheng nien-p'u ch'ang-pien ch'u-kao* (Taipei, 1958), II, 581, cited by Y. C. Wang, "Free Enterprise in China: The Case of a Cigarette Concern, 1905-1953," *Pacific Historical Review*, 29.4 (November 1960), 389n18; Ai Wu, *Wo-ti ch'ing-nien shih-tai* (Shanghai, 1949), p. 51.

46. Provision 6, contract concerning the reorganization of Nanyang, July 17, 1919, *Nanyang*, p. 136.

47. Minutes of the meeting of the Nanyang board of directors, January 18, 1933, *Nanyang*, pp. 250-251.

48. Summary of newspaper reports, *Wu-hu*, p. 23a.

49. *Millard's Review*, 10.10 (November 8, 1919), 417-418; *NCH*, May 24, 1919, p. 507; Marie-Claire Bergère, "Le mouvement du 4 mai 1919 en Chine," *Revue historique*, 241 (June 1969), 324; Joseph Chen, pp. 98-99. On the reorganization of the Sun and Sincere companies in response to the May Fourth boycott, see *Millard's Review*, 10.10 (November 8 and 22, 1919), 417 and 504. On the Commercial Press' decision to buy out its Japanese stockholders between 1912 and 1914 in response to criticism from political nationalists (including the Chinese owners of its rival, the Chung-hua Book Store), see Alice Herman, "The Early Commercial Press and the Episode over Foreign Investment, 1897-1914," seminar paper, Cornell University, August 1975. And on Sun Yat-sen's and Yuan Shih-k'ai's resistance to Japanese efforts to gain partial control over the Hanyehping Coal and Iron Company, Ltd., in 1912, see Feuerwerker, "China's Nineteenth Century Industrialization" in Cowan, ed., *The Economic Development*, p. 91.

6. The Postwar Golden Age

1. See Huang Ch'eng-ching, "Ts'ung Nan-yang," p. 35; Ho Ping-yin, *The Foreign Trade of China* (Shanghai, 1935), pp. 55-57; Hou, *Foreign Investment*, pp. 80 and 86-87; Wang, *Chinese Intellectuals and the West*, pp. 475-476; Chou Hsiu-luan, *Ti-i-tz'u shih-chieh ta-chen shih-ch'i Chung-kuo min-tsu kung-yeh ti fa-chan* (Shanghai, 1929), pp. 93-100; Lieu, *The Growth and Industrialization of Shanghai*, pp. 22-24; Kikuchi, pp. 243-252. For statistics on Nanyang's sales and profits, see Appendix, table 6.

2. Feuerwerker, *The Chinese Economy, 1912-1949*, pp. 21-22; Marie-Claire Bergère, "La première guerre mondiale et les développements des industries nationales en Chine," paper read at the 29e Congrès International des Orientalistes, Paris, 1973; Bergère, "La bourgeoisie chinoise et les problèmes de développement économique, 1917-1923," *Revue d'histoire modern et contemporaine*, 16.2 (April-June 1969), 246-267. On China's lack of producer industries, see Rawski, "The Growth of Producer Industries," in Perkins, ed., *China's Modern Economy*, p. 220.

3. Some of the other developments abroad which affected China's economy in the postwar period besides those I have already mentioned include changes in Western demand for Chinese goods, rising price levels in Japan, and fluctuations in the value of gold in the West.

4. See FO 371/3701, no. 180832, Thornton to Hood, February 19, 1920.

5. FO 371/3701, no. F334, Allen of BAT (which refers to Duke's personal conversation with Geddes) to Geddes, December 6, 1920; FO 371/11688, no. F1593, Rickards of BAT to the Colonial Office, November 4, 1925. As noted in Chapter 5, Chiang K'ung-yin claimed that BAT had paid between ¥8 and ¥9 million in dividends in Chinese stockholders as early as 1918.

6. An official in the Foreign Office noted in likelihood that BAT formed

BAT (China), Ltd., to avoid paying taxes in England. See FO 371/9205, no. F127, unsigned comment on the letter from Rickards to the Colonial Office, December 21, 1922. The historian Ōi Senzō has interpreted this change in BAT's legal status as the turning point in its history in China, marking the transition from "a period of consolidation" (1902-1919) to "a period of self-sufficiency" (1919-1932). From the standpoint of business history, this periodization seems inappropriate because the change in the legal status of BAT's head office in China from a branch to a subsidiary does not seem to have had any direct or significant effect on the company's organization or operation as a business in China. Cf. Ōi, pp. 23-24.

7. FO 371/9205, no. F127, Rickards to the Colonial Office, December 21, 1922.

8. Paul A. Cohen, *Between Tradition and Modernity* (Cambridge, Mass., 1974), pp. 256-257.

9. Cobbs to Thomas, November 17, 1920, Thomas Papers.

10. HKRS 111/4/263, Rose to Fraser, March 14, 1922.

11. Pettitt to Thomas, April 5, 1922, Thomas Papers.

12. Thomas, *Pioneer*, p. 24.

13. Ch'eng et al., *Chüan-yen t'ung-shui shih*, pt. 2, p. 5. On BAT's agreement to pay likin in 1905, see FO 371/180, no. 18884, Keily to Warren, January 14, 1905. On BAT's lobbying efforts in 1917, see Chien Chao-nan to Chien Yü-chieh, June 17 and October 18, 1917, *Nanyang*, pp. 114 and 119. For a copy of one of BAT's schemes to give it control over tax collection, see "A Proposal for the Cooperation of the Chinese Government and the British-American Tobacco Company, Ltd., for the Improvement of the Tobacco Industry in China," December 30, 1916, enclosed in FO 371/2659, no. 264342. According to the same file, the British government informed BAT of its opposition to the scheme on January 15, 1917.

14. Hollington K. Tong, "Japan Seeking China's Tobacco Monopoly," *Millard's Review*, 5 (June 18, 1918), 49-52; London *Times*, May 31, 1918; New York *Times*, June 1, 1918; FO 371/5300, no. F1331, Williams quoted by Lampson to Curzon, April 14, 1920; FO 371/9211, no. F619, report by Williams to the Wine and Tobacco Monopoly, October 21, 1920; Thomas to Bruce, November 5, 1920, Thomas Papers. Williams' uneventful career in the Wine and Tobacco Monopoly contrasts sharply with the record of his contemporary, Sir Richard Dane, who, though also a foreigner, instituted a series of bold reforms in the Chinese Salt Administration. Cf. S. A. M. Adshead, *The Modernization of the Chinese Salt Administration, 1900-1920* (Cambridge, Mass., 1970), pp. 82-117 and passim.

15. Ch'eng et al., *Yen chiu shui shih*, II, ch. 7, sec. 2, p. 1; FO 371/6670, no. F4838, Alston to Curzon, November 4, 1921; Chu Hsieh, *Chung-kuo tzu-shui wen-t'i* (Shanghai, 1936), pp. 484-488.

16. Hollington K. Tong, "Development of China's Wine and Tobacco Administration," *Millard's Review*, 8 (April 5, 1919), 198-199.

17. FO 371/6670, no. F4830, Alston to Curzon, November 4, 1921. Enclosed is a copy of BAT's English translation of the agreement. For a Chinese summary of the agreement, see Ch'eng et al., *Chüan-yen t'ung-shui shih*, pt. 2, pp. 5-7.

18. *Hsiang-tao*, 38 (August 29, 1923), 288, translated into English by Stuart

R. Schram in Schram, ed. and trans., *The Political Thought of Mao Tse-tung* (New York, 1969), pp. 209-210.

19. At this time Sun Yat-sen not only refused to accept the new tax but promulgated his own 20 percent tax on foreign wine and tobacco in Canton. See FO 371/6667, file 3652; FO 371/8011, file 182; FO 371/8024, file 473 (documentation from October through November 1924). On implementation of the agreement, see FO 371/8036, no. F1633, Rose to Lampson, March 16, 1922; FO 371/9217, no. F1433, Rose, "Memorandum on the Practical Working of Special Arrangement," March 19, 1923; Thomas to Allen, October 15, 1921, Thomas Papers.

20. Nanyang's annual reports for 1923 and 1924, *Nanyang*, p. 222; NA 693.1112/104, Cunningham to State, July 20, 1925. The provinces' rebellions against Peking's centralized tobacco tax which spread to more than half of China's provinces between 1923 and 1925 is described in Sanford, pp. 27-39.

21. Research is needed on local taxation in Republican China. For brief references to it which suggest that local taxes proliferated after World War I, see Jerome Ch'en, "Historical Background," in Jack Gray, ed., *Modern China's Search for a Political Form* (London, 1969), p. 30; Sheridan, *Chinese Warlord*, pp. 24-26; Donald G. Gillin, *Warlord: Yen Hsi-shan in Shansi Province, 1911-1949* (Princeton, 1967), p. 55.

22. See Appendix, tables 3 and 8. On the opening of BAT's second factory in 1917, see *Chan-tou*, p. 13.

23. Appendix, table 7; FO 371/3701, no. 171045, Hood of BAT to the Colonial Office, January 6, 1920; *North China Daily News*, January 21, 1916, p. 7; Thomas, "Selling and Civilization," p. 949; The British-American Tobacco Company, *The Record in China*, p. 10; Thomas to Pettitt, June 15, 1920, Thomas Papers; *Nanyang*, p. 407; *Chinese Economic Bulletin*, 119 (August 30, 1924), 1-2; Chin-ling ta-hsueh nung-hsueh yuan nung-yeh ching-chi hsi, comp., "Ho-nan sheng ch'an yen-yeh ch'ü chih tiao-ch'a pao-kao," in *Yü O Wan Kan ssu-sheng nung-ts'un ching-chi tiao-ch'a ch'u-pu pao-kao*, 6 (June 1934), 6; Ming Chieh, "Ying-Mei yen-ts'ao kung-ssu ho Yü-chung nung-min," in Chung-kuo nung-ts'un ching-chi yen-chiu hui, ed., *Chung-kuo nung-ts'un tung-t'ai* (Shanghai, 1937), pp. 1-2; Crow, p. 58. Photographs of the Tientsin factory appear in the *B.A.T. Bulletin*, 17.73 (May 1926), 30-31. On BAT's investments in print shops and equipment, see Ch'en Chen, col. 1, p. 125.

24. Thomas to DuPont, November 5, 1924, Thomas Papers. The quotation is from W. H. Jansen, "Eight Years Pioneering in China," *American Cinematographer*, 11 (February 1931), 11, quoted in Jay Leyda, *Dianying: An Account of Films and Film Audiences in China* (Cambridge, Mass., 1972), p. 44.

25. Financial control over Cheng's company seems to have shifted into BAT's hands between 1919 and 1921. See Ōi Senzō, p. 24; Ch'en Chen, col. 2, pp. 95 and 133; Bassett to Cobbs, November 17, 1919, Thomas Papers; Wang Hsi, p. 83.

26. Li Hsien-wei, *The Tobacco in China* (Tientsin, 1934), p. 43; Ch'en Chen, col. 1, p. 95; Thomas to Jeffress, July 11 and August 6, 1923, Thomas Papers; Thomas to Flowers, March 5, 1928, Thomas Papers. On compradors in the nineteenth century, cf. Hao, chs. 5 and 6.

27. H. L. Kung (Kung Ho-ch'ien) to Thomas, October 5, 1920; Thomas to Jeffress, April 1, 1922; Kung to Thomas, December 5, 1926 and March 25, 1927; Thomas to Kung, January 3, 1927, all in Thomas Papers. Wang Ching-yü, p. 222; The British-American Tobacco Company, *The Record in China*, p. 32; Hutchison, pp. 309-310; *B.A.T. Bulletin*, 11.3 (July 1920), 41-42. Illustrative of Ts'ui's success at distribution are the following lines from a poem composed by J. H. Pritchett, a BAT manager who lamented that he could not supply Ts'ui and other Chinese agents in North China fast enough in 1921:

When old Tsui drops in crying 'bout his "Pinhead" [a popular BAT brand],
Says he's lost an even thousand "bones" today,
There is nothing left to do but just to ask him
To wait on Shanghai's shipment *one more day*.
When a shipment comes with just eleven "Pinhead,"
The dealers here bought ten, the thing is done,
Then you sit that night and wonder what's to happen,
When your country dealers find there's only one.

B.A.T. Bulletin, 12.13 (May 1921), 290.

28. FO 371/10297, F3405, Rose of BAT to the British minister in Peking, August 7, 1924; Crow, pp. 57-58. BAT directors made similar comments. See Wang Hsi, p. 84; and Thomas, *Pioneer*, p. 107.

29. Cobbs to Cheang Park Chew (Cheng Po-chao), May 11, 1919, Thomas Papers. Cf. Wang Hsi, p. 83.

30. Hutchison, p. 30.

31. Thomas to Jeffress, June 17, 1920, and Pettitt to Thomas, October 5, 1920, Thomas Papers; NA 693.1112/104, Cunningham to State, July 20, 1925; Appendix, table 1. I have assumed that all exports from the United States, Great Britain, and Canada and none from Hong Kong (where Nanyang manufactured large numbers of cigarettes for export to China) belonged to BAT.

32. China, Inspectorate General of Customs, *Decennial Reports, 1922-31* (Shanghai, 1933), p. 361, estimated production in BAT's Tientsin factory at 30 billion per year, but Thomas Wiens has pointed out that in light of the 60 cigarette rolling machines available there (as mentioned in the same report) 6 billion is a more probable estimate of the factory's capacity. Wiens, "The British American Tobacco Company and the Nanyang Brothers Tobacco Company," seminar paper, Harvard University, 1966, pp. 11-12.

33. FO 371/10940, no. 5505, Cunliffe-Owen to the Foreign Office, November 13, 1925; Chien Chao-nan to Chou Shou-chen, February 26, 1922, *Nanyang*, p. 425. Unfortunately, there are no available figures on the value of BAT's annual sales between 1916 and 1924. See Appendix, table 3.

34. Wolsiffer to Thomas, July 24, 1923, Thomas Papers.

35. Sun Chia-chi, p. 59; The British-American Tobacco Company, *The Record in China*, p. 22.

36. Pearl S. Buck, "China the Eternal," *The Living Age*, 324.4205 (February 7, 1925), 325.

37. Wang Li (Wang Liao-i), *Lung-ch'ung ping tiao chai so yü* (Shanghai, 1949), p. 82.

38. Arnold et al., *Commercial Handbook of China*, no. 84, II, p. 391.

39. William Jansen, p. 11, quoted in Leyda, p. 42; *B.A.T. Bulletin*, 15.53 (September 1924), 150.

40. Leyda, p. 44; Ch'eng Chi-hua, *Chung-kuo tien-ying fa-chan shih*, I (Peking, 1963), 122-124. Ch'eng estimates that BAT's initial investment in its movie division was about ¥50,000.

41. William Jansen, p. 11; *Chinese Economic Monthly*, 11 (August 1924), 3. For examples of BAT films, see *NCH*, April 25, 1925, p. 132; September 5, 1925, p. 306; July 10, 1926, p. 70.

42. William Jansen, p. 12; Sanger, p. 78.

43. Cheng Po-chao started construction of the Ao-ti-an (Odeon) Theater in 1923 and opened it in Shanghai. See Shang-hai t'ung-she, ed., *Shang-hai yen-chiu tzu-liao hsu-chi* (Shanghai, 1939), p. 543; and Thomas to Kingsbury, October 23, 1923, Thomas Papers. Thomas' company, the Peacock Motion Picture Corporation, was capitalized at $5 million. Its promoters included five directors of the DuPont Company, which was deeply involved in the motion picture business in the United States; a former Minister of Finance, Chou Tzu-ch'i; an admiral from the Chinese navy, Ts'ai T'ing-kan; and twelve other prominent Chinese. Thomas believed that his movies could serve as "an excellent advertising medium" and urged BAT to advertise in conjunction with his films. Thomas to Cunliffe-Owen, February 7, 1923; and Thomas to Page, May 10, 1923, both in the Thomas Papers. See also Leyda, p. 39.

44. Ho Wei-hsing, "Chung-kuo tien-ying chieh ti chueh-tai wei-chi," *Ying-hsi ch'un-ch'iu* (April 5, 1925), quoted in Ch'eng Chi-hua, I, 125-126.

45. The British-American Tobacco Company, *The Record in China*, pp. 12 and 17-21; *NCH*, August 14, 1926, p. 304; "Labour Conditions and Labour Regulation in China," *International Labour Review*, 10.6 (December 1924), 1018; *Chan-tou*, pp. 13-17.

46. "Labour Conditions and Labour Regulation in China," p. 1018; Chesneaux, *The Chinese Labor Movement*, pp. 72 and 95; Wang Hsi, p. 89. Cf. Chan Ming Kou, "Labor and Empire," Ph.D. dissertation, Stanford University, 1975, pp. 145-153; and Christopher Howe, *Wage Patterns and Wage Policies in Modern China, 1919-1972* (Cambridge, England, 1973), p. 22, table 11.

47. "Labour Conditions and Labour Regulation in China," p. 1018; *NCH*, August 14, 1926, p. 304. Cf. John Gittings, *The World and China, 1922-1972* (New York, 1974), p. 23. From the vantage point of the People's Republic in the 1950s, several workers bitterly recalled the insecurity that they had felt while working at BAT in the 1920s and 1930s. See *Chan-tou*, pp. 6-14 and passim.

48. *Chan-tou*, pp. 13-17; *Shen-pao*, June 7, 1919, and *Hsin-wen pao*, June 9, 1919, both in *Wu-ssu yun-tung tsai Shang-hai hsuan-chi*, pp. 311 and 337; *NCH*, June 14, 1919, p. 721.

49. Chesneaux, *The Chinese Labor Movement*, ch. 8; *Chan-tou*, pp. 20-22; C. Martin Wilbur and Julie Lien-ying How, eds., *Documents on Communism, Nationalism, and Soviet Advisers in China, 1918-1927* (New York, 1956), p. 55.

50. *Chan-tou*, pp. 20-28. During this period, the name of the tobacco workers' union was changed to the Tobacco Workers' Club (*Yen-ts'ao kung-jen chü-lo-pu*).

51. *NCH*, October 21, 1922, p. 154; and August 28, 1922, p. 217; FO 371/

9217, no. F851, Goffe to Clive, December 9, 1922 and January 22, 1923; FO 228/3529, Porter to Palairet, June 4, 1925; Chesneaux, *The Chinese Labor Movement*, pp. 323, 401, and 522n33. The full text of the agreement has been published in the *Far Eastern Review*, 19.2 (February 1923), 91-92.

52. *Chan-tou*, pp. 31-40; Chesneaux, *The Chinese Labor Movement*, pp. 195 and 537n117; *NCH*, November 11, 1922, p. 377; November 18, 1922, p. 450; November 25, 1922, p. 529.

53. FO 371/9217, no. F851, report of the BAT strike prepared by BAT. Whether the Nanyang company actively publicized BAT's labor problems or not, BAT's workers appear to have been convinced that the Chiens were sympathetic to their cause, for they approached the Chiens for help against BAT's management during the strike. The Chiens' accommodating attitude toward labor was apparently typical of management in Chinese-owned factories during the big wave of strikes in 1921 and 1922. See Bèrgere, "La bourgeoisie chinoise et les problèmes du développement économique," pp. 264-265.

54. FO 371/9217, no. F851, Goffe to Clive, January 22, 1923. The Chinese labor movement as a whole also suffered setbacks in 1923 and 1924. See Chesneaux, *The Chinese Labor Movement*, pp. 206-233.

55. *Chan-tou*, pp. 20-28, 34-36, and 39 (including a cartoon picturing the conspiracy). Cf. Chuan-hua Lowe, *Facing Labor Issues in China* (Shanghai, 1933), pp. 66-67; and Nym Wales (Helen Foster Snow), *The Chinese Labor Movement* (New York, 1945), pp. 195-196.

56. FO 371/3180, no. 2564, Kennett of BAT to Jordan, October 22, 1917; Ming Chieh, pp. 1-2; W. Y. Swen, "Types of Farming, Costs of Production and Annual Labor Distribution in Wei hsien county, Shantung, China," *Chinese Economic Journal*, 3.2 (August 1928), 658.

57. Ōi, p. 23; Ch'en Chen, col. 2, p. 133; *Who's Who in China*, I, 241; Ming Chieh, p. 2. Rose is quoted by Wang Hsi, p. 82. Shen also recruited members of his family to work for BAT. For example, he recommended his cousin, a Cambridge University graduate, to BAT to help it with "diplomatic questions . . . for instance the cigarette taxes, etc. [because] he is very well connected with all the officials both at the capital and the different Ports." Shen to Thomas, March 26, 1924, Thomas Papers.

58. Ramon H. Myers, *The Chinese Peasant Economy* (Cambridge, Mass., 1970), p. 349n11. BAT also conducted experiments at growing bright tobacco in Kwangtung, Kiangsi, and Kirin provinces. See Ch'en Chen, col. 1, p. 731; and Chang Yu-i, pp. 152-153, 503, and 508.

59. Chang Yu-i, pp. 202 and 225; Chen Han Seng, p. 93; *Chinese Economic Bulletin*, 119 (June 2, 1923), 2. The land in Shantung yielded about 400 *chin* (535 pounds) of tobacco per *mou* of land, and BAT paid between ¥.10 and ¥.30 for every *chin*. At this rate, a reporter for *Ta-kung pao* estimated that the value of tobacco purchased by BAT at Fangtze in 1918 totaled ¥5 million. (See Chang Yu-i, pp. 225 and 244.) This estimate correlates with Wu Ch'eng-lo's later calculations that in 1922 and 1923 ¥16 million was paid for tobacco in Shantung, of which he estimates ¥9 million was profit. Cf. Wu, II, 71.

60. Swen, pp. 658-659; U.S. Department of Commerce, *Economic Development of Shantung, 1912-21*, Trade Information Bulletin no. 70, p. 8, quoted by

Allen and Donnithorne, *Western Enterprise in Far Eastern Economic Development*, p. 170; Thomas, *Pioneer*, pp. 44-45.

61. The quotations are from Myers, *The Chinese Peasant Economy*, pp. 293 and 294. My brief summary does not do justice to the complexity of these approaches. For more detailed and sophisticated discussions of them, see Myers, ibid., ch. 2; and Carl Riskin, "Surplus and Stagnation in Modern China," in Perkins, ed., *China's Modern Economy*, pp. 56-64.

62. Chen Han Seng, pp. 15, 25-26, and passim.

63. Myers, *The Chinese Peasant Economy*, p. 16.

64. Chen Han Seng, p. 62; Swen, pp. 650-653 and 661; Hattori Mitsue, "Hoku-Shi ni okeru hatabako saibai fukyū irai no nōgyō keiei no henka," *Mantetsu chōsa geppō*, 20.20 (December 1941), 96-99. According to another source, cultivating bright tobacco in Wei-hsien during the late 1920s and early 1930s was still less profitable than during the early 1920s (when the study by Swen was conducted). See Lung An, "Rival Auxiliary Occupations in Weihsien, Shantung," *I-shih pao*, May 12, 1934, translated into English in Institute of Pacific Relations, comp., *Agrarian China* (London, 1939), p. 232.

65. Thomas Wiens makes this point incisively in his review of Myers' book in *Modern Asian Studies*, 9.2 (April 1975), 281.

66. FO 371/3180, no. 2564, Kennett of BAT to Jordan, October 22, 1917; Chang Yu-i, p. 502; Ch'en Chen, col. 2, pp. 141-142; Huang I-feng, "Kuan-yü chiu Chung-kuo," pp. 103 and 107; Chen Han Seng, p. 11; Thomas, *Pioneer*, pp. 66-69.

67. Hattori, p. 99. Myers has analyzed why Chinese adopted a variety of cash crops (one of which was bright tobacco) between 1890 and 1937 in "The Commercialization of Agriculture in Modern China," in Willmott, ed., *Economic Organization*, pp. 173-191.

68. The average prices have been computed on the basis of Chen Han Seng, p. 94; and U.S. Department of Commerce, Bureau of the Census, *Stocks of Tobacco Leaf*, Bulletin no. 155 (Washington, D.C., 1924), p. 36.

69. Pettitt to Thomas, March 14, 1921, Thomas Papers.

70. See Appendix, table 8. On BAT's direct purchasing system, see Tilley, pp. 307-308 and 331-332.

71. Allen to Thomas, February 18, 1920, Thomas Papers; Hu Kuang-piao, *Po-chu liu-shih-nien* (Hong Kong, 1964), p. 227.

72. Chien Chao-nan to Chien Yü-chieh et al., February 6, 1921, *Nanyang*, pp. 424-425. This letter and all documents published in *Nanyang*, pp. 423-429, bear the date (in their published form) of 1922. Comparing them with unpublished correspondence in the Thomas Papers has convinced me that these documents were written one year earlier and misdated in the process of publication. Jeffress' unpublished correspondence indicates that he became deeply involved in negotiations with Nanyang in Shanghai during February and March of 1921, and the Chiens' published correspondence clearly indicates that Jeffress participated in the negotiations in question. Since I have found no evidence in unpublished correspondence to suggest that Jeffress participated in negotiations with the Chiens in 1922, I have proceeded on the assumption that the *Nanyang* documents on this subject were written in 1921, not 1922, and I have dated them as such in my footnotes.

Among Chien Chao-nan's allies on the Nanyang board of directors who favored a merger with BAT was Li Yuan-hung, former president of the Republic. See Thomas to Allen, March 27, 1920, Thomas Papers; and *Nanyang*, p. 444.

73. Chien Yü-chieh et al., to Chien Chao-nan, February 5 and 11, 1921, *Nanyang*, pp. 423-424 and 426-427; Chien Chao-nan (quoting from an earlier letter by Chien Yü-chieh) to Chien Yü-chieh, February 6, 1921, *Nanyang*, p. 425.

74. Chien Chao-nan and Ch'en Ping-ch'ien to Chien Yü-chieh et al., February 4 and 6, 1921, *Nanyang*, pp. 423-425.

75. Chien Yü-chieh et al., to Chien Chao-nan, February 10, 1921, *Nanyang*, pp. 426-427; Chien Chao-nan to Chien Yü-chieh and Chien Chao-nan to Chien Ying-fu, both February 11, 1921, *Nanyang*, p. 427.

76. Chien Chao-nan to Chien Yü-chieh, February 14, 1921, and Chien Yü-chieh to Chien Chao-nan, February 20, 1921, both in *Nanyang*, p. 427; Jeffress to Thomas, March 13, 1921, Thomas Papers.

77. Thomas to Jeffress, March 26 and August 21, 1921, Thomas Papers. As a token of their friendship, Liang made Thomas godfather to his daughter, and Thomas acted *in loco parentis* for her while she was in the United States. See Violet C. Liang to Thomas, October 2, 1922 and July 11, 1924, Thomas Papers.

78. *NCH*, April 13, 1918, p. 84; the quotation is from Thomas to Jeffress, September 1, 1921, Thomas Papers. On Liang's rise to prominence and power, see Stephen R. MacKinnon, "Liang Shih-i and the Communications Clique," *Journal of Asian Studies*, 29.3 (May 1970), 581-602.

79. Thomas to Liang Shih-i, August 18, 1921; and Thomas to Jeffress, September 1, 1921, Thomas Papers.

80. Thomas to Pettitt, June 8, 1921, Thomas Papers.

81. Pettitt to Thomas, December 17, 1921; Thomas to Pettitt, February 4, 1922, both in Thomas Papers. Pettitt noted that BAT's policy of all-out commercial warfare would be used not only to stop Nanyang but also to bar a newcomer from China, Liggett and Myers Tobacco Company, which was trying to gain a foothold in China at this time.

82. Fairley to Thomas, December 24, 1919, Thomas Papers.

83. Thomas to Rustard, March 23, 1921, Thomas Papers.

84. Kong Sook Wing (Chiang Shu-ying) to Morris, April 11, 1922, Thomas Papers.

85. Allen to Thomas, March 7, 1922, Thomas Papers. Chiang Shu-ying came to Thomas recommended by Thomas' friend, Liang Shih-i, and according to a mutual friend of Thomas and Chiang, H. C. Tan, Thomas had "quite a talk" with Chiang, but there is no conclusive evidence to show whether Thomas approved of Chiang's strategy. Liang to Thomas, March 22, 1922; Tan to Thomas, June 7, 1922, Thomas Papers.

86. Kan Chiu Nam (Chien Chao-nan) to Thomas, July 7, 1922; and Thomas to Morris, August 26 and October 10, 1922, Thomas Papers.

87. Pettitt to Thomas, April 18, 1921, Thomas Papers.

88. Ezra Vogel has summarized the stereotype of Canton as "the home of the small independent trader and craftsman" by contrast with Shanghai as "a center of large commercial and industrial enterprises" in *Canton under Communism* (Cambridge, Mass., 1969), p. 21. On this distinction between Cantonese and Shanghainese businessmen in contemporary Hong Kong, see Marjorie Topley,

"The Role of Savings and Wealth among Hong Kong Chinese," in Jarvie, ed., *Hong Kong: A Society in Transition*, pp. 202 and 221-222.

89. Minutes of the meetings of the Nanyang board of directors, March 17 and December 30, 1921, and January 25, 1922; interviews with Nanyang staff members, all in *Nanyang*, pp. 443, 445-446, and 753. It seems plausible to infer that Chien's son and Ch'en's daughter were married by 1919 from the fact that they had a six year old son in 1925. See Thomas to Herbsman, July 11, 1925, Thomas Papers.

90. Minutes of the meetings of the Nanyang board of directors, August 27, 1920, March 15, 1921, and January 31, 1923, *Nanyang*, p. 137. Chien Chao-nan held 26.9 percent of the stock, Chien K'ung-chao 26.9, Chien Yü-chieh 7.8, Chien Ying-fu 2.1, and Chien Chao-nan's two sons 2.2. Not counting the Chiens, 107 individuals held 11.9 percent of the stock, and small stockholders (owning less than 500 shares and generally less than 100 shares) held 27.5 percent.

91. Minutes of the meetings of the Nanyang board of directors, August 30, October 27, and December 21, 1922, December 17, 1923, June 28, 1924; Chien Ying-fu to Chien Yü-chieh, February 22, 1923; Chien Ying-fu to Lin Shih, July 27, 1923; Chien Ying-fu to T'ien Ju, November 14, 1923; Chien Ying-fu to Chien Ching-shan, November 22, 1923, all in *Nanyang*, pp. 492-495.

92. Minutes of the meeting of the Nanyang board of directors, December 21, 1922, *Nanyang*, p. 446; Thomas to MacNaughten, April 20, 1921 and Thomas to Pettit, October 27, 1921, Thomas Papers; FO 371/9217, Rose, "Memorandum," March 19, 1923; Nanyang's annual report for 1922, *Nanyang*, p. 221. When provincial governments resumed taxation in the mid-1920s, Nanyang's board of directors considered working jointly with provincial cigarette monopolies (*ho-tso shih-yeh*) and resolved to discuss this possibility with the Kiangsu provincial government, but apparently no such arrangement was every made. See minutes of the meetings of the Nanyang board of directors, August 9, 1924, and June 2, 1927, *Nanyang*, p. 457.

93. Appendix, table 7 (which shows Nanyang's capital in U.S. dollars rather than *yuan*); minutes of the meetings of the Nanyang board of directors, January 25 and December 21, 1922, *Nanyang*, p. 446; interview with Nanyang employees, *Nanyang*, p. 445; contract between the Chiens and Ch'en Ping-ch'ien, February 2, 1923, *Nanyang*, p. 447. The Japanese historian, Shibaike Yasuo, has suggested that in granting Ch'en this power the Chiens took a step "about as undesirable [for them] as being forced to open the gates and hand over the fortress." But I have seen no evidence that they suffered from the consequences. Cf. Shibaike, "1930 nendai no keizai kikika ni okeru Chūgoku minzoku shihon kigyo no jittai," *Shodai ronshu*, 24.1-3 (June 1972), 54-56. Ch'en remained with Nanyang until 1929 when he apparently resigned in protest against the Chiens' decision to close one of their Shanghai factories. See minutes of the meetings of the Nanyang board of directors, April 4, May 28, and July 8, 1929, *Nanyang*, pp. 449-450.

94. Hu Kuang-piao, pp. 227 and 91; interview with former Nanyang staff members, April 22, 1958, *Nanyang*, p. 322; Allen to Thomas, February 18, 1920, Thomas Papers. Hu summarizes these ideas in a chapter from whose title I have taken this quotation, pp. 105-110. He also mentions (on p. 91) that Ch'en Ch'i-

chün was part of the group of Chinese exposed to these ideas in the late 1910s.

95. Frederick Winslow Taylor, *The Principles of Scientific Management* (New York, 1911), p. 11. Published in New York in 1911, this book was translated into Chinese in 1916 by the industrialist Mu Hsiang-yueh. A copy of his translation may be found in the Frederick Winslow Taylor Collection at the Stevens Institute of Technology in Hoboken, New Jersey, along with other translations of this highly influential book into Dutch, Esperanto, French, German, Hungarian, Italian, Japanese, Lettish, Russian, Spanish, and Swedish. See Elizabeth Gardner Hayward, *A Classified Guide to the Frederick Winslow Taylor Collection* (Hoboken, 1951), pp. 4-6.

96. Taylor, pp. 10 and 25. Cf. Hu Kuang-piao, pp. 105-110; and Samuel Haber, *Efficiency and Uplift: Scientific Management in the Progressive Era, 1890-1920* (Chicago, 1964), pp. ix-x.

97. *Kung-yeh tsa-chih*, 10.11 (November 1922), *Nanyang*, pp. 145-146.

98. Minutes of the meetings of the Nanyang board of directors, March 15, 1921, and March 15, 1922, *Nanyang*, pp. 144 and 146; *Nanyang*, p. 486; Thomas to Jeffress, May 15, 1920, Thomas Papers. The quotation is from *Min-kuo jih-pao*, September 12, 1919, *Nanyang*, p. 56.

99. Interviews with former Nanyang staff members and workers, January 21 and April 22, 1958, *Nanyang*, pp. 149 and 322-323; *Min-kuo jih-pao*, September 12, 1919, *Nanyang*, p. 56. The investment rose from ¥589,000 in 1920 to ¥1,276,000 in 1921 to ¥1,732,000 in 1922. See *Nanyang*, pp. 144-145. On Nanyang's earlier preference for Japanese machinery, see *NCH*, May 13, 1916, p. 354; and interviews with former staff and workers from Nanyang's Shanghai factory, January 21, 1958, *Nanyang*, p. 149.

100. Interviews with former Nanyang staff members and workers, January 21, 1958, *Nanyang*, p. 149. I use the word "appears" here because some Nanyang managers at the time said that American-educated managers (like Ch'en) were not popular with workers, but such allegations were contradicted in the interviews cited above and were, therefore, presumably false charges made by enemies of the American-educated members of Nanyang's managerial staff. Cf. *Min-kuo jih-pao*, November 9, 1922, *Nanyang*, p. 324.

101. Chien Yü-chieh to Hsu Tsou-yun, September 23, 1921; and two posters used in Nanyang's Hong Kong factory, September 1921, *Nanyang*, pp. 326-327; *Min-kuo jih-pao*, November 8 and 9, 1922, *Nanyang*, pp. 323-325; *South China Morning Post*, September 1, 1923. For the agreement between management and labor, see *Nanyang*, pp. 325-326. The militant labor leaders at BAT aspired to extend their organization to Nanyang's Shanghai factory but were unable to do so. See *Chan-tou*, p. 27n1.

102. More than 75 percent of the strikes in Shanghai by businesses between 1918 and 1933 were against foreign companies. See Shang-hai she-hui chü, comp., *Chin shih-wu-nien lai Shang-hai chih pa-kung t'ing-yeh* (Shanghai, 1933), pp. 41-42. The number of Nanyang's workers was noted in a report by Chien Chao-nan to Nanyang's board of directors, March 15, 1921, a copy of which was given to me by Y. L. Kan.

103. Interviews with Wu Chiu-chang, Wu Ping-hou, Ch'en Wan-ts'ai, Wu San-mei, Sung A-kuei, Wang Yung-sheng, Shih Hsiao-feng, Lu San-yuan, and

other former Nanyang staff members and workers from the Shanghai factory, 1957-58, *Nanyang*, pp. 293-294 and 304-317. For details on wages and conditions, see Pei-p'ing she-hui tiao-ch'a, ed., *Ti-i tse Chung-kuo lao-tung nien-chien* (Peiping, 1928), p. 25, *Nanyang*, p. 299; *Min-kuo jih-pao*, September 12, 1919, *Nanyang*, p. 56; *Kung shang pan-yueh-k'an*, 1.1 (January 1, 1929), *Nanyang*, pp. 299-300 and 311; Chu Pang-hsing et al., *Shang-hai ch'an-yeh yü Shang-hai chih-kung* (Shanghai, 1939), pp. 517-518 and 527-529, *Nanyang*, pp. 310-311 and 294-295; *Nanyang*, pp. 21-22; "The Nanyang Bros. Tobacco Co.," *Chinese Economic Journal*, 3.5 (November 1928), 989.

104. Yü Ta-fu, "Chün-feng ch'en-tsui ti wan-shang," in T'ung Wen, ed., *Chung-kuo hsien-tai tuan-pien hsiao-shuo hsuan* (Hong Kong, 1963), I, 187-201, especially 192-193. Regulations in Nanyang's plants tend to confirm that workers were at the mercy of their supervisors. See posters dated October 4, 1915, and June 16, 1916, listing factory regulations in *Nanyang*, pp. 23-24. Undated regulations (which, according to the *Nanyang* compilers, applied before 1931) empowered the employer to fire a worker without notice or compensation and terminate a worker automatically if he or she failed to appear on the job for three consecutive days. See regulations 9, 40, 44, 46, and 47 in *Nanyang*, pp. 290 and 296-297.

105. "Labour Conditions and Labour Regulation in China," p. 1018; *Chantou*, pp. 13-17.

106. On Nanyang's improvements of conditions, see Chan Ming Kou, pp. 115-116 and 129.

107. H. C. Tan to Thomas, August 20, 1920; Pettitt to Thomas, December 17, 1921; Thomas to Pettitt, February 4, 1922, Thomas Papers. Pettitt was disappointed that Wu had joined Nanyang and thus failed to remain "loyal" after being well treated by BAT, but Thomas felt that Wu regretted leaving BAT and would not betray it. Thomas assured Pettitt that "Wu can still be handled, if necessary, without much trouble."

108. Nanyang's annual report for 1921, *Nanyang*, p. 221; Thomas to Jeffress, March 26, 1921, Thomas Papers; *NCH*, June 18, 1921, p. 833.

109. Cobbs to Thomas, November 17, 1920, Thomas Papers.

110. Fairley to Thomas, December 24, 1919, Thomas Papers.

111. Nanyang's annual report for 1921, *Nanyang*, p. 221.

112. Cobbs to Thomas, November 17, 1920, Thomas Papers.

113. Minutes of the meeting of the Nanyang board of directors, April 30, 1918, *Nanyang*, p. 211; Chien Ying-fu to Chien Yü-chieh, February 22, 1923, *Nanyang*, pp. 211-212. On P'an, see Rhoads, pp. 109, 182, 184, 197, 211.

114. Sun Chia-chi, pp. 61-62 and 71.

115. Chien Yü-chieh to Nanyang's branch in Canton, August 4, 1922, *Nanyang*, p. 248.

116. Y. P. Wang, *The Rise of the Native Press in China* (New York, 1924), p. 45. Another study of advertising in the 1920s similarly ranked Nanyang second only to BAT among advertisers of all products in China. See C. A. Bacon, "Advertising in China," *Chinese Economic Journal*, 5.3 (September 1929), 756-758 and 764-766.

117. *North China Standard*, December 4, 1920; *NCH*, July 26, 1924, p. 130; minutes of the meeting of the Nanyang board of directors, October 20, 1920,

Nanyang, p. 249. I am grateful to David Strand for calling my attention to the first of these items.

118. Kuo-huo shih-yeh ch'u-pan she, ed., *Chung-kuo kuo-huo*, p. 8; *NCH*, August 13, 1921, p. 476. Bōrisat Yāsūp Nānyāng Čhamkat, Bangkok, *Thawāi phraphonchai Phrabāt Somdet Phra Čhao Yūhūa ratchakān thi 7* (Bangkok, 1925), pp. 7-8. I wish to thank David Wyatt for recommending this last item and Nantaga Kanjanapan for translating selected passages in it from Thai into English.

119. On Shantung, see the *Chinese Economic Monthly*, II, 6 (March 1925), 32-33; on Hupeh, ibid., 9 (June 1925), 6; on Anhwei, Chang Yu-i, p. 203. The third largest tobacco purchaser was the Japanese Imperial Tobacco Monopoly. Its representatives were active around Mukden in Manchuria and to a lesser extent in Shantung and Hupeh. See China, Inspectorate General of Customs, *Returns of Trade, 1919*, p. 88; *Chinese Economic Monthly*, 97 (January 31, 1925), 2; 119 (June 2, 1923), 2; 184 (August 30, 1924), 3; 2.9 (June 1925), 6; 173 (July 31, 1926), 48; Chang Yu-i, pp. 503-504. A few Chinese companies besides Nanyang also entered the leaf market but not on a large scale.

120. *Chinese Economic Journal*, 10.1 (January 1932), 41; Chang Yu-i, pp. 511-512; *The Weekly Review*, November 1922, p. 160.

121. Nanyang financial records compiled by the Li Hsin accountants, *Nanyang*, p. 203. The evidence on BAT's use of similar practices is contradictory. Two sources, both based on Chinese eyewitness accounts from the mid-1930s, reach opposite conclusions, one saying that BAT used "light" scales and the other maintaining that BAT "gives no better treatment to the peasants than any other collecting agency . . . [but] this concern strives for strict standardization without falsification of weights, and will not allow any argument or bargaining [whereas the other tobacco companies] tolerate negotiation but also resort to deceit in matters of weight and classification." For the evidence that BAT used "light" scales, see Wang Hsi, p. 92; for the evidence that it did not, see Institute of Pacific Relations, comp., *Agrarian China*, p. 174.

122. Chen Han Seng, p. 43; Huang I-feng, "Kuan-yü chiu Chung-kuo," p. 103; Thomas to Jeffress, June 26, 1923, Thomas Papers.

123. Tilley, pp. 307-308, 331-334, and 426-427. I am indebted to Nannie May Tilley not only for her fine book but also for permitting me to read her research notes on Gravely.

124. Minutes of the meetings of the Nanyang board of directors, April 3, 1924, January 3 and October 13, 1925, March 13, 1926, *Nanyang*, pp. 204-208; Thomas to Jeffress, December 1, 1919 and May 15, 1920, Thomas Papers; Tilley, pp. 283 and 333. As early as 1919, Thomas suspected that J. P. Taylor would buy American tobacco grown in China and export it to the West, but I have found no evidence that such a scheme was ever carried out. When Yuille visited Shanghai to confer with the Chiens in 1921, Thomas remarked to him about the keenness of the competition for Nanyang's business among American leaf suppliers. Soon thereafter, Pemberton and Penn, Inc., of Danville, Virginia, another leaf supplier, also approached Nanyang. Thomas to Jeffress, December 2, 1919; Thomas to Yuille, April 15, 1921; Pemberton and Penn to Thomas, July 9, 1921, all in Thomas Papers.

125. The annual exports of this commodity to the United Kingdom and

China far exceeded exports of it to all other countries of the world combined. See U.S. Department of Commerce, Bureau of the Census, *Stocks of Tobacco Leaf, 1923*, Bulletin no. 155 (Washington, D.C., 1924), p. 36; and U.S. Department of Agricultural Economics, *First Annual Report on Tobacco Statistics*, Statistical Bulletin no. 58 (Washington, D.C., 1937), p. 109.

126. For an excellent analysis of the typical stages of development from the small, owner-operated business to the professionally managed big business, see Alfred D. Chandler, Jr., *Strategy and Structure* (Cambridge, Mass., 1962).

127. For celebrations of Duke's rise from rags to riches, see Jenkins; and Watson S. Rankin, *James Buchanan Duke (1865-1925): A Great Pattern of Hard Work, Wisdom and Benevolence* (New York, 1952). For a lively journalistic characterization, see Winkler. And for a critical assessment which confirms the authenticity of Duke's Horatio Alger story, see Porter.

For celebrations of Chien Chao-nan's career, see Liu Huo-kung, "Chien Chao-nan," in Chu Hsiu-hsia, ed., *Hua-ch'iao ming-jen chuan* (Taipei, 1955), I, 127-133; A. Sy-hung Lee, pp. 158-159; Shu-lun Pan, *The Trade of the United States with China* (New York, 1924), p. *v* and ch. 12 (a book dedicated to Chien Chao-nan out of gratitude for the Nanyang scholarship program which put Pan through Columbia Graduate School); *Who's Who in China* (1926), p. 83; and the six volumes of eulogies cited in n. 137 below.

On Chinese mobility and myths of mobility, see Ping-ti Ho, *The Ladder of Success in Imperial China* (New York, 1962). The social mobility of merchants in Republican China has not been closely analyzed, but scattered examples suggest that they rose more readily following the revolution of 1911 (as the Chiens did) than in imperial times. For examples, see the biographical sketches of the Soong family in Boorman and Howard, III, 137-153; and of various Yunnanese merchants in Yung-teh Chow, *Social Mobility in China* (New York, 1966), pp. 157-158, 190-194, 220-225, 230-231, 236-238.

128. *NCH*, September 15, 1923, p. 772.

129. Fitzgerald, pp. 145-146; Reavis Cox, *Competition in the American Tobacco Industry, 1911-1912* (New York, 1933), p. 73.

130. *NCH*, September 15, 1923, p. 772.

131. The British-American Tobacco Company, *The Record in China*, p. 10; Hutchison, p. 280. Photographs of BAT's factory at Tientsin appear in the *B.A.T. Bulletin*, 17.73 (May 1926), 30-31.

132. Hutchison, p. 280. See also U.S. Department of Commerce, *Tobacco Markets and Conditions Abroad*, 26 (January 5, 1926), 5.

133. All quotations and other material in this paragraph are from Hutchison, pp. 280-281.

134. Ibid., p. 281.

135. See Appendix, table 3.

136. Appendix, table 6; Chou Hsiu-luan, p. 58; Chesneaux, *The Chinese Labor Movement*, pp. 215-217; Feuerwerker, *The Chinese Economy, 1912-1949*, pp. 21-22; *Min-kuo jih-pao*, September 27, 1924, *Nanyang*, p. 337; *NCH*, May 9, 1925, p. 240.

137. These and many other eulogies written in commemoration of Chien Chao-nan appear in a compilation edited by Chu Hsiao-tsang (Chien Yao-teng),

Chien chün Chao-nan ai-wan lu (n.p., n.d.), 6 vols.

138. Untitled clippings from English language newspapers published in China, enclosed in Kingsbury to Thomas, December 21, 1923, Thomas Papers.

139. Minutes of the meeting of the Nanyang board of directors, July 30, 1924, *Nanyang*, p. 281. The higher salaries had been fixed at the board meeting of September 1, 1920; see minutes in *Nanyang*, p. 281.

140. Quoted from a presentation given by Chien Jih-lin, *Shang-hai hsin-wen jih-pao*, November 28, 1955, *Nanyang*, pp. 461-462; Chung-wai ch'u-pan she, *Chung-kuo hao-men* (April 1949), p. 34, *Nanyang*, p. 461.

141. Minutes of the meetings of the Nanyang board of directors, October 13, 1925, and March 13, April 25, and December 28, 1926, *Nanyang*, pp. 205-207 and 210; Nanyang's annual reports, 1925-1930, *Nanyang*, pp. 222-223; *NCH*, February 4, 1930, p. 180; and February 11, 1930, p. 218. The full text of the contract, signed by Gravely and Chien Yü-chieh on March 22, 1916, is in *Nanyang*, pp. 208-210.

142. Minutes of the meetings of the Nanyang board of directors, July 26, September 8, and October 15, 1926, *Nanyang*, pp. 485-486, 463 and 468; personal reminiscences of members of the Nanyang staff, *Nanyang*, p. 462; interview with Wu Ting-chu, assistant manager of Nanyang, May 24, 1957, *Nanyang*, pp. 467-468.

143. Chien Chao-nan to Chien Yü-chieh, January 28, 1919, *Nanyang*, p. 422; Lou T'ing-chen, "Tsai Nan-yang 25-nien ti hui-i" (March 1957), *Nanyang*, p. 462; and interview with the former comptroller of Nanyang, April 24, 1957, *Nanyang*, p. 261.

144. P'an Hsu-lun of the Li-hsin accountants to Hsu Tsou-yun, based on an audit conducted in February 1929, *Nanyang*, pp. 462-463.

145. Hsu Yung-tso and P'an Hsu-lun of the Li-hsin accountants to the Shanghai Municipal Government, January 1929, *Nanyang*, pp. 157-158. For an example of embezzlement by one of Nanyang's regional distributors, see minutes of the meeting of the Nanyang board of directors, October 13, 1928, *Nanyang*, pp. 463-465.

7. Business and Politics

1. For stimulating papers and discussions which have shaped my understanding of the relationship between business and politics, I am grateful to participants in a conference held at Cornell University, Ithaca, New York, in October 1975. For a summary of the proceedings, see my report, "Business and Politics in Republican China," *Chinese Republican Studies Newsletter*, 1.2 (February 1976), 13-15.

2. See Appendix, table 6.

3. C. Martin Wilbur, *Sun Yat-sen: Frustrated Patriot* (New York, 1976), p. 52. On the continuing rivalry between moderate and militant leaders of the labor movement, see Chesneaux, *The Chinese Labor Movement*, pp. 249-253.

4. *Min-kuo jih-pao*, September 11 and 13, 1924, *Nanyang*, pp. 329-330; *Chung-kuo kung-jen*, 2 (November 1924), 57-58, *Nanyang*, p. 328.

5. *Min-kuo jih-pao*, September 11, 13, and 14, 1924, *Nanyang*, pp. 329-

332; *Chung-kuo kung-jen,* 2 (November 1924), 57-58, *Nanyang,* p. 328. Between 1922 and 1924, the club appears to have become increasingly disenchanted with the Nanyang managers who initially dominated its leadership. According to one source, in February 1924 (seven months before the strike) most of these were replaced by leaders drawn from the ranks of the workers. See *Chung-kuo kung-jen,* 2 (November 1924), 55, *Nanyang,* p. 322.

6. *Min-kuo jih-pao,* September 13, 1924, *Nanyang,* p. 332; interviews with former Nanyang workers Wu Ping-hou and Liu Lin, April 16 and 17, 1958, *Nanyang,* p. 333.

7. Minutes of the meeting of the Nanyang board of directors, September 23, 1924, *Nanyang,* p. 332; *Min-kuo jih-pao,* September 21 and 24, 1924, *Nanyang,* pp. 334 and 345; *Hsiang-tao chou-pao,* 85 (October 1, 1924), 695, *Nanyang,* p. 339. The Chiens' eagerness for arbitration belies their contention that they were prepared to carry on business as usual within two weeks after the beginning of the strike, but the exact number of workers who were back on the job at this time is unknown. Estimates vary from 1,000 (by a source sympathetic to the strike) to 5,000 (by one unsympathetic to it). See *Min-kuo jih-pao,* September 26 and 27, 1924, *Nanyang,* pp. 338-339; *Hsiang-tao chou-pao,* 85 (October 1, 1924), 695, *Nanyang,* p. 339; minutes of the meeting of the Nanyang board of directors, September 23, 1924, *Nanyang,* p. 337.

8. James P. Harrison discusses the significance of this journal in *The Long March to Power* (New York, 1972), pp. 38 and 65.

9. *Hsiang-tao chou-pao,* 83 and 84 (September 17 and 24, 1924), 672-673 and 685, *Nanyang,* pp. 335-336.

10. Ibid., 83-85 (September 17 and 24 and October 1, 1924), 672-673, 685-686, and 695, *Nanyang,* pp. 340-345. On the treatment of strikers by police and the courts, see also *Min-kuo jih-pao,* September 13, 16, and 24, 1924, *Nanyang,* pp. 333-334; interviews with former Nanyang workers Wu Ping-hou and Liu Lin, *Nanyang,* p. 333. On Yeh Ch'u-tsang, see Boorman and Howard, IV, 27-29.

11. *Hsiang-tao chou-pao,* 84 (September 24, 1924), 685-686, *Nanyang,* p. 344; Chesneaux, *The Chinese Labor Movement,* pp. 245-250 and 503n115; Chan Ming Kou, p. 56.

12. *Hsiang-tao chou-pao,* 96 (December 24, 1924), 807, *Nanyang,* pp. 341-342; Li Ming, "Yang Yin t'ung-chih chuan-lueh," *Lieh-shih chuan* (August 1949), p. 251, *Nanyang,* p. 329; interviews with Wu Ping-hou and Liu Lin, *Nanyang,* p. 346; minutes of the meetings of the Nanyang board of directors, January 15, 1925, *Nanyang,* p. 347; NCH, January 27, 1924, p. 531; January 10, 1925, p. 51. The provisions of the proposal were conveyed in a letter from Liao Chung-k'ai and Yang Yin to Nanyang which was quoted in the minutes of the meeting of the Nanyang board of directors, January 15, 1925, *Nanyang,* p. 346.

13. Minutes of the meeting of the Nanyang board of directors, January 15, 1925, *Nanyang,* p. 347; interviews with Wu Chiu-chang, Ma Kuei-sheng, and Wu Ping-hou, April 1958, *Nanyang,* pp. 306-308. For the Federation's endorsement of the strike, see *Min-kuo jih-pao,* September 13, 24, 26, 27, and 28, 1924, *Nanyang,* pp. 337-338 and 345; on the Federation's withdrawal of its endorsement, see *Hsiang-tao chou-pao,* 96 (December 24, 1924), 807, *Nanyang,* p. 342. Chesneaux characterizes the Federation in *The Chinese Labor Movement,* pp. 223-227.

14. *Chan-tou*, pp. 43-46; Chesneaux, *The Chinese Labor Movement*, pp. 262-264, 269, and 506n11. Historians have noted the significance of the May Thirtieth movement for certain segments of China's urban population, but a full-scale study is needed. On the role of labor, see Chesneaux, *The Chinese Labor Movement*, ch. 11. Concerning the movement's impact on writers, see Leo Lee, *The Romantic Generation of Modern Chinese Writers* (Cambridge, Mass., 1973), p. 177.

15. On the role of this union in the May Thirtieth movement, see Chesneaux, *The Chinese Labor Movement*, ch. 11; Liang Hsiao-ming, *Wu-sa yun-tung* (Peking, 1956), pp. 17, 41-42 and passim; Hsu Shih-hua and Ch'iang Chung-hua, *Wu-sa yun-tung* (Peking, 1956), pp. 25-26 and 56.

16. *Chan-tou*, pp. 46-49 and 55-60. According to BAT, Hsu received $30,000 from the Soviet diplomat Leo M. Karakhan to finance the strike at this time. FO 228/3529, Porter to Palairet, June 4, 1925; *Chan-tou*, pp. 43-46; Chesneaux, *The Chinese Labor Movement*, pp. 262-264, 323, 401, 506nn11 and 33; *NCH*, June 6, 1925, p. 428. Shang-hai-shih she-hui chü, *Chin shih-wu-nien lai Shang-hai pa-kung t'ing-yeh* (Shanghai, 1933), appendix 1, p. 22.

17. The unions raised ¥350,000, and the General Chamber of Commerce (through which Nanyang made its contribution) added ¥470,000 for a total of ¥820,000. See Chesneaux, *The Chinese Labor Movement*, pp. 266 and 508n43; and FO 405/F4325/194/10, Brett to Chamberlain, July 24, 1925, enclosed in Department of Overseas Trade to Foreign Office, September 1, 1925. I am indebted to Nicholas Clifford for the latter reference.

18. Shang-hai-shih she-hui chü, *Chin shih-wu-nien lai Shang-hai pa-kung t'ing-yeh*, appendix 1; Ta Chen, "Analysis of Strikes in China, from 1918 to 1926," *Chinese Economic Journal*, 1.10 (December 1927), 1088; *Chan-tou*, p. 61; *NCH*, August 29, 1925, p. 249; October 10, 1925, p. 57; October 17, 1925, p. 107.

19. Chan Ming Kou, p. 338; NA 693.1112/104, Cunningham to State, July 20, 1925. British consular reports from Yentai, Shanghai, Kiukiang, Ningpo, Amoy, and Swatow, indicated that Nanyang was making every effort to sabotage BAT in these cities as well. FO 228/3529, Palairet's summary of consular reports, September 1, 1925; and Jamieson to Palairet, September 11, 1925. See also *NCH*, August 22 and 29, 1925, pp. 221 and 268.

20. FO 228/3529, Wallace to Kerr, August 3, 1925; King to Palairet, June 27, 1925; Hunt to Palairet, August 14, 1925; Palairet's summary of consular reports, September 1, 1925. In support of the assertion "that the Nanyang Tobacco Co. are more responsible than any other single agency in the promotion of the Boycott movement," BAT agents showed proof that information on brands, trademarks, etc., contained in the contraband lists, pamphlets, and other anti-BAT propaganda was very specific and extensive. Such information, BAT persuasively argued, "could only be gotten from the Trademark Journals or from our competitors' files." See FO 228/3529, BAT to British Consulate General, July 29, 1925.

21. Minutes of the meeting of the Nanyang board of directors, February 2, 1925, *Nanyang*, pp. 248-249.

22. FO 228/3529, Palairet's summary of consular reports, September 1, 1925; Jamieson to Palairet, September 11, 1925.

23. *Tien-ying ch'un-ch'iu*, 11 (May 10, 1925); *Tien-ying tsa-chih*, 13 (September 1925); *Yin-teng*, 1 (April 24, 1926); *Ming-hsing t'e-k'an*, 3 (July 27, 1925), all in Ch'eng Chi-hua, I, 124 and 127; Kinglu S. Chen, "Chinese Papers as Advertising Mediums," *China Weekly Review* (September 1, 1928), p. 15; *Chinese Economic Monthly*, 11 (August 1924), 3; Leyda, p. 52; *Chan-tou*, p. 51.

24. FO 228/3529, BAT to the British Consulate General, July 29, 1925; Liang Hsiao-ming, *Wu-sa yun-tung*, pp. 32 and 36-37. On Yü's investments in shipping, see Hao, pp. 101, 124, and 126, table 15. Communists at the time and recently Jean Chesneaux have analyzed Yü's behavior in terms of his bourgeois class interests. Chesneaux has berated Yü as typical of the "tepid revolutionaries" who were willing to compromise too readily with Westerners at the expense of workers, students, and shopkeepers. On June 20, 1925, at the height of the May Thirtieth movement, Chesneaux has noted, the Communist-led Shanghai General Union accused Yü of accepting bribes from Western interests. I have insufficient evidence to demonstrate what motivated Yü to make the decisions I have described, but from their consequences and from the bitter BAT complaints about them, it seems that any efforts BAT might have made to bribe or woo Yü during the movement were unsuccessful. Cf. Chesneaux, *The Chinese Labor Movement*, pp. 267 and 508n54.

25. *Chan-tou*, pp. 51-52; *NCH*, August 8, 1925, p. 123; FO 228/3529, King to Palairet, June 27, 1925. For evidence of a similar incident in Hangchow, see *NCH*, August 15, 1925, p. 160; September 19, 1925, p. 381; and September 26, 1925, p. 426.

26. FO 228/3529, Walker to Kerr, August 3, 1925; BAT to British Consulate General, July 29, 1925; King of Palairet, June 27, 1925; Davidson to Palairet, August 11 and September 18, 1925; Jamieson to Palairet, September 11, 1925; Palairet's summary of consular reports, September 1, 1925; *Chan-tou*, pp. 49-51.

27. NA 693.1112/104, Cunningham to State, July 20, 1925; NA 693.1112/84, Bowling of BAT, affidavit, June 13, 1925; Bowling to Kellogg, June 15, 1925; Hackworck memo of conversation with Bowling, June 15, 1925.

28. NA 693.1112/104, Cunningham to State, July 20, 1925; NA 693.1112/112, MacMurray to Kellogg, September 22, 1925, and the following enclosures: Gauss to MacMurray, September 1, 1925; Cunningham to MacMurray, July 14, 1925; undated BAT advertisement from the *Chinese Peking and Tientsin Times*.

29. FO 228/3529, BAT to the British Consulate General, July 29, 1925; FO 228/3529, Walker to Kerr, August 3, 1925; Palairet's summary of consular reports, September 1, 1925; NA 693.1112/95, Stanton to Mayer, June 19, 1925; U.S. Department of Commerce, *Tobacco Markets and Conditions Abroad*, 50 (June 22, 1926), 8. The cities covered in the reports included: Tientsin, Anyang, Yentai, Tsingtao, and Tsinan in North China; Kalgan and Kueihua in Northwest China; Shanghai, Ningpo, Hangchow, Chinkiang, Nanking, and Wuhu in the Lower Yangtze region; Chiukiang, Hankow, Nanchang, Ichang, and Shasi in the Middle Yangtze; Chungking in the Upper Yangtze; Foochow in the Southeast; Swatow and Canton in the South; Kunming in the Southwest.

30. FO 228/3529, King to Palairet, June 27, 1925; Jamieson to Palairet, September 11, 1925; Archer to Palairet, August 8, 1925; BAT to British Consulate General, July 29, 1925; FO 228/3520, summary of consular reports on the

anti-British boycott as of February 24, 1926; Hewlett to Macleay, March 26, 1926; Palairet's summary of consular reports, September 1, 1925; Huang I-feng, "Wu-sa yun-tung chung ti ta tzu-ch'an chieh-chi," *Li-shih yen-chiu*, 3 (1965), 19. BAT's division manager in Changsha estimated that total losses were still greater because he had originally anticipated a 10 percent increment in sales in 1925. FO 371/12417, F1216, Joyner of BAT to Grant-Jones, October 7, 1926.

31. Appendix, table 5; FO 228/3529, Palairet's summary of consular reports, September 1, 1925; Hunt to Palairet, August 14, 1925; Archer to Palairet, August 8, 1925; Huang, "Wu-sa yun-tung," p. 19; minutes of the meeting of the Nanyang board of directors, June 28, 1925, *Nanyang*, pp. 146-147. Cf. Appendix, table 6.

32. *South China Morning Post*, August 27, 1926, p. 6. Former Nanyang workers later recalled that the May Thirtieth movement brought a rise in Nanyang's work force to 7,000 or 8,000 and raises in wages (which had been denied during the strike in the previous year). Interviews with Ma Kuei-sheng and Wu San-mei, April 1958, *Nanyang*, pp. 307 and 313-314.

33. Interview with a former Nanyang employee, *Nanyang*, p. 148; Chesneaux, *The Chinese Labor Movement*, pp. 267-268 and 508n56.

34. *Min-kuo jih-pao*, January 5, 1929; and personal recollections of two former Nanyang staff members, April 1958; minutes of the meeting of the Nanyang board of directors, June 28, 1925, all in *Nanyang*, pp. 147-148.

35. Minutes of the meeting of the Nanyang board of directors, July 29, 1925, *Nanyang*, p. 147; Nanyang's annual report for 1925, *Nanyang*, pp. 222-223. Cf. Wiens, pp. 22-23.

36. Minutes of the meeting of the Nanyang board of directors, August 20, 1925, *Nanyang*, pp. 147-148. Between 1925 and 1927 the Chiens invested over ¥700,000 to acquire a site, erect a four-story building, and install machinery for a new factory in Hankow, but floods and warfare inhibited construction, and it did not begin to operate until 1934. See minutes of the meetings of the Nanyang board of directors, August 20, 1925, January 18, 1926, and July 4, 1927, *Nanyang*, pp. 146-149; and note 89 below.

37. *NCH*, September 26, 1925, p. 422. On the settlement of the strike against BAT in Shanghai, see *Chan-tou*, pp. 60-61 and 64; Lowe, p. 67; Ta Chen, *The Labor Movement in China* (Honolulu, 1927), p. 7; *NCH*, October 10, 1925, p. 57; October 17, 1925, p. 107.

38. Minutes of the meetings of the Nanyang board of directors, December 30, 1925, and April 25, 1926, *Nanyang*, pp. 148-149, 206, and 489-490; Nanyang's annual reports for 1925 and 1926, *Nanyang*, pp. 222-223; Huai Shu, p. 105. For a hyperbolic but revealing characterization of BAT's distributing system and the carriers with which it worked in the mid-1920s (" . . . rail, steamer, junk, sampan, mule, camel, donkey, wheelbarrow, coolie, motor—in fact by all ways and means, except, as yet, an 'air route.' "), see *B.A.T. Bulletin*, 16.69 (January 1926), 268-270.

39. Nanyang's annual report for 1925, *Nanyang*, p. 222. See Appendix, table 6.

40. FO 371/11632, F4275, Rose to Mounsey, October 11, 1926; Sanford, p. 40. Cf. David A. Wilson, "Principles and Profits: Standard Oil Responds to Chi-

nese Nationalism, 1925-1927," unpublished paper, 1976.

41. On the Merchant Volunteer uprising, see Hsiang-kang Hua-tzu jih-pao, comp., *Kuang-tung k'ou-chieh ch'ao* (Hong Kong, 1924). On the events of March 1926, see Harold R. Isaacs, *The Tragedy of the Chinese Revolution*, 2nd rev. ed. (Stanford, 1961), pp. 89-110; Tien-wei Wu, "Chiang Kai-shek's March Twentieth Coup d'Etat of 1926," *Journal of Asian Studies*, 27 (May 1968), 585-602.

42. *South China Morning Post*, December 19, 1925, p. 9; August 26, 1926, p. 10. Cf. Chan Ming Kou, pp. 348-349; Chesneaux, *The Chinese Labor Movement*, pp. 349 and 529n31.

43. The role of BAT workers (and Nanyang workers as well) in these three "righteous risings" (*ch'i-i*) against warlords in Shanghai is described on the basis of workers' recollections in *Chan-tou*, pp. 68-86, and in interviews with three former Nanyang workers, *Nanyang*, pp. 351-353. BAT managers' complaints about the strikes, confiscation of cigarettes, and harassment of foreigners in Shanghai and Yangtze River ports may be found in FO 371/12478, F3905, Morris to BAT, April 15, 1927; FO 371/12482, F5074, Lampson to FO, May 28, 1927; FO 371/12469, F8319, Lampson to FO, August 29, 1927; FO 371/12507, F6832, Prideaux-Brune to Lampson, June 22, 1927. On bargaining between BAT and the unions with the Kuomintang present for the negotiations, see FO 371/11661, F5184, Cousins and Emery to BAT, November 27, 1926; F5234, Goffe to FO, December 3, 1926; FO 371/11662, F5276 and F5315, Goffe to FO, December 5 and 6, 1926; FO 371/12416, F8, Goffe to FO, January 1, 1927; FO 371/12407, F6568, Rose to Lampson, June 22, 1927; FO 371/12416, Goffe to FO; F7423, F8010, F8464, Lampson to FO, September 8, October 12, and November 1, 1927; Chesneaux, *The Chinese Labor Movement*, pp. 327 and 522n48.

44. Chen Han Seng, p. 28; Ming Chieh, pp. 1-2. BAT lost an estimated 15 million pounds of tobacco in the episode at Hsuchang. (*NCH*, March 6, 1926, p. 152.) On the sharp decline of total tobacco production at this time, see Appendix, table 8.

45. Diary of Julean H. Arnold, April 15, 1927, Arnold Papers; FO 228/3529, Coates to Macleay, February 10, 1926; FO 228/3530, summary of consular reports as of February 15, 1926; letters published in *Shih-shih kung-pao*, cited in FO 228/3530, Dandley to Macleay, October 6, 1926; FO 371/12417, F1216, correspondence between Lo Cheng of the Kuomintang and Grant-Jones, October 23 and 25 and November 3 and 8, 1926; Joyner to Grant-Jones, October 7 and November 8, 1926; FO 371/12416, F8464, Lampson to FO, November 6, 1927; FO 371/12482, Lampson to FO, May 28, 1927; FO 371/12469, F8319, Lampson to FO, August 29, 1927; FO 371/12514, F9244, Bailey to Cunliffe-Owen, December 14, 1927; FO 371/13226, file 1475 on correspondence between BAT and FO, March-April, 1928; FO 371/13227, F5859, Fitzmaurice to Lampson, August 21, 1928.

46. FO 371/12416, F7423, Lampson to FO, September 8, 1927; *NCH*, December 17, 1927, p. 490; December 31, 1927, p. 567; Chen Han Seng, p. 44; Ming Chieh, pp. 1-2. Though Nanyang benefited from the change, its profits were less than they might have been for two reasons. In the late 1920s new official exactions were imposed in Hsuchang (as they had not been in the early 1920s), and the quality of tobacco leaf grown around Hsuchang degenerated after

the departure of BAT (as less American seed was imported, more Chinese seed was mixed in, and crops thus became "mongrelized" between 1926 and 1934). On exactions, see the audit of Nanyang's records by the Li-hsin accountants, *Nanyang*, pp. 203-204, especially n4. On "mongrelization," see Chin-ling ta-hsueh, "Ho-nan sheng-ch'an yen-yeh ch'ü chih tiao-ch'a pao-kao," pp. 6 and 12-13.

47. *NCH*, January 28, 1928, p. 132.

48. FO 228/3520, Dandley to Macleay, October 6, 1926.

49. Diary of Julean H. Arnold, April 13, 1927, Arnold Papers; U.S. Department of Commerce, *Tobacco Markets and Conditions Abroad*, 119 (October 17, 1927), 6; Appendix, table 6.

50. *Shen-pao*, January 1, 4, 6, 7, and 8, February 17, and April 8, 1927; interview with former Nanyang worker Chuang Yü-ch'ing, May 12, 1958, all in *Nanyang*, pp. 347-351; *NCH*, February 26, 1927, p. 317. On strikes at Nanyang in 1926, see Chesneaux, *The Chinese Labor Movement*, pp. 281, 384, and 526n187.

51. Isaacs, *The Tragedy of the Chinese Revolution*, chs. 8-9; Parks M. Coble, "The Shanghai Financial, Commercial, and Industrial Elite and the Kuomintang, 1927-1937," unpublished paper, 1974, p. 11.

52. On the Kuomintang's priorities and perceptions of the urban economy in this period, see Lloyd E. Eastman, *The Abortive Revolution* (Cambridge, Mass., 1974), especially pp. 226-240.

53. Chiang probably removed Soong because he suspected that Soong sympathized with Chiang's rival at Wuhan, the Kuomintang "left" government, in which Soong's sister, Soong Ching-ling, was a major leader. Coble, p. 10.

54. *Chinese Economic Bulletin*, 9 (November 13, 1926), 286-288; Stanley Wright, pp. 608-622. These accounts emphasize that new tobacco taxes levied during the Northern Expedition varied from locale to locale.

55. Coble, p. 11.

56. *NCH*, July 2, 1927, p. 4; FO 371/10284, F485, Pratt to Macleay, December 10, 1923. During the mid-1920s (prior to the Northern Expedition), BAT's directors claimed that provincial taxes were exorbitant, but the Peking government argued that official tax rates of 25 to 80 percent were low by international standards and that in actual practice BAT paid less than the official rates—rarely more than 1 or 2 percent. FO 371/9217, F1573, Ministry of Foreign Affairs (Wai-chiao pu) to Macleay, 1923; FO 371/10284, F3651, Macleay to MacDonald, August 20, 1924; FO 371/10933, F1439, Palairet to Chamberlain, March 12, 1925; FO 371/11647, F115, memo from the Chinese government for the tariff conference of November 1925; and F113, BAT to Stewart, November 12, 1925 and Cunliffe-Owen to FO, January 8, 1926. See also Sanford, pp. 38-39 and 43-46.

57. See Appendix, table 5. I have used these figures rather than another set (also given in table 5) because I found them (and not the others) confirmed in more than one survey conducted at the time. (See sources cited in table 5.) For examples of Chinese cigarette manufacturers' complaints, see "Pronouncement of the Chinese Tobacco Factories," *China Times*, May 14, 1927, enclosed in NA 893.61331/10, Arthur to Gauss, May 16, 1927; *NCH*, December 31, 1929, p. 549; Wang Chung-fang (Paul K. Whang), "Tobacco and Cigarettes in China," *China*

Weekly Review, 61 (August 6, 1932), 358-359.

58. For details on these new firms, see Shang-hai-shih she-hui chü, *Shang-hai chih kung-yeh* (Taipei reprint, 1970), pt. 1, pp. 102-105; Kung Chün, *Chung-kuo hsin kung-yeh fa-chan shih ta-kang* (Shanghai, 1933), pp. 225-226. The largest of these new firms was the Chinese Hua-ch'eng Tobacco Company (*Chung-kuo Hua-ch'eng yen-ts'ao kung-ssu*). On it, see Kuo-huo shih-yeh ch'u-pan she, comp., *Chung-kuo kuo-huo kung-ch'ang shih-lueh*, pp. 10-11; Ch'en and P'ing, pp. 77-78; Chen Han Seng, p. 42. Nanyang's management lamented the effects of this new competition on its sales. See Nanyang's annual report for 1927, *Nanyang*, p. 223.

59. *NCH*, June 25, 1927, p. 541; July 2, 1927, p. 4; July 9, 1927, pp. 46-47, January 12, 1929, p. 71; February 4, 1930, p. 180.

60. *NCH*, February 11, 1930, p. 218.

61. Ibid. The Kuomintang's Ministry of Finance replied to these charges with a denial that its tax policies were the cause of Nanyang's financial trouble. NCH, March 25, 1930, p. 477.

62. Chien Yü-chieh and Ch'en Lien-po to Lao Ching-hsiu, January 10, 1929, *Nanyang*, pp. 460-461.

63. Ibid.; minutes of the meeting of the Nanyang board of directors, December 10, 1934; investigative department of Nanyang's Hong Kong branch to the management (with enclosure), November 21, 1935, all in *Nanyang*, pp. 458-461.

64. Interview with former Nanyang worker, Wang Yung-sheng, January 21, 1958; *Min-kuo jih-pao*, January 5, 6, and 7, 1929, all in *Nanyang*, pp. 357-360; *NCH*, January 12, 1929, p. 71; March 1, 1929, p. 365.

65. *NCH*, February 11, 1930, p. 218.

66. *Min-kuo jih-pao*, January 4, 5, 6, and 30 and February 1 and 7, 1929; *Chung-kuo kung-jen*, no. 5, pp. 5-8, and no. 6, pp. 71-76; minutes of the meetings of the Nanyang board of directors, December 11, 1928, January 31, 1929, February 18 and July 5, 1930; Wen Chih to Nanyang, February 1934; interviews with former Nanyang staff members and workers, 1957-1958, all in *Nanyang*, pp. 357-369, 372-374, and 468-469; U.S. Department of Commerce, *Tobacco Markets and Conditions Abroad*, 260 (July 1, 1930), and 277 (October 28, 1930).

67. *Min-kuo jih-pao*, January 9 and 11 and February 1, 1929, *Nanyang*, pp. 366-369; minutes of the meetings of the Nanyang board of directors, February 18 and July 5, 1930, *Nanyang*, pp. 372-373; *NCH*, February 9, 1929, p. 236; February 11, 1930, p. 218; September 16, 1930, p. 435; September 30, 1930, p. 507; *China Weekly Review*, 52.14 (May 31, 1930), 535.

68. *NCH*, March 2, 1929, p. 365; September 30, 1930, p. 507; U.S. Department of Commerce, *Tobacco Markets and Conditions Abroad*, 277 (October 28, 1930), 1. For another example of the Chiens complying with orders from the Kuomintang to rehire workers in late 1929, see minutes of the meetings of the Nanyang board of directors, November 15 and December 17, 1929, *Nanyang*, pp. 370-372. Though following the government's orders, the Chiens used their own techniques for breaking down the workers' resistance to their offers. For example, they paid some Nanyang workers to lobby among others workers for acceptance of relations between management and labor on the Chiens' terms. See

Wen Chih to Nanyang's management, February 1934, and interviews with former Nanyang workers, February 1958, *Nanyang*, pp. 373-374.

69. Appendix, table 6; *Nanyang*, p. 486; minutes of the meetings of the Nanyang board of directors, January 31, February 26, and April 4, 1929, and January 23 and May 13, 1930; Ch'en Lien-po to Lao Ching-hsiu, February 19, 1929, *Nanyang*, pp. 450-452. In 1935 the Chiens noted that they had not been able to maintain payments according to the schedule established in the original agreement with the Hong Kong and Shanghai Banking Corporation. Minutes of the meeting of the Nanyang board of directors, January 19, 1935, *Nanyang*, p. 452.

70. FO 371/12407, F6568, Rose to Lampson, June 22, 1927. American businessmen, unlike American missionaries and diplomats, called for Western military intervention in the Chinese revolution in April 1927, two months before Rose wrote this letter. See Dorothy Borg, *American Policy and the Chinese Revolution, 1925-1928* (New York, 1968), pp. 342-344.

71. Chesneaux, *The Chinese Labor Movement*, pp. 349-350 and 370. According to *Chan-tou* (p. 95), later in the year during the autumn of 1927, two hundred men from the British army and navy were used to try to break the BAT strike in Shanghai.

72. FO 371/12415, F8923, Lampson to FO, October 3, 1927. For expressions of the Chiens' resentment over BAT's not paying these taxes, see minutes of the meeting of the Nanyang board of directors, January 10, 1928 and the board's report to Nanyang stockholders for 1927, both in *Nanyang*, p. 383.

73. *Chan-tou*, pp. 90-96. See also Lowe, pp. 56 and 67; and *NCH*, September 30, 1927, p. 192; FO 371/13218, F370, Lampson to FO, January 25, 1928; FO 371/13157, F3181, Barton to Lampson, April 13, 1928. Shang-hai-shih she-hui chü, *Chin shih-wu-nien lai Shang-hai pa-kung t'ing-yeh*, appendix 1, p. 59.

74. NA 893.504/40, Lockhart to MacMurray, December 31, 1926; U.S. Department of Commerce, *Tobacco Markets and Conditions Abroad*, 109 (August 19, 1927), 7; 116 (September 27, 1927), 7; 119 (October 17, 1927), 5; 124 (November 22, 1927), 8; 127 (December 13, 1927), 4; 128 (December 20, 1927), 2-3; 172 (October 23, 1928), 8; and note 43 above.

75. FO 371/13156, F1697, correspondence between various BAT carriers and Maritime Customs officials, July 1927-February 1928; Thomas to Rustard, August 31, 1927 and Thomas to Cobbs, November 11, 1927, Thomas Papers.

76. Thomas to Bassett, March 13, 1928, Thomas Papers; John Carter Vincent, *The Extraterritorial System in China: Final Phase* (Cambridge, Mass., 1970), pp. 44-45; Wang, *Chinese Intellectuals and the West*, pp. 434-435 and 460-462. The figure of $1,000,000 is from FO 371/13218, F370, Lampson to FO, January 25, 1928. Higher figures are quoted in two other sources: $3,000,000 in *Chan-tou*, p. 97, and $5,000,000 in FO 371/13195, F293 and F304, Senior British Naval Officer at Shanghai to the Admiralty, January 17 and 18, 1928. English translations of the private agreement appear in FO 371/13157, F3181, Barton to Lampson, April 23, 1928; and *NCH*, January 21, 1928, p. 97. An English translation of the related agreement between the management and the BAT union is in Lowe, pp. 67-68. The quotations are from *NCH*, April 28, 1928, p. 147.

77. U.S. Department of Commerce, *Tobacco Markets and Conditions*

Abroad, 212 (July 30, 1929), 2; 237 (January 21, 1930); 222 (October 8, 1929). See also enclosure to the notice from the Nanyang office in Hong Kong, July 25, 1931, *Nanyang,* pp. 439-442. For a comment on BAT's billboard advertising at this time, see Frank G. Carpenter, *China* (New York, 1928), p. 140.

78. Appendix, table 3. The recovery of BAT's manufacturing and distributing systems was reflected in the increase of the amount of tobacco imported from the United States in 1928 as compared with 1927. See Appendix, table 8.

79. U.S. Department of Commerce, *Tobacco Markets and Conditions Abroad,* 222 (October 8, 1929). A study by Chinese reached the same conclusion. See Chao Ching-yuan, "The Present Crisis in Chinese Industry," trans. Hoh Chieh-hsiang, *China Weekly Review,* 53.13 (August 30, 1930), 490.

80. *NCH,* February 18, 1928, p. 247. For Other Chinese complaints about Kuomintang taxes that gave BAT advantages over its Chinese competition, see *NCH,* December 1, 1928, p. 339; and August 3, 1929, p. 173.

81. Cheng Yu-kwei, p. 61.

82. Wu Ch'eng-lo, II, 74-75; FO 228/1663, Smith to Jordan, March 27, 1907; FO 228/1628, Giles to Jordan, November 10, 1906; FO 371/3822, no. 165579, Brooks of BAT to Coales, October 28, 1919; Paton to Alston, November 4, 1919; Remer, *A Study of Chinese Boycotts,* chs. 5-9 and 11-14; Kikuchi, passim. For a summary of numerous complaints from BAT to the Foreign Office about Japanese competition in Manchuria, see FO 371/3189, no. 126013, Rickards to Grey, July 17, 1918.

83. Liu Ta-chün, "Industry," *Chinese Year Book, 1935* (Shanghai, 1935), pp. 1126-27; John Ahlers, "China's Cigarette Industry," *China Weekly Review,* 93.3 (June 15, 1940), 88-89; Allen and Donnithorne, *Western Enterprise in Far Eastern Economic Development,* p. 172; Chen Han Seng, pp. 33-35. On Tōa's tobacco purchasing in the 1910s and 1920s, see ch. 5, n119.

84. See Appendix, table 6.

85. This problem plagued Chinese cigarette companies, including Nanyang, as early as 1930. See Nanyang's annual report for 1930, *Nanyang,* pp. 223-224; *Kung shang pan-yueh k'an,* 2.8 (April 15, 1930), *Nanyang,* p. 160; *China Weekly Review,* 51.5 (January 4, 1930), 198; 51.10 (February 8, 1930), 351; 53.2 (July 14, 1930), 68.

86. Minutes of the meeting of the Nanyang board of directors, August 2, 1933, *Nanyang,* pp. 457-458.

87. Y. L. Kan to me, June 6, 1974. Cf. Wang, "Free Enterprise," p. 407. A copy of the agreement, signed by the Chiens and by Soong's representatives on March 1, 1937, may be found in *Nanyang,* pp. 502-503.

88. C. C. Kwong (Kuang Ch'ao-ch'i) to me, April 7, 1974. Nanyang's accounting records; *Chen-pao,* May 15, 1944; contract between Nanyang and the Great Southeastern Tobacco Company (*Ta tung-nan yen kung-ssu*), all in *Nanyang,* pp. 514-516. This "subcontracting" was an extension of a practice that Nanyang had used before the war. See minutes of the meeting of the Nanyang board of directors, October 27, 1933, with enclosed contract between Nanyang and the Liao-ning Tobacco Company, *Nanyang,* pp. 179-181.

89. Appendix, table 6. Minutes of the meetings of the Nanyang board of directors, February 21, 1934, and March 7, 1936; Chien Ch'ien-man to the company, November 5, 1935; a history of the Hankow factory by the workers in the

factory (October 1957); Nanyang's Hankow branch to the company's headquarters, April 4 and October 21, 1931, and July 2 and 6 and October 4, 1938, all in *Nanyang*, pp. 176-177 and 522-525.

90. Kōain, *Jūkei seifu no seinan keizai kensetsu jōkyō* (Tokyo, 1940), p. 93. On Nanyang's introduction of bright tobacco into Yunnan and its purchasing there in the early 1940s, see Ch'u Shou-chuang, *Yun-nan yen-ts'ao shih-yeh* (Kunming, 1947), p. 179; and minutes of the Nanyang board of directors, December 28, 1939, May 13, 1940, and April 22, 1941, *Nanyang*, pp. 375-397.

91. Nanyang's accounting records, *Nanyang*, p. 528; Ch'en Lien-po to Chien Yü-chieh and Chien Ying-fu, June 27, 1930; Ch'en Po-ying to the Nanyang office in Hong Kong, May 31, 1932; materials from Nanyang's Hong Kong files, *Nanyang*, pp. 186-187 and 534; *Nanyang*, p. 531; meeting of the Nanyang board of directors, May 13, 1940, *Nanyang*, p. 619.

92. Wang Hsi, p. 85; materials from the Foreign Affairs Office of Shanghai City (*Shang-hai-shih wai-shih ch'u*), *Nanyang*, p. 410.

93. Hsi Ch'ao, "Ying-Mei yen kung-ssu tui-yü Chung-kuo kuo-min ching-chi ti ch'in-shih," in *Chung-kuo ching-chi lun-wen chi* (Shanghai, 1936), I, 93.

94. Ibid., pp. 96-97; Chen Han Seng, p. 20.

95. Chen Han Seng, pp. 37-41; Wang, *Chinese Intellectuals and the West*, pp. 434-435; Wang, "Free Enterprise," p. 404; and Hou, *Foreign Investment*, pp. 148-149. For extensive documentation on these taxes, see *Nanyang*, pp. 375-397.

96. Wang Hsi, pp. 77 and 92. Cf. Ch'en Chen, col. 2, pp. 98-99; HKRS 111/4/263, Price to British Consulate-General, Shanghai, January 11, 1937.

97. Ch'en Chen, col. 2, pp. 94 and 135; Allen and Donnithorne, *Western Enterprise in Far Eastern Economic Development*, pp. 172-173; Dobson, pp. 171-172 and 196-198; Ahlers, p. 89; Wang Hsi, pp. 85, 88, and 93; Cheng Yu-kwei, p. 40; Michael Shapiro, *Changing China* (London, 1958), p. 19.

98. Minutes of the meeting of the Nanyang board of directors, July 11, 1947; Love to Cousins, April 1, 1946; I-chung (BAT) file no. 0302/27; Love to Terrell, May 20, 1947, all in *Nanyang*, pp. 608-611; Appendix, table 6.

99. Y. L. Kan to me, June 6, 1974; editors' introduction, *Nanyang*, p. 14; Shapiro, pp. 25-26; interview with B. G. Pearson, London, September 8, 1971. See Appendix, table 3 (including note h) and Chapter 8, n42. The full text of the agreement between Nanyang and the government can be found in *Nanyang*, pp. 654-658. On the negotiations that led to this agreement, see Wang, "Free Enterprise," pp. 409-411.

100. Chou-mo pao-she, ed., *Hsin Chung-kuo jen-wu chih* (Hong Kong, 1950), II, 148-150; Boorman and Howard, I, 365-366. Even before 1949, a biographer said of Chien Yü-chieh that his "life could be viewed as the embodiment of the history of Chinese national industry (*min-tsu kung-yeh*)." See *Ching-chi tao-pao*, 24 (1947), in Ch'en Chen, col. 1, pp. 469-490.

101. Chien Yü-chieh, "Wo-ti hsi-yueh ho an-wei," *Chung-kuo hsin-wen*, 721 (February 16, 1956), *Nanyang*, p. 684.

8. Conclusion: Imperialism, Nationalism, and Entrepreneurship

1. Richard Caves, *American Industry* (Englewood Cliffs, N.J., 1967), pp. 30-31.

2. See Dwight H. Perkins, *Agricultural Development in China, 1368-1968* (Chicago, 1969), p. 216; and Perkins, "Growth and Changing Structure of China's Twentieth Century Economy," pp. 122-123. The quotation is from the latter.

3. See Appendix, table 9.

4. Cohen, *Between Tradition and Modernity*, pp. 196-208; Hou, *Foreign Investment*, pp. 93-94, 131, 254; Wang Hsi, pp. 92-94.

5. Wang Hsi, p. 93. Admitting that his figures may not be representative, Hou has calculated that 56.3 percent of foreign manufacturing firms reinvested 30 percent or more of their profits in China between 1872 and 1936. Cf. his *Foreign Investment*, p. 102.

6. David S. Landes, "Some Thoughts on the Nature of Economic Imperialism," *The Journal of Economic History*, 21.4 (1961), 499-500.

7. Wang Hsi, p. 89; Gittings, p. 23.

8. Ernest Mandel, *Marxist Economic Theory*, trans. Brian Pearce (New York, 1968), II, 459-465.

9. Figures for bright tobacco are from Chen Han Seng, p. 20. Percentages of land occupied by bright tobacco have been estimated on the basis of figures in Perkins, *Agricultural Development*, p. 236, table B.14; and p. 262, table C.17.

10. Cf. Wilkins, *Emergence*, pp. 215-216; and Wilkins, *Maturing*, p. 92.

11. Stephen Lyon Endicott, *Diplomacy and Enterprise: British China Policy, 1933-1937* (Toronto, 1975), p. 177.

12. For application of this argument to the cigarette industry, see Wang Hsi, pp. 78-80 and 85-89. As Hou Chi-ming has pointed out, Mao Tse-tung, Chiang Kai-shek, Sun Yat-sen, and other leaders have also made this argument. See Hou, *Foreign Investment*, pp. 1-3.

13. Appendix, table 3. Cf. Hou, *Foreign Investment*, pp. 112-118; and Tennant, p. 5.

14. Several historians have argued that the unfavorable effects of imperialism in China were primarily political and psychological rather than economic. For a review article that makes this point and cites the major works on the subject, see Andrew J. Nathan, "Imperialism's Effects on China," *Bulletin of Concerned Asian Scholars*, 4.4 (December 1972), 3-8.

15. On BAT's brief rivalries during the 1920s with these American-owned companies in China, see Thomas to Pettitt, January 9 and May 9, 1922; Thomas to Allen, April 10, 1922; and Pettitt to Thomas, March 13, 1922, all in the Thomas Papers; NA 693.1112/104, Cunningham to State, July 20, 1925; *NCH*, October 29, 1927, p. 206; Cox, pp. 300-301; Crow, pp. 60-61.

16. Hou, *Foreign Investment*, pp. 134 and 217; Cf. Bergère, "The Role of the Bourgeoisie," p. 249.

17. Y. L. Kan to me, June 6, 1974.

18. For an additional example of Nanyang borrowing from a foreign bank —this time the Dutch-owned Netherlandsche Handel-Maatschappij—see the promissory note signed by Chien Ying-fu as Nanyang's representative, November 24, 1924, in *Nanyang*, pp. 455-456.

19. James E. Sheridan, *China in Disintegration* (New York, 1975), p. 153. See also Hou, *Foreign Investment*, p. 152; and Remer, *A Study of Chinese Boycotts*, p. 247.

20. Shang-hai-shih she-hui chü, *Chin shih-wu-nien lai Shang-hai pa-kung t'ing-yeh,* appendix 1. There were 45 strikes against Naigai Wata and 38 against the Japan-China Spinning and Weaving Company.

21. Y. L. Kan (Chien Jih-lin) to me, June 6, 1974; "The Nanyang Bros. Tobacco Co.," *Chinese Economic Journal,* 3.5 (November 1928), 990.

22. Hou, *Foreign Investment,* pp. 130-134.

23. Jerome Ch'en has pointed out that Mao, in making the distinction between national and comprador bourgeoisie, reached a formulation that was parallel to Stalin's theory of the "two sections" within the national bourgeoisie (the revolutionary section and the compromising section). See Ch'en, "The Development and Logic of Mao Tse-tung's Thought, 1928-49," in Chalmers Johnson, ed., *Ideology and Politics in Contemporary China* (Seattle, 1973), pp. 81-83.

24. Hao, chs. 5 and 6, especially pp. 111-112.

25. Bergère, "The Role of the Bourgeoisie," pp. 249-250. For her application of a similar formulation to the early 1920s, see Bergère, "National Bourgeoisie and Bourgeois Nationalism," prepared as a draft of chapter 7 for *The Cambridge History of China,* xii-xiii, and presented to the Working Conference on Republican China at Harvard University, Cambridge, Mass., August 15-20, 1976.

26. Chien Chao-nan to Chien Yü-chieh, February 22, 1917, *Nanyang,* pp. 104-105; Wang Hsi, p. 88.

27. A Chinese-owned firm outside the cigarette industry that faced this dilemma, for example, was the Chinese-owned Yung-li Chemical Company which the British-owned Brunner, Mond and Company tried repeatedly to buy out in the 1920s. See Hou Teh-pang, "The Yung-li Company: Pioneer of Chemical Industry in China," *China Reconstructs,* 4.3 (March 1955), 22.

28. Joseph A. Schumpeter, "The Creative Response in Economic History," *Journal of Economic History,* 8 (November 1947), 151.

29. See Chapter 1, nn12-16.

30. Schumpeter, *The Theory of Economic Development,* p. 66.

31. Cf. Hou, *Foreign Investment,* pp. 123-124; and Ragnar Nurkse, *Problems of Capital Formation in Underdeveloped Countries* (New York, 1953), pp. 25-29, cited by Hou.

32. Interview with B. G. Pearson of BAT, London, September 8, 1971.

33. See Chapter 1, nn12-13.

34. Hou Chi-ming has outlined definitions of "oppression" that various writers have used. See his "The Oppression Argument on Foreign Investment in China, 1895-1937," *Journal of Asian Studies,* 20.4 (August 1961), 435-436.

35. Hou, *Foreign Investment,* pp. 134-136; Dernberger, "The Role of the Foreigner," pp. 39-47.

36. See Appendix, tables 3, 5, 6, and 7. Cf. Hou, *Foreign Investment,* pp. 138-139.

37. Feuerwerker, *The Chinese Economy, 1912-1949,* p. 17.

38. For example, Thomas Rawski has suggested to me that if BAT paid a higher percentage of its sales revenue to residents in China than Nanyang did, then it had a legitimate right to claim that it was contributing more to China's welfare than Nanyang was. (As far as I know, this example is purely hypothetical, for available evidence does not show whether BAT paid a higher percentage of its sales revenue to residents of China than Nanyang did.)

39. For a historiographical discussion of the "business in history" approach which emphasizes the interaction between businesses and other institutions in history (as I have tried to do in this book), see James P. Baughman, "New Directions in American Economic and Business History," in George Athan Billias and Gerald N. Grob, eds., *American History* (New York, 1971), pp. 302-313.

40. See Appendix, table 9.

41. Jay Mathews, "Between Puffs," *Washington Post*, September 6, 1978, p. A17. On the growth of cigarette production and sales in China during the 1950s, see Dwight H. Perkins, *Market Control and Planning in Communist Society* (Cambridge, Mass., 1966), p. 250, table 32. On further growth of the market in the 1960s and 1970s, see Hugh C. Kiger, "Prospects for Tobacco Trade between the United States and Mainland China," *Foreign Agriculture*, 10.19 (May 8, 1972), 2-3. For travelers' reports commenting on the pervasiveness of cigarette smoking in China during the mid- and late 1970s, see Orville Schell, *In the People's Republic* (New York, 1977), pp. 124 and 228-229; Michael Pertschuk, "Smoking in China," *Washington Post*, October 17, 1976, pp. C1 and C5; and Susan Brownmiller, "A Former Fan's Notes," *The Village Voice*, 22.49 (December 5, 1977), 28-29.

42. Interview with Y. L. Kan (Chien Jih-lin), Hong Kong, June 28, 1977; United Nations Conference on Trade and Development, *Marketing and Distribution of Tobacco* (United Nations, 1978), pp. 38-42; *Forbes*, March 20, 1978, p. 79; *Asian Wall Street Journal*, March 8, 1979, p. 3.

43. Quoted by Charles P. Kindleberger in his *American Business Abroad* (New Haven, 1969), p. 16.

Bibliography

A Ying 阿英 (Ch'ien Hsing-ts'un 錢杏邨), comp. *Fan-Mei Hua-kung chin-yueh wen-hsueh chi* 反美華工禁約文學集 (A collection of anti-American literature relating to the exclusion of Chinese laborers). Peking: Chung-hua shu-chü, 1960.

Adshead, S. A. M. *The Modernization of the Chinese Salt Administration, 1900–1920*. Cambridge, Mass.: Harvard University Press, 1970.

Ai Wu 艾蕪 (pseud.). *Wo-ti ch'ing-nien shih-tai* 我的青年時代 (My youth). 2nd ed. Shanghai: K'ai-ming shu-tien, 1949.

Alford, B. W. E. *W. D. & H. O. Wills and the Development of the U. K. Tobacco Industry, 1786–1965*. London: Methuen, 1973.

Allen, G. C., and Audrey G. Donnithorne. *Western Enterprise in Far Eastern Economic Development*. London: Allen & Unwin, 1954.

————*Western Enterprise in Indonesia and Malaya*. London: Allen & Unwin, 1957.

Anderson, William Ashley. *The Atrocious Crime (of Being a Young Man)*. Philadelphia: Dorrance, 1973.

Arnold, Julean H. Papers. Hoover Institution, Stanford University, Stanford.

————comp. *China through an American Window*. Shanghai: The American Chamber of Commerce, 1932.

————et al. *China: A Commercial and Industrial Handbook*, U.S. Department of Commerce, Bureau of Foreign and Domestic Commerce, Trade Promotion Series no. 38. Washington, D. C.: Government Printing Office, 1926.

————*Commercial Handbook of China*, U.S. Department of Commerce, Bureau of Foreign and Domestic Commerce, Miscellaneous Series no. 84. Washington, D.C.: Government Printing Office, 1920. 2 volumes.

Asia. New York: The Asia Society, 1918–1930.

Baba Masao and Masahiro Tatemoto. "Foreign Trade and Economic Growth in Japan: 1858–1937." In Lawrence Klein and Kazushi Ohkawa, eds., *Economic Growth: The Japanese Experience since the Meiji Era*. Homewood, Illinois: R.D. Irwin, 1968. Pp. 162–196.

Bagnall, Dorothy, ed. *Cigarette Card News*. Brentford, England, vols. 1–24 (1933–1965).

Bain, Joe S. *Barriers to New Competition: Their Character and Consequences in Manufacturing Industries*. Cambridge, Mass.: Harvard University Press, 1956.

Baker, Hugh. *A Chinese Lineage Village: Sheung Shui*. Stanford: Stanford University Press, 1968.

Bank of China, comp. *Chinese Government Foreign Loan Obligations*. Shanghai, 1935.

Barnet, Richard J., and Ronald E. Müller. *Global Reach: The Power of the Multinational Corporations*. New York: Simon and Schuster, 1974.

Barth, Gunther. *Bitter Strength: A History of the Chinese in the United States, 1850–1870*. Cambridge, Mass.: Harvard University Press, 1964.

Baughman, James P. "New Directions in American Economic and Business History." In George Athan Billias and Gerald N. Grob, eds., *American History: Retrospect and Prospect*. New York: Free Press, 1971. Pp. 271–314.

Beale, Howard K. *Theodore Roosevelt and the Rise of America to World Power*. Baltimore: Johns Hopkins Press, 1956.

Bell, H. T. Montague, and H. G. Woodhead. *China Yearbook*. London: G. Routledge and Sons, 1913–1930.

Bergère, Marie-Claire. "La bourgeoisie chinoise et les problèmes du développement économique (1917–1923)," *Revue d'histoire moderne et contemporaine*, 16 (April–June, 1969), 246–267.

———"Le mouvement du 4 mai 1919 en Chine: La conjoncture économique et le rôle de la bourgeoisie nationale," *Revue historique*, 241 (June 1969), 309–326.

———"La première guerre mondiale et les développements des industries nationales en Chine." Paper read at the 29e Congrès International des Orientalistes. Paris, 1973.

———"National Bourgeoisie and Bourgeois Nationalism." Draft of chapter 7 for *The Cambridge History of China*, XII–XIII, presented to the Working Conference on Republican China, Harvard University, Cambridge, Mass., August 15–20, 1976.

———"The Role of the Bourgeoisie." In Mary C. Wright, ed., *China in Revolution*. New Haven: Yale University Press, 1968. Pp. 229–295.

Bernstein, Barton J., ed. *Towards a New Past: Dissenting Essays in American History*. New York: Vintage Books, 1967.

Blick, Judith. "The Chinese Labor Corps in World War I," *Harvard Papers on China*, 9 (1955), 111–146.

Blum, John Morton. *The Republican Roosevelt*. Cambridge: Harvard University Press, 1954.

Boorman, Howard L., and Richard C. Howard, eds. *Biographical Dictionary of Republican China*. New York: Columbia University Press, 1967–1971. 4 volumes.

Borg, Dorothy. *American Policy and the Chinese Revolution, 1925–1928*. New York: Octagon Books, 1968.

Bŏrisat Yāsūp Nānyāng Čhamkat, Bangkok (Nanyang Tobacco Company, Ltd., Bangkok). *Thawāi phraphonchai Phrabāt Somdet Phra Čhao Yūhūa ratchakān thi 7* (Felicitations to His Majesty the seventh king [of the Bangkok Dynasty]). Bangkok, 1925.

British–American Tobacco Company, Ltd. *The Record in China of the British–American Tobacco Company, Limited.* Shanghai, 1925 (?).

———ed. *B.A.T. Bulletin.* London, vols. 1–20 (1915–1930).

Brown, William H. Papers. Sterling Memorial Library, Yale University, New Haven, Connecticut.

Brownmiller, Susan. "A Former Fan's Notes," *The Village Voice,* 22.49 (December 5, 1977), 1, 23–24, 26, and 28–29.

Buck, Pearl S. "China the Eternal," *The Living Age,* 324.4205 (February 7, 1925), 324–330.

Bureau of Corporations. *Report of the Commissioner of Corporations on the Tobacco Industry.* Washington, D.C.: Government Printing Office. Pt. 1, 1909; pt. 2, 1911.

Bureau of Social Affairs of Shanghai. *The Index Numbers of Earnings of the Factory Laborers in Greater Shanghai, July–December, 1928.* Shanghai, 1929.

Campbell, Charles S., Jr. *Special Business Interests and the Open Door.* New Haven: Yale University Press, 1951.

Carlson, Ellsworth C. *The Kaiping Mines (1877–1912).* Cambridge, Mass.: Harvard University, East Asian Research Center, 1957.

Carpenter, Frank G. *China.* New York: Doubleday, Doran and Company, 1925.

The Cartophilic Society of Great Britain, Ltd., comp. *The British–American Tobacco Company Booklet.* London, 1952.

Castles, Lance. *Religion, Politics and Economic Behavior in Java: The Kudus Cigarette Industry.* New Haven: Southeast Asian Studies, Yale University, 1967.

Caves, Richard. *American Industry: Structure, Conduct, Performance.* Englewood Cliffs, N. J.: Prentice-Hall, 1964.

Chan Ming Kou. "Labor and Empire: The Chinese Labor Movement in the Canton Delta, 1895–1927." Ph.D. dissertation, Stanford University, 1975.

Chan, Wellington K. K. *Merchants, Mandarins, and Modern Enterprise in Late Ch'ing China.* Cambridge, Mass.: Harvard University East Asian Research Center, 1977.

Chandler, Alfred D., Jr. *Strategy and Structure: Chapters in the History of the American Industrial Enterprise.* Cambridge, Mass.: M.I.T. Press, 1962.

———*The Visible Hand: The Managerial Revolution in American Business.* Cambridge, Mass.: Harvard University Press, 1977.

Chang, John K. *Industrial Development in Pre-Communist China.* Chicago: Aldine, 1969.

Chang Ts'un-wu 張存武. *Kuang-hsu sa-i-nien Chung-Mei kung-yueh feng-ch'ao* 光緒卅一年中美工約風潮 (The crisis of the Sino-American dispute over the labor agreement of 1905). Taipei: Chung-kuo hsueh-shu chu-tso chiang-tsu wei-yuan-hui, 1965.

Chang Yen-shen 張雁深. *Jih-pen li-yung so-wei "ho-pan shih-yeh" ch'in-Hua ti li-shih* 日本利用所謂 "合辦事業" 侵華的歷史 (The history of the Japanese use of so-called "joint management enterprises" to exploit China). Peking: San-lien shu-tien, 1958.

Chang Yu-i 章有義, comp. *Chung-kuo chin-tai nung-yeh shih tzu-liao, ti erh-chi,*

1912–1927 中國近代農業史資料第二輯, 1912–1927 (Historical materials on modern Chinese agriculture, second collection, 1912–1927). Peking: San-lien shu-tien, 1957.

Chao Kang. "The Chinese–American Textile Trade, 1830–1930." Paper presented at the Conference on the History of American-East Asian Economic Relations," Mt. Kisco, New York, June 1976.

————*The Development of Cotton Textile Production in China.* Cambridge, Mass.: Harvard University Press, 1977.

Ch'en Chen 陳眞, comp. *Chung-kuo chin-tai kung-yeh shih tzu-liao* 中國近代工業史資料 (Historical materials on modern Chinese industry). Peking: San-lien shu-tien, first collection, 1957, 2 volumes; second collection, 1958, 2 volumes; fourth collection, 1961, 2 volumes.

Ch'en Chin-miao 陳金淼. "T'ien-chin chih mai-pan chih-tu" 天津之買辦制度 (The comprador system in Tientsin), *Ching-chi hsueh-pao* 經濟學報 (Journal of economics). Peking: Yen-ching ta-hsueh ching-chi hsueh-hui, I (1940), 27–69.

Chen Han Seng. *Industrial Capital and Chinese Peasants: A Study of the Livelihood of Chinese Tobacco Cultivators.* Shanghai: Kelly and Walsh, 1939.

Ch'en, Jerome. "Historical Background." In Jack Gray, ed., *Modern China's Search for a Political Form.* London: Oxford University Press, 1969. Pp. 1–40.

————"The Development and Logic of Mao Tse-tung's Thought, 1928–49." In Chalmers Johnson, ed., *Ideology and Politics in Contemporary China.* Seattle: University of Washington Press, 1973. Pp. 78–114.

Ch'en Kung Ping-ch'ien fu-kao 陳公炳謙訃告 (Obituaries of Mr. Ch'en Ping-ch'ien). Macao, 1938.

Chen, Joseph T. *The May Fourth Movement in Shanghai.* Leiden: Brill, 1971.

Chen Ta. *The Labor Movement in China.* Honolulu: Peking Leader Press, 1927.

Ch'en Tzu-ch'ien 陳子謙 and P'ing Chin-ya 平襟亞. "Hsiang-yen hsiao-shih" 香煙小史 (A short history of cigarettes). In Shang-hai jen-min ch'u-pan she 上海人民出版社 (Shanghai People's Publishing House), ed., *Shang-hai ching-chi shih-hua, ti-i chi* 上海經濟史話第一輯 (A history of Shanghai's economy, collection 1). Shanghai: Jen-min ch'u-pan she, 1963. Pp. 72–80.

Ch'eng Chi-hua 程季華. *Chung-kuo tien-ying fa-chan shih* 中國電影發展史 (The history of the development of Chinese movies). Peking: Chung-kuo tien-ying ch'u-pan she, 1963.

Ch'eng Shu-tu 程叔度 et al., comps. *Chüan-yen t'ung-shui shih* 捲菸統稅史 (A history of the consolidated tobacco tax). Shanghai, 1929.

————*Yen chiu shui shih* 菸酒稅史 (A history of tobacco and wine taxes). Nanking, 1929. 2 volumes.

Cheng Yu-kwei. *Foreign Trade and Industrial Development of China: An Historical and Integrated Analysis through 1948.* Washington, D.C.: The University Press of Washington, D.C., 1956.

Chesneaux, Jean. *The Chinese Labor Movement, 1919–1927.* Translated from the French by H. M. Wright. Stanford: Stanford University Press, 1968.

————"The Labour Force in the First Part of the Twentieth Century." In C. D. Cowan, ed., *The Economic Development of China and Japan: Studies in Economics, History and Political Economy.* New York: Praeger, 1964. Pp. 111–127.

Chia I-chün 賈逸君. *Chung-kuo ming-jen chuan* 中國名人傳 (Eminent Chinese). Peiping, 1940.

Chien Chao-liang 簡朝亮 et al., comps. *Yueh-tung Chien-shih ta-t'ung-p'u* 粵東簡氏大同譜 (A genealogy of the Chien family of Kwangtung province). N.p., 1928.

Ch'ien I-shih 錢亦石. *Chin-tai Chung-kuo ching-chi shih* 近代中國經濟史 (A history of the Chinese economy). Chungking: Sheng-huo shu-tien, 1939.

Chin Tsu-hsun 金祖勛. "1905-nien Kuang-tung fan-Mei yun-tung ti p'ien-tuan hui-i" 1905年廣東反美運動的片斷回憶 (Reminiscences of the anti-American movement of 1905 in Kwangtung). *Chin-tai shih tzu-liao* 近代史資料 (Materials on modern history), 5 (October 1958), 52–55.

China, Inspectorate General of Customs. *Decennial Reports*. Shanghai, for 1892–1901 (1906); for 1902–1911 (1913); for 1912–1921 (1924); for 1922–1931 (1933).

———*Returns and Reports of Trade*. Shanghai, annual, 1890–1930.

The China Weekly Review. See *Millard's Review of the Far East*.

Chinese Economic Bulletin. Peking, 1919–1927.

Chinese Economic Journal. Peking and Shanghai, 1927–1930 (Successor to *Chinese Economic Bulletin*).

Chinese Economic Monthly. Peking, 1924–1926.

Chinese Year Book. Shanghai, 1935–1945.

Chin-ling ta-hsueh nung-hsueh yuan nung-yeh ching-chi hsi 金陵大學農學院農業經濟系 (The department of agriculture and economics in the institute of agricultural science at Chin-ling University), comp. "Ho-nan sheng ch'an yen-yeh ch'ü chih tiao-ch'a pao-kao" 河南省產菸葉區之調查報告 (Report of investigators of the tobacco region in Honan province), *Yü O Wan Kan ssu-sheng nung-ts'un ching-chi tiao-ch'a ch'u-pu pao-kao* 豫鄂皖贛四省農村經濟調查初步報告 (Preliminary report on investigations into the village economies of the four provinces of Honan, Hupeh, Anhwei, and Kiangsi), 6 (June 1934).

Ch'iu Shih 秋石 (pseud.). "T'an kung-ch'ang shih pien-hsieh chung ti chi-ko wen-t'i" 談工廠史編寫中的幾個問題 (Some problems in writing factory history), *Hsueh-shu yueh-k'an* 學術月刊 (The scholarship monthly), 2.26 (February 10, 1959), 57–59.

Chou Chih-hua 周志驊. *Chung-kuo chung-yao shang-p'in* 中國重要商品 (Major Chinese articles of commerce). Shanghai: Hua-t'ung shu-chü, 1931.

Chou Hsiu-luan 周秀鸞. *Ti-i-tz'u shih-chieh ta-chan shih-ch'i Chung-kuo min-tsu kung-yeh ti fa-chan* 第一次世界大戰時期中國民族工業的發展 (The development of China's national industry during World War I). Shanghai: Jen-min ch'u-pan she, 1958.

Chou-mo pao-she 週末報社, ed. *Hsin Chung-kuo jen-wu chih* 新中國人物誌 (Directory of people in new China). Hong Kong: Chou-mo pao-she, 1950. 2 volumes.

Chow Tse-tsung. *The May Fourth Movement*. Cambridge, Mass.: Harvard University Press, 1960.

Chow Yung-tê. *Social Mobility in China*. New York: Atherton Press, 1966.

Chu Ch'ang 竹廠 (pseud.). "Shang-tien chao-p'ai" 商店招牌 (Shop signs),

Chung-ho yueh-k'an 中和月刊 (Chung-ho monthly), 3.7 (July 1942), 25–27.

Chu Hsiao-tsang 朱孝臧 (Chien Yao-teng 簡耀登), comp. *Chien chün Chao-nan ai-wan lu* 簡君照南哀輓錄 (Funeral scrolls in memory of the honorable Chien Chao-nan). N.p., n.d. 6 volumes.

Chu Hsieh 朱偰. *Chung-kuo tsu-shui wen-t'i* 中國租稅問題 (China's taxation problem). Shanghai: Shang-wu yin-shu kuan, 1936.

Chu Hsin-fan 朱新繁. *Chung-kuo tzu-pen chu-i chih fa-chan* 中國資本主義之發展 (The development of capitalism in China). Shanghai: Lien-ho shu-tien, 1929.

Chu Hsiu-hsia 祝秀俠, ed. *Hua-ch'iao ming-jen chuan* 華僑名人傳 (Eminent Overseas Chinese). Taipei: Chung-hua wen-hua ch'u-pan shih-yeh wei-yuan-hui, 1955.

Chu, Samuel C. *Reformer in Modern China: Chang Chien, 1853–1926.* New York: Columbia University Press, 1965.

Chu Shih-chia 朱士嘉. *Mei-kuo p'o-hai Hua-kung shih-liao* 美國迫害華工史料 (Historical materials on American persecution of Chinese laborers). Peking: Chung-hua shu-chü, 1959.

Ch'u Shou-chuang 褚守莊. *Yun-nan yen-ts'ao shih-yeh* 雲南菸草事業 (The tobacco business in Yunnan province). Kunming: Hsin Yun-nan ts'ung-shu she, 1947.

Ch'u T'ung-tsu. *Local Government in China under the Ch'ing.* Cambridge, Mass.: Harvard University Press, 1962.

Chung-kuo k'o-hsueh yuan li-shih yen-chiu ti-san-so 中國科學院歷史研究第三所 (Historical research institute no. 3 in the Chinese academy of sciences), comp. *Wu-ssu ai-kuo yun-tung tzu-liao* 五四愛國運動資料 (Materials on the May Fourth patriotic movement). Peking: K'o-hsueh ch'u-pan she, 1959.

Chung-kuo k'o-hsueh yuan Shang-hai ching-chi yen-chiu so Shang-hai she-hui k'o-hsueh yuan ching-chi yen-chiu so 中國科學院上海經濟研究所上海社會科學院經濟研究所 (The Shanghai Economic Research Institute of the Chinese Academy of Sciences and the Economic Research Institute of the Shanghai Academy of Social Sciences), comps. *Nan-yang hsiung-ti yen-ts'ao kung-ssu shih-liao* 南洋兄弟煙草公司史料 (Historical materials on the Nanyang Brothers Tobacco Company). Shanghai: Jen-min ch'u-pan she, 1958.

Chung-kuo kung-ch'an tang 中國共產黨 (Chinese Communist Party). *Chan-tou ti wu-shih-nien—Shang-hai chüan-yen i ch'ang kung-jen tou-cheng shih-hua* 戰鬥的五十年—上海卷菸一廠工人鬥爭史話 (Fighting for fifty years—a history of the workers' struggle based on interviews in Shanghai Cigarette Factory No. 1). Shanghai: Jen-min ch'u-pan she, 1960.

Chung-kuo nien-chien 中國年鑑. (China Yearbook). Vol. I (1923). Shanghai: Shang-wu yin-shu kuan.

Clementi, Cecil. *Cantonese Love Songs.* Oxford: Clarendon Press, 1904.

Clubb, O. Edmund. *Twentieth Century China.* New York: Columbia University Press, 1964.

Coble, Parkes M. "The Shanghai Financial, Commercial, and Industrial Elite and the Kuomintang, 1927–1937." Unpublished paper, 1974.

Cochran, Sherman G. "Business and Politics in Republican China," *Chinese Republican Studies Newsletter*, 1.2 (February 1976), 13–15.

Cohen, Paul A. *Between Tradition and Modernity: Wang T'ao and Reform in*

Late Ch'ing China. Cambridge, Mass.: Harvard University Press, 1974.

——*China and Christianity: The Missionary Movement and the Growth of Chinese Antiforeignism, 1860–1870*. Cambridge, Mass.: Harvard University Press, 1963.

Corina, Maurice. *Trust in Tobacco: The Anglo-American Struggle for Power*. New York: St. Martin's Press, 1975.

Cowan, C. D., ed. *The Economic Development of China and Japan*. New York: Praeger, 1964.

Cox, Reavis. *Competition in the American Tobacco Industry, 1911–1932*. New York: Columbia University Press, 1933.

Crist, Raymond F., and Harry R. Burrill. *Trade with China*, U.S. Department of Commerce and Labor, Bureau of Manufactures. Washington, D.C.: Government Printing Office, 1906.

Crow, Carl. *Foreign Devils in the Flowery Kingdom*. New York and London: Harper and Brothers, 1940.

Dennett, Tyler. *Americans in Eastern Asia*. New York: The Macmillan Company, 1922.

Dernberger, Robert F. "The Role of the Foreigner in China's Economic Development, 1840–1949." In D. H. Perkins, ed., *China's Modern Economy in Historical Perspective*. Pp. 19–47.

DesForges, Roger V. *Hsi-liang and the Chinese National Revolution*. New Haven: Yale University Press, 1973.

Dingle, Edwin J. *Across China on Foot: Life in the Interior and the Reform Movement*. New York: H. Holt and Company, 1911.

Dobson, Richard P. *China Cycle*. London: Macmillan and Company, 1946.

Doré, Henry, S. J. *Researches into Chinese Superstitions*. Translated from the French by M. Kennelly, S. J. Shanghai: T'usewei Printing Press, 1918. Volume 5.

Durden, Robert F. *The Dukes of Durham, 1865–1929*. Durham, North Carolina: Duke University Press, 1975.

——"Tar Heel Tobacconist in Tokyo, 1899–1904," *North Carolina Historical Review*, 53.4 (October 1976), 347–363.

Eastman, Lloyd E. *The Abortive Revolution: China under Nationalist Rule, 1927–1937*. Cambridge, Mass.: Harvard University Press, 1974.

——"The Kwangtung Anti-foreign Disturbances during the Sino-French War," *Papers on China*, 13 (1959), 1–31. Harvard University, East Asian Research Center.

Easton, Robert. *Guns, Gold and Caravans*. Santa Barbara: Capra Press, 1978.

Endicott, Stephen Lyon. *Diplomacy and Enterprise: British China Policy, 1933–1937*. Vancouver, B.C.: University of British Columbia Press, 1975.

Esherick, Joseph. "Harvard on China: The Apologetics of Imperialism," *Bulletin of Concerned Asian Scholars*, 4.4 (December 1972), 9–16.

Estes, John M. *Tobacco: Instructions for Its Cultivation and Curing*, U.S. Department of Agriculture, Farmers Bulletin no. 6. Washington, D.C.: Government Printing Office, 1892.

Evans, I. O. *Cigarette Cards and How to Collect Them*. London: H. Jenkins, Ltd., 1937.

Fairbank, John K., Alexander Eckstein, and Yang Lien-sheng. "Economic Change in Early Modern China: An Analytic Framework," *Economic Development and Cultural Change*, 9.1 (October 1960), 1–26.

Fan Yin-nan 樊蔭南 *Tang-tai Chung-kuo ming-jen lu* 當代中國名人錄 (Eminent figures of contemporary China). Shanghai: Liang-yu t'u-shu yin-tse kung-ssu, 1931.

Fang Hsien-t'ing 方顯廷 (H. D. Fong). *Chung-kuo chih mien-fang-chih yeh* 中國之棉紡織業 (China's cotton textile industry). Shanghai: Shang-wu yin-shu kuan, 1934.

The Far Eastern Review. Shanghai, 1904–1941.

Faung Lih-chung. "A Chinese Corporation." In C. F. Remer, ed., *Readings in Economics for China*. Shanghai: Commercial Press, Ltd., 1922. Pp. 310–314.

Feuerwerker, Albert. *China's Early Industrialization: Sheng Hsuan-huai (1844–1916) and Mandarin Enterprise*. Cambridge, Mass.: Harvard University Press, 1958.

——"China's Modern Economic History in Communist Chinese Historiography." In Feuerwerker, ed., *History in Communist China*. Cambridge, Mass.: Harvard University Press, 1968. Pp. 216–246.

——"China's Nineteenth Century Industrialization: The Case of the Hanyehping Coal and Iron Company Limited." In C. D. Cowan, ed., *The Economic Development of China and Japan*. Pp. 79–110.

——*The Chinese Economy ca. 1870–1911*. Ann Arbor: University of Michigan Center for Chinese Studies, 1969.

——*The Chinese Economy, 1912–1949*. Ann Arbor: University of Michigan Center for Chinese Studies, 1968.

——*The Foreign Establishment in China in the Early Twentieth Century*. Ann Arbor: University of Michigan Center for Chinese Studies, 1976.

——"Industrial Enterprise in Twentieth-Century China: The Chee Hsin Cement Co." In Feuerwerker, Rhoads Murphey, and Mary C. Wright, eds., *Approaches to Modern Chinese History*. Berkeley: University of California Press, 1967. Pp. 304–341.

Fewsmith, Joseph. "Yü Hsia-ch'ing and the Evolution of Kuomintang–Merchant Relations, 1924–1930." Paper presented at the convention of the American Historical Association, December 29, 1973.

Field, Frederick V. *American Participation in the China Consortiums*. Chicago: University of Chicago Press, 1931.

Field, Margaret. "The Chinese Boycott of 1905," *Papers on China*, 11 (1957), 63–95. Harvard University, East Asian Research Center.

Fitzgerald, Patrick. *Industrial Combination in England*. London: Sir I. Pitman and Sons, Ltd., 1927.

Freedman, Maurice. *Chinese Lineage and Society: Fukien and Kwangtung*. London: Athlone Press, 1966.

——"The Growth of a Plural Society in Malaya," *Pacific Affairs*, 33 (1960), 158–168.

——*Lineage Organization in Southeastern China*. London: Athlone Press, 1958.

——"Immigrants and Associations: Chinese in Nineteenth Century Singapore," *Comparative Studies in Society and History*, 3 (1960), 25–48.

Friedman, Edward. *Backward toward Revolution: The Chinese Revolutionary Party*. Berkeley: University of California Press, 1974.

————"The Center Cannot Hold: The Failure of Parliamentary Democracy in China from the Chinese Revolution of 1911 to the World War of 1914." Ph.D. dissertation, Harvard University, 1968.

————"The Failure to Modernize China's Most Modern City, Canton, 1911–1914." Paper presented at the Conference on Urban Society and Political Development in Modern China, St. Croix, Virgin Islands, December 1968–January 1969.

————and Mark Selden, eds. *America's Asia*. New York: Pantheon Books, 1971.

Gaimushō jōhōbu 外務省情報部 (Foreign Office, Intelligence Bureau). *Gendai Chūka minkoku Manshūkoku jimmei kan* 現代中華民國滿洲國人名鑑 (Biographical dictionary of contemporary Republic of China and Manchukuo). Tokyo, 1932.

Gibb, George Sweet. *The Saco-Lowell Shops: Textile Machinery Building in New England, 1813–1949*. Cambridge, Mass.: Harvard University Press, 1950.

Gillin, Donald G. *Warlord: Yen Hsi-shan in Shansi Province, 1911–1949*. Princeton: Princeton University Press, 1967.

Gittings, John. *The World and China, 1922–1972*. London: Methuen, 1974.

Golas, Peter J. "Early Ch'ing Guilds." In G. William Skinner, ed. *The City in Late Imperial China*. Stanford: Stanford University Press, 1977. Pp. 555–580.

Goodrich, L. Carrington. "Early Prohibitions of Tobacco in China and Manchuria," *Journal of the American Oriental Society*, 58 (1938), 648–657.

Gracey, Samuel L., "Suggestions for Labels and Trade Marks in China," *Scientific American*, 78 (January 12, 1898), 51.

Great Britain. Records of the British Foreign Office. Public Records Office, London.

Great Britain, Foreign Office. *British and Foreign State Papers, 1921*, vol. 114. London, 1924.

————*British Diplomatic and Consular Reports, China, Report on the Trade of Hankow, 1913*. London, 1914.

————*British Diplomatic and Consular Reports, Foreign Trade of China, 1906*, no. 3943. London, 1907.

————*British Diplomatic and Consular Reports, Shanghai, 1908*, no. 4366. London, 1909.

Great Britain, Parliamentary Command Papers. *Foreign Import Duties, 1904*. London, 1905.

Greene, Graham. *A Sort of Life*. New York: Simon and Schuster, 1971.

Gregory, Richard Henry. Papers. Manuscript Department, William R. Perkins Library, Duke University, Durham, N. C.

Haber, Samuel. *Efficiency and Uplift: Scientific Management in the Progressive Era, 1890–1920*. Chicago: University of Chicago Press, 1964.

Hao Yen-p'ing. *The Comprador in Nineteenth Century China: Bridge between East and West*. Cambridge, Mass.: Harvard University Press, 1970.

Harrison, James P. *The Long March to Power*. New York: Praeger, 1972.

Hattori Mitsue 服部滿江. "Hoku-Shi ni okeru hatabako saibai fukyū irai no nōgyō keiei no henka" 北支に於ける葉煙草栽培普及以來の農業經營の變化

(The transformation of farm management since the spread of tobacco cultivation in North China), *Mantetsu chōsa geppō* 滿鐵調查月報 (South Manchurian Railway Company research department monthly), 21.20 (December 1941), 81–106.

Hauser, Ernest O. *Shanghai: City for Sale*. New York: Harcourt, Brace and Company, 1940.

Hayward, Elizabeth Gardner. *A Classified Guide to the Frederick Winslow Taylor Collection*. Hoboken, N. J.: Stevens Institute of Technology, 1951.

Henderson, Dan Fenno. *Foreign Enterprise in Japan: Laws and Policies*. Chapel Hill: University of North Carolina Press, 1973.

Herman, Alice. "The Early Commercial Press and the Episode over Foreign Investment, 1897–1914." Seminar paper, Cornell University, 1975.

Ho Ping-ti. *The Ladder of Success in Imperial China: Aspects of Social Mobility, 1368–1911*. New York: Columbia University Press, 1962.

————*Studies on the Population of China, 1368–1953*. Cambridge, Mass.: Harvard University Press, 1959.

————and Tang Tsou, eds. *China in Crisis*, vol. I, bk. 1. Chicago: University of Chicago Press, 1968.

Ho Ping-yin. *The Foreign Trade of China*. Shanghai: Commercial Press, 1935.

Ho Tso 和作 (pseud.). "1905-nien fan-Mei ai-kuo yun-tung" 1905年反美愛國運動 (The anti-American patriotic movement of 1905), *Chin-tai shih tzu-liao* 近代史資料 (Historical materials on modern history), I (1956), 1–90.

Hoku-Shi jimukyoku chōsashitsu 北支事務局調查室 (Research division of the bureau for North China affairs). *Kō-Zai ensen ni okeru kōshoku hatabako seisan jōkyō chōsa* 膠濟沿線ニ於ケル黄色葉煙草生產狀況調查 (A survey of the production conditions for yellow leaf tobacco along the Tsinan–Tsingtao Railway). Tientsin, 1938.

The Hong Kong Daily Press, comp. *The Directory and Chronicle for China, Japan, Corea, Indo-China, Straits Settlements, Malay States, Siam, Netherlands India, Borneo, the Philippines, etc. for the Year 1906*. London: Hong Kong Daily Press Office, 1906.

Hornbeck, Stanley K. Papers. Hoover Institution, Stanford University, Stanford, California.

Hou Chi-ming. *Foreign Investment and Economic Development in China, 1840–1937*. Cambridge, Mass.: Harvard University Press, 1965.

————"The Oppression Argument on Foreign Investment in China, 1895–1937," *Journal of Asian Studies*, 20.4 (August 1961), 435–448.

Hou Teh-pang. "The Yung-li Company: Pioneer of Chemical Industry in China," *China Reconstructs*, 4.3 (March 1955), 21–24.

Howe, Christopher. *Wage Patterns and Wage Policies in Modern China, 1919–1972*. Cambridge, England: Cambridge University Press, 1973.

Hsi Ch'ao 希超 (pseud.). "Ying-Mei yen kung-ssu tui-yü Chung-kuo kuo-min ching-chi ti ch'in-shih 英美煙公司對於中國國民經濟的侵蝕 (The invasion of the Chinese national economy by the British-American Tobacco Company)," Chung-kuo ching-chi ch'ing-pao she 中國經濟情報社 (Chinese Economic Intelligence Bureau), eds., *Chung-kuo ching-chi lun-wen chi* 中國經濟

論文集 (Essays on the Chinese economy). Shanghai: Sheng-huo shu-tien, 1936, I, 91–99.

Hsiang-kang Hua-tzu jih-pao 香港華字日報 (Hong Kong Chinese language daily), comp. *Kuang-tung k'ou hsieh ch'ao* 廣東扣械潮 (The weapon-impounding crisis in Kwangtung). Hong Kong, 1924.

Hsu Shih-hua 許世華 and Ch'iang Chung-hua 強重華 *Wu-sa yun-tung* 五卅運動 (The May Thirtieth movement). Peking: Kung-jen ch'u-pan she, 1956.

Hsu Yen-cho 許衍灼, ed. *Chung-kuo kung-i yen-ko shih-lueh* 中國工藝沿革史略 (A history of technological change in China). Shanghai: Hsin-hsueh-hui she ts'ang-pan, 1917.

Hu Kuang-piao 胡光麃. *Po-chu liu-shih nien* 逐波六十年 (Sixty years of turbulence). Hong Kong: Hsin-wen t'ien-ti she, 1964.

Hu Pang-hsien 胡邦憲. *Huang-kang hsien yen-yeh mao-i tiao-ch'a chi* 黃岡縣菸葉貿易調查記 (A report of an investigation on the tobacco leaf trade in Huang-kang hsien [in Hupeh province]). Nanking: Chin-ling ta-hsueh nung-hsueh-yuan nung-yeh ching-chi hsi, 1934.

Hua Kang 華岡. *Wu-ssu yun-tung shih* 五四運動史 (A history of the May Fourth movement). Shanghai: Hai-yen shu-tien, 1951.

Huai Shu 懷庶 (pseud.). *Chung-kuo ching-chi nei-mo* 中國經濟內幕 (Behind the scenes in the Chinese economy). Hong Kong: Hsin-min chü ch'u-pan she, 1948.

Huang Ch'eng-ching 黃澄靜. "Ts'ung Nan-yang hsiung-ti yen-ts'ao kung-ssu lai k'an min-tsu tzu-ch'an chieh-chi ti hsing-k'o" 從南洋兄弟煙草公司來看民族資產階級的性格 (Insights into the character of the national capitalist class based on the Nanyang Brothers Tobacco Company), *Hsueh-shu yueh-k'an* 學術月刊 (The scholarship monthly), 10.22 (October 10, 1958), 34–39.

Huang I-feng 黃逸峰 "Kuan-yü chiu Chung-kuo mai-pan chieh-chi ti yen-chiu" 關於舊中國買辦階級的研究 (A study of the comprador class in old China), *Li-shih yen-chiu* 歷史研究 (Historical research), 87.3 (June 15, 1964), 89–116.

———"Wu-sa yun-tung chung ti ta tzu-ch'an chieh chi" 五卅運動中的大資產階級 (The big bourgeoisie in the May Thirtieth movement), *Li-shih yen-chiu* 歷史研究 (Historical research), 3 (1965), 11–24.

Huang Yueh-po 黃于鮑 et al., eds. *Chung-wai tiao-yueh hui-pien* 中外條約彙編 (A collection of China's treaties with foreign countries). Shanghai: Shang-wu yin-shu kuan, 1935.

Hunt, Michael H. *Frontier Defense and the Open Door: Manchuria in Chinese–American Relations, 1895–1911.* New Haven: Yale University Press, 1973.

Hutchison, James Lafayette. *China Hand.* Boston: Lothrop, Lee and Shepard Company, 1936.

Institute of Pacific Relations, comp. *Agrarian China: Selected Source Materials from Chinese Authors.* London: Allen and Unwin, 1939.

Iriye, Akira. "Public Opinion and Foreign Policy: The Case of Late Ch'ing China." In Albert Feuerwerker et al., eds., *Approaches to Modern Chinese History.* Berkeley: University of California Press, 1967. Pp. 216–238.

Isaacs, Harold. *Five Years of Kuomintang Reaction.* Shanghai: China Forum Publishing Company, 1932.

——*The Tragedy of the Chinese Revolution.* 2nd rev. ed. Stanford: Stanford University Press, 1961.

Jansen, Marius B. *The Japanese and Sun Yat-sen.* Cambridge, Mass.: Harvard University Press, 1954.

Jansen, William H. "Eight Years Pioneering in China," *American Cinematographer*, 11 (February 1931), 11 and 26.

Jenkins, John Wilbur. *James B. Duke: Master Builder.* New York: George H. Doran Company, 1927.

Jones, Susan Mann. "Finance in Ningpo: The 'Ch'ien Chuang,' 1750–1880." In W. E. Willmott, ed., *Economic Organization in Chinese Society.* Pp. 47–77.

Kao Chen-pai 高貞白. "Chiang t'ai-shih chih 'ku'" 江太史之 '古' (The "story" of Chiang, a second-class compiler of the Hanlin Academy), *Ta-ch'eng* 大成 (Parorama magazine), 24 (November 1975), 12–19.

Kennan, George F. *American Diplomacy, 1900–1950.* Chicago: University of Chicago Press, 1951.

Kiger, Hugh C. "Prospects for Tobacco Trade between the United States and Mainland China," *Foreign Agriculture*, 10.19 (May 8, 1972), 2–3.

Kikuchi Takaharu 菊池貴晴 *Chūgoku minzoku undō no kihon kōzō: Taigai boikotto no kenkyū* 中國民族運動の基本構造：對外ボイコットの研究 (The structure of Chinese nationalism: a study of anti-foreign boycotts). Tokyo: Daian, 1966.

Kindleberger, Charles P. *American Business Abroad: Six Lectures on Direct Investment.* New Haven: Yale University Press, 1969.

Ko Kung-chen 戈公振. *Chung-kuo pao-hsueh shih* 中國報學史 (A history of Chinese journalism). Shanghai: Shang-wu yin-shu kuan, 1927.

Kōain seimubu 興亞院政務部 (Asia Development Board, Political Affairs Bureau). *Jūkei seifu no seinan keizai kensetsu jōkyō* 重慶政府の西南經濟建設狀況 (Southwestern economic development under the Chungking government). Tokyo: Daian, 1940.

Kolko, Gabriel. *The Triumph of Conservatism: A Reinterpretation of American History, 1900–1916.* New York: Quadrangle Books, 1963.

Kuang-tung kuo-huo chiu-cheng hui 廣東國貨糾正會 (Society for the protection of national goods in Kwangtung province), ed. *Wu-hu Nan-yang hsiung-ti yen-ts'ao kung-ssu chih hei-mo* 嗚呼南洋兄弟煙草公司之黑幕 (An exposé of the swindle by the Nanyang Brothers Tobacco Company). Canton, 1919.

Kung Chün 龔駿. *Chung-kuo hsin kung-yeh fa-chan shih ta-kang* 中國新工業發展史大綱 (An outline history of the development of China's new industries). Shanghai: Shang-wu yin-shu kuan, 1935.

K'ung Hsiang-hsi. "The Reminiscences of K'ung Hsiang-hsi (1880–   )" as told to Julie Lien-ying How. Manuscript. Special Collections Library, Butler Library, Columbia University, 1961.

Kuo-huo shih-yeh ch'u-pan she 國貨事業出版社 (Publishing house for national goods industries), comp. *Chung-kuo kuo-huo kung-ch'ang shih lueh* 中國國貨工廠史略 (Short histories of Chinese national goods companies). Shanghai, 1935.

"Labour Conditions and Labour Regulation in China," *International Labour Review*, 10.6 (December 1924), 1005–1028.

LaFeber, Walter. *The New Empire: An Interpretation of American Expansion, 1860-1898*. Ithaca, N. Y.: Cornell University Press, 1963.

Landes, David S. "Some Thoughts on the Nature of Economic Imperialism," *The Journal of Economic History*, 21.4 (December 1961), 496–512.

Landon, Kenneth P. *The Chinese in Thailand*. London and New York: Oxford University Press, 1941.

Laws and Regulations of the Government Monopoly of Japan. N.p., 1906.

Lawton, James. *Shoptalk: Papers on Historical Business and Commercial Records in New England*. Boston: Boston Public Library, 1975.

Lee, A. Sy-hung. "The Romance of Modern Chinese Industry," *China Review* (November 1923), pp. 158–160 and 200.

Lee, Leo Ou-fan. *The Romantic Generation of Modern Chinese Writers*. Cambridge, Mass.: Harvard University Press, 1973.

Lethbridge, Henry J. "Hong Kong under Japanese Occupation: Changes in Social Structure." In I. C. Jarvie, ed., *Hong Kong: A Society in Transition*. London: Routledge and Kegan Paul, 1969. Pp. 77–127.

LeFevour, Edward. *Western Enterprise in Late Ch'ing China: A Selective Survey of Jardine, Matheson & Company's Operations, 1842–1895*. Cambridge, Mass.: Harvard University, East Asian Research Center, 1968.

Levy, Marion J., Jr. "Contrasting Factors in the Modernization of China and Japan," *Economic Development and Cultural Change*, 2.3 (October 1953), 161–197.

————and Shih Kuo-heng. *The Rise of the Modern Chinese Business Class*. New York: Institute of Pacific Relations, 1949.

Leyda, Jay. *Dianying: An Account of Films and Film Audiences in China*. Cambridge, Mass.: M.I.T. Press, 1972.

Li Hsien-wei. *The Tobacco in China*, Economic Studies no. 1 in the Hautes Etudes Industrielles et Commerciales. Tientsin, 1934.

Li Wen-chih 李文治, comp. *Chung-kuo chin-tai nung-yeh shih tzu-liao, ti-i-chi, 1840–1911* 中國近代農業史資料, 第一輯, 1840–1911 (Historical materials on modern Chinese agriculture, first collection, 1840–1911). Peking: San-lien shu-tien, 1957.

Liang Hsiao-ming 梁曉明. *Wu-sa yun-tung* 五卅運動 (The May Thirtieth movement). Peking: T'ung-shu tu-wu ch'u-pan she, 1956.

Lieu, D. K. (Liu Ta-chün). *China's Industries and Finance*. Peking: The Chinese Government Bureau of Economic Information, 1927.

————劉大鈞 *Chung-kuo kung-yeh tiao-ch'a pao-kao* 中國工業調查報告 (Investigation and report on Chinese industry). Nanking: Chung-kuo ching-chi t'ung-chi yen-chiu so, 1937.

————*The Growth and Industrialization of Shanghai*. Shanghai: Institute of Pacific Relations, 1936.

Lin Pin 林斌. "Kuang-tung kuai-chieh Chiang Hsia-kung t'ai-shih i-shih" 廣東怪傑江霞公太史軼事 (Biographical notes on the extraordinary Cantonese second class compiler of the Hanlin Academy, Chiang Hsia-kung [Chiang K'ung-yin]), *I-wen chih* 藝文誌 (Annals of literary pursuits), 30 (March 1, 1968), 18–20 and 22.

Liu Huo-kung 劉豁公 "Nan-yang yen-ts'ao kung-ssu Chien Chao-nan ch'uang yeh shih" 南洋菸草公司簡照南創業史 (The history of Chien Chao-nan's founding of the Nanyang Tobacco Company), *I-wen chih* 藝文誌 (Annals of literary pursuits), 6 (December 1965), 14–18.

Liu Kwang-ching. *Anglo-American Steamship Rivalry, 1862–1874.* Cambridge, Mass.: Harvard University Press, 1962.

———"British-Chinese Steamship Rivalry in China." In C. D. Cowan, ed., *The Economic Development of China and Japan.* Pp. 49–78.

Liu Shou-lin 劉壽林. *Hsin-hai i-hou shih-ch'i nien chih-kuan nien-piao* 辛亥以後十七年職官年表 (Tables of officials by year during the 17 years following the revolution of 1911). Peking: Chung-hua shu-chü, 1966.

Liu Ta-chung and Yeh Kung-chia. *The Economy of the Chinese Mainland: National Income and Economic Development, 1933–1959.* Princeton: Princeton University Press, 1965.

Liu Wu-chi. *Su Man-shu.* New York: Twayne Publishers, 1972.

Lo Hui-min. "Some Notes on Archives on Modern China." In Donald D. Leslie, Colin Mackerras, and Wang Gungwu, eds., *Essays on the Sources for Chinese History.* Canberra: Australian National University Press, 1973. Pp. 203–220.

Lo Tun-wei 羅敦偉. *Wu-shih nien hui-i lu* 五十年回憶錄 (Personal recollections of the last fifty years). Taipei: Chung-kuo wen-hua kung-ying she, 1952.

Lockwood, William W. *The Economic Development of Japan: Growth and Structural Change, 1868–1938.* Princeton: Princeton University Press, 1954.

Lorence, James J. "Business and Reform: The American Asiatic Association and the Exclusion Laws," *Pacific Historical Review*, 39 (November 1970), 421–438.

Lowe, Chuan-hua. *Facing Labor Issues in China.* London: Allen and Unwin, 1934.

Lunt, Carroll. *Some Builders of Treatyport China.* Los Angeles, 1965.

McAleavy, Henry. *Su Man-shu (1884–1918): Sino-Japanese Genius.* London: China Society, 1960.

McCormick, Thomas J. *China Market: America's Quest for Informal Empire, 1893–1901.* Chicago: Quadrangle Books, 1967.

MacKinnon, Stephen R. "Liang Shih-i and the Communications Clique," *Journal of Asian Studies*, 29.3 (May 1970), 581–602.

MacNair, Harley F. *The Chinese Abroad: Their Position and Protection.* Shanghai: Commercial Press, 1925.

———"The Relation of China to Her Nationals Abroad," *The Chinese Social and Political Science Review*, 7.1 (January 1923), 23–43.

Mandel, Ernest. *Marxist Economic Theory.* Translated from the French by Brian Pearce. New York: Monthly Review Press, 1968. 2 volumes.

Mao Cho-ting. "Taxation and Accelerated Industrialization—with Special Reference to the Chinese Tax System during 1928–1936." Ph.D. dissertation, Northwestern University, 1954.

Marshall, Byron K. *Capitalism and Nationalism in Prewar Japan.* Stanford: Stanford University Press, 1967.

Martin, Albro. "Towards a More Precise Definition of Entrepreneurship." Paper presented to the Columbia University Seminar in Economic History, March 4, 1976.

Matchi kōgyō hōkokusho 燐寸工業報告書 (Report on the match industry). Nanking, 1940.

Mathews, Jay. "Between Puffs," *Washington Post*, September 6, 1978, p. A17.

Maugham, W. Somerset. *On a Chinese Screen*. New York: George H. Doran Company, 1922.

Mayers, William F., comp. *Treaties between the Empire of China and Foreign Powers*. Shanghai: North China Herald, 1906.

Medhurst, C. Spurgeon. "Report on Land Tenure in Shantung," *Journal of the China Branch of the Royal Asiatic Society for the Year 1888*, 23 (1889), 85–89.

Medhurst, Walter H. *The Foreigner in Far Cathay*. New York: Scribner, Armstrong and Company, 1873.

Metzger, Thomas A. "The Organizational Capabilities of the Ch'ing State in the Field of Commerce: The Liang-huai Salt Monopoly, 1740–1840." In Willmott, ed., *Economic Organization in Chinese Society*. Pp. 9–45.

Millard's Review of the Far East (1917–May 28, 1921); name changed to *The Weekly Review of the Far East* (June 4, 1921–July 1, 1922); name changed to *The Weekly Review* (July 8, 1922–June 16, 1923); name changed to *The China Weekly Review* (June 23, 1923–April 7, 1947). Shanghai.

Ming Chieh 明潔 (pseud.). "Ying-Mei yen-ts'ao kung-ssu ho Yü-chung nung-min" 英美煙草公司和豫中農民 (The British-American Tobacco Company and the peasants of Central Honan). In Chung-kuo nung-ts'un ching-chi hui 中國農村經濟會 (Society on China's rural economy), ed. *Chung-kuo nung-ts'un tung-t'ai* 中國農村動態 (China's rural situation). Pp. 1–17.

Morse, Hosea Ballou. *The Gilds of China*. London: Longmans, Green and Company, 1909.

Munro, Dana G. "American Commercial Interests in Manchuria," *The Annals of the American Academy of Political and Social Science*, 39.128 (January 1912), 154–168.

Murphey, Rhoads. *The Treaty Ports and China's Modernization: What Went Wrong?* Ann Arbor: University of Michigan Center for Chinese Studies, 1970.

Myers, Ramon H. *The Chinese Peasant Economy: Agricultural Development in Hopei and Shantung, 1890–1949*. Cambridge, Mass.: Harvard University Press, 1970.

————"The Commercialization of Agriculture in Modern China." In W. E. Willmott, ed., *Economic Organization in Chinese Society*. Pp. 173–191.

Nagano, Akira. *Development of Capitalism in China*. Tokyo: Japan Council of the Institute of Pacific Relations, 1931.

Nan-yang hsiung-ti yen-ts'ao kung-ssu 南洋兄弟煙草公司 (Nanyang Brothers Tobacco Company). *Nan-yang hsiung-ti yen-ts'ao ku-fen yu-hsien kung-ssu k'uo-ch'ung kai-tsu chao-ku chang-ch'eng* 南洋兄弟煙草股份有限公司擴充改組招股章程 (Regulations governing the expansion and reorganization of the Nanyang Brothers Tobacco Company, Limited). Shanghai, 1919.

————comp. *Chen tsai chi* 賑災記 (Record of famine). Canton (?), 1918.

Nathan, Andrew J. "Imperialism's Effects on China," *Bulletin of Concerned Asian Scholars*, 4.4 (December 1972), 3–8.

————*Peking Politics, 1918–1923: Factionalism and the Failure of Constitutionalism*. Berkeley: University of California Press, 1976.

Navin, Thomas R. *The Whitin Machine Works since 1831: A Textile Machinery*

Company in an Industrial Village. Cambridge, Mass.: Harvard University Press, 1950.

Nelliest, George F., ed. *Men of Shanghai and North China.* Shanghai: The University Press, 1933.

Nicholls, William H. *Price Policies in the Cigarette Industry: A Study of "Concerted Action" and Its Social Control, 1911–50.* Nashville: Vanderbilt University Press, 1951.

Nien Ch'eng 念澄 (pseud.). *Huang-pu chiang-pan hua tang-nien* 黃浦江畔話當年 (Years along the banks of the Whangpoo River). Hong Kong: Chih-ch'eng ch'u-pan she, 1971.

Nurkse, Ragnar. *Problems of Capital Formation in Underdeveloped Countries.* New York: Oxford University Press, 1953.

Ōi Senzō 大井專三, "Shina ni okeru Ei-Bei tabako torasuto no keiei keitai, zai-Shi gaikoku kigyō no hatten to baiben soshiki no ichikōsatsu" 支那に於ける英美煙草トラストの經營形態—在支外國企業の發展と買辦組織の一考察 (The form of administration of the British–American Tobacco Company in China, a study of the development of foreign enterprises in China and of the comprador system), *Tōa kenkyū shohō* 東亞研究所報 (Report of the East Asia Institute), 26 (February 1944), 1–47.

Pan Shu-lun. *The Trade of the United States with China.* New York: China Trade Bureau, Inc., 1924.

Parker, Lee, and Ruth Dorval Jones. *China and the Golden Weed.* Ahoskie, N. C.: Herald Publishing Company, 1976.

Parrish, Edward J. Papers. Manuscript Department, William R. Perkins Library, Duke University, Durham, N. C.

Patterson, Don D. *The Journalism of China.* Columbia, Mo.: University of Missouri Bulletin, 1922.

Payer, Cheryl. "Harvard on China II: Logic, Evidence, and Ideology," *Bulletin of Concerned Asian Scholars*, 6.2 (April–August 1974), 62–68.

P'eng Tse-i 彭澤益, comp. *Chung-kuo chin-tai shou-kung-yeh shih tzu-liao, 1840–1949* 中國近代手工業史資料 1840–1949 (Historical materials on modern Chinese handicraft industries, 1840–1949). Peking: San-lien shu-tien, 1957. 4 volumes.

Perkins, Dwight H. *Agricultural Development in China, 1368–1968.* Chicago: Aldine Publishing Company, 1969.

———"Growth and Changing Structure of China's Twentieth Century Economy." In Perkins, ed., *China's Modern Economy in Historical Perspective.* Pp. 115–165.

———"Introduction: The Persistence of the Past." In Perkins, ed., *China's Modern Economy in Historical Perspective.* Pp. 1–18.

———*Market Control and Planning in Communist China.* Cambridge, Mass.: Harvard University Press, 1966.

———ed. *China's Modern Economy in Historical Perspective.* Stanford: Stanford University Press, 1975.

Pertschuk, Michael. "Smoking in China," *Washington Post*, October 17, 1976, pp. C1 and C5.

Pien Chieh 边洁. "'Nan-yang hsiung-ti yen-ts'ao kung-ssu shih-liao' p'ing-chieh"

"南洋兄弟煙草公司史料" 評介 (A review of "Historical materials on the Nanyang Brothers Tobacco Company"), *Hsueh-shu yueh-k'an* 學術月刊 (Scholarship monthly), 6.30 (June 10, 1959), 62–63.

Porter, Patrick G. "Origins of the American Tobacco Company," *Business History Review*, 43.1 (Spring 1969), 59–76.

Potter, Jack M. *Capitalism and the Chinese Peasant: Social and Economic Change in a Hong Kong Village*. Berkeley: University of California Press, 1968.

Pun, Tsze E. "Modern Development in the Chinese Cigarette Trade and Industry," *Tobacco* (April 24, 1924), pp. 19 and 21.

Rankin, Watson S. *James Buchanan Duke (1865–1925): A Great Pattern of Hard Work, Wisdom, and Benevolence*. New York: The Newcomen Society in North America, 1952.

Rawski, Thomas G. "The Growth of Producer Industries, 1900–1931." In D. H. Perkins, ed., *China's Modern Economy in Historical Perspective*. Pp. 203–234.

———"Producer Goods and Industrialization in Twentieth Century China." Manuscript, May 1977.

Reader, W. J. *Imperial Chemical Industries: I, The Forerunners, 1870-1926; II, The First Quarter Century*. London: Oxford University Press, 1970 and 1975.

REA's Far Eastern Manual. Shanghai, 1923.

Remer, C. F. *Foreign Investments in China*. New York: Macmillan Company, 1933.

———*Readings in Economics for China*. Shanghai: Commercial Press, 1922.

———*A Study of Chinese Boycotts*. Baltimore: Johns Hopkins Press, 1933.

Rhoads, Edward J. M. *China's Republican Revolution: The Case of Kwangtung, 1895–1913*. Cambridge, Mass.: Harvard University Press, 1975.

Riskin, Carl. "Surplus and Stagnation in Modern China." In D. H. Perkins, ed., *China's Modern Economy in Historical Perspective*. Pp. 49–84.

Rosenbaum, Arthur Lewis. "China's First Railway: The Imperial Railways of North China, 1880–1911." Ph. D. dissertation, Yale University, 1972.

Sanford, James C. "Tobacco Taxation on the Eve of Tariff Autonomy." Seminar paper, Harvard University, 1971.

Sanger, J. W. *Advertising Methods in Japan, China, and the Philippines*, U.S. Department of Commerce, Special Agent Series no. 209. Washington, D.C.: Government Printing Office, 1921.

Schell, Orville. *In the People's Republic: An American's Firsthand View of Living and Working in China*. New York: Random House, 1977.

Scherer, Frederic M. *Industrial Market Structure and Economic Performance*. Chicago: Rand McNally, 1970.

Schiffrin, Harold Z. *Sun Yat-sen and the Origins of the Chinese Revolution*. Berkeley: University of California Press, 1968.

Schram, Stuart R., ed. and trans. *The Political Thought of Mao Tse-tung*. New York: Praeger, 1969.

Schumpeter, Joseph A. *Capitalism, Socialism, and Democracy*. New York: Harper and Brothers, 1947.

———"The Creative Response in Economic History," *Journal of Economic History*, 8 (November 1947), 149–159.

———*The Theory of Economic Development: An Inquiry into Profits, Capital,*

Credit, Interest, and the Business Cycle. Translated from the German by Redvers Opie. Cambridge, Mass.: Harvard University Press, 1959.

Shang-hai jen-min ch'u-pan she 上海人民出版社 (People's publishing house of Shanghai), ed. *Shang-hai ching-chi shih-hua* 上海經濟史話 (A history of Shanghai's economy). Shanghai: Shang-hai jen-min ch'u-pan she, 1964.

Shang-hai she-hui k'o-hsueh-yuan li-shih yen-chiu so 上海社會科學院歷史研究 所 (The historical research institute of the Shanghai social science academy), comp. *Wu-ssu yun-tung tsai Shang-hai shih-liao hsuan-chi* 五四運動在上海 史料選輯 (Selected historical materials on the May Fourth movement in Shanghai). Shanghai: Jen-min ch'u-pan she, 1960.

Shang-hai-shih cheng-fu she-hui chü 上海市政府社會局 (Bureau of social affairs of the government of Shanghai), comp. *Chin shih-wu-nien lai Shang-hai chih pa-kung t'ing-yeh* 近十五年來上海之罷工停業 (Strikes and lockouts in Shanghai since 1918). Shanghai: Chung-hua shu-chü, 1933.

Shang-hai-shih she-hui chü 上海市社會局 (Bureau of social affairs of Shanghai), comp. *Shang-hai chih chi-chih kung-yeh* 上海之機製工業 (Mechanized industries of Shanghai). Shanghai: Chung-hua shu-chü, 1933.

——— *Shang-hai chih kung-yeh* 上海之工業 (Industries of Shanghai). Taipei; Hsueh-hai ch'u-pan she, 1970.

Shang-hai t'ung-she 上海通社 (United publishing house of Shanghai), ed. *Shang-hai yen-chiu tzu-liao hsu-chi* 上海研究資料續集 (Supplementary research materials on Shanghai). Shanghai: Chung-hua shu-chü, 1939.

Shapiro, Michael. *Changing China*. London: Lawrence and Wishart, 1958.

Sheridan, James E. *China in Disintegration: The Republican Era in Chinese History, 1912–1949*. New York: Free Press, 1975.

———*Chinese Warlord: The Career of Feng Yü-hsiang*. Stanford: Stanford University Press, 1966.

Shibaike Yasuo 芝池晴夫. "1930 nendai no keizai kikika ni okeru Chūgoku minzoku shihon kigyo no jittai" 1930 年代の經濟危機下における中國民族 資本企業の實態 (The true situation of Chinese national capitalist enterprise during the economic crisis of the 1930s), *Shōdai ronshū* 商大論集 (Shōdai journal), 24.1–3 (June 1972), 44–61.

Shih Ch'eng-chih 史誠之. "Shih-lun 'ssu-shih' yü 'wen-ko'" 試論 '四史' 與 '文革' (A tentative discussion of the 'Four Histories' and the 'Cultural Revolution'), *Ming-pao yueh k'an* 明報月刊 (Ming-pao monthly), 72 (December 1971), 5–17; and 73 (January 1972), 37–43. An English translation of this article appears in *Chinese Sociology and Anthropology*, 4.3 (Spring 1972), 175–233.

Skinner, G. William. *Chinese Society in Thailand: An Analytical History*. Ithaca, N. Y.: Cornell University Press, 1957.

———"Marketing and Social Structure in Rural China," *Journal of Asian Studies*, 24.1 (November 1964), 3–43; and 2 (February 1965), 195–227.

———"Regional Urbanization in Nineteenth Century China." In Skinner ed., *The City in Late Imperial China*. Stanford: Stanford University Press, 1977. Pp. 211–249.

Spence, Jonathan D. "Opium Smoking in Ch'ing China." In Frederic Wakeman, Jr., and Carolyn Grant, eds., *Conflict and Control in Late Imperial China*. Berkeley: University of California Press, 1975. Pp. 143–173.

Stern, Michael. "British–American Tobacco's Big Move," *New York Times*, June 23, 1973, p. F3.

Sumiya Mikio 隅谷三喜男. *Dai-Nihon teikoku no shiren* 大日本帝國の試錬 (Tribulations of the Japanese empire). Vol. XXII of *Nihon no rekishi* 日本の歴史 (History of Japan). Tokyo, 1966.

Sun Chia-chi. "Cigarette Cards," translated from the Chinese by Robert Christensen, *Echo of Things Chinese*, 6.4 (January 1977), 58–67, 71, and 74.

Sun Yü-t'ang 孫毓棠, comp. *Chung-kuo chin-tai kung-yeh shih tzu-liao, ti-i-chi* 中國近代工業史資料 第一集 (Historical materials on modern Chinese industry, first collection). 2 volumes. Peking: K'o-hsueh ch'u-pan she, 1957.

————et al. *Mei ti-kuo chu-i ching-chi ch'in-Hua shih lun-ts'ung* 美帝國主義經濟侵華史論叢 (Essays on the history of American imperialist economic invasion of China). Peking, 1963.

Suzuki Tsutomu 鈴木勤, ed *Nis-Shin Nichi-Ro sensō* 日清日露戰爭 (The Sino-Japanese War and the Russo-Japanese War). Vol. XIX in *Nihon rekishi shirizu* 日本歷史シリーズ (A series in Japanese history). Tokyo: Sekai bunka sha, 1967.

Ta Ch'ing Kuang-hsu hsin-fa-ling 大清光緒新法令 (New laws and ordinances of the Kuang-hsu period). Shanghai: Shang-wu yin-shu kuan, 1909.

"T'an-pa-ku ts'ai chih fa" 淡芭菰栽製法 (Techniques for cultivating and producing tobacco), *Nung-hsueh ts'ung-shu* 農學叢書 (Collected works on agricultural science), second collection, no. 24, pp. 1a-4b.

Taylor, Frederick Winslow. *The Principles of Scientific Management*. New York: Harper and Brothers, 1911.

Teng Chung-hsia 鄧中夏. *Chung-kuo chih-kung yun-tung chien-shih (1919–1926)*. 中國職工運動簡史 (1919–1926) (A brief history of the Chinese labor movement, 1919–1926). Shanghai: Hsin-hua shu-tien, 1949.

Tennant, Richard B. *The American Cigarette Industry: A Study in Economic Analysis and Public Policy*. New Haven: Yale University Press, 1950.

Thomas, James A. Papers. Manuscript Department, William R. Perkins Library, Duke University, Durham, N. C.

————*A Pioneer Tobacco Merchant in the Orient*. Durham, N. C.: Duke University Press, 1928.

————"Selling and Civilization," *Asia*, 23.12 (December 1923), 896–899 and 948–950.

Tilley, Nannie May. *The Bright Tobacco Industry, 1860–1929*. Chapel Hill: University of North Carolina Press, 1948.

Ting Wen-chiang 丁文江, ed. *Liang Jen-kung hsien-sheng nien-p'u ch'ang-pien ch'u-kao* 梁任公先生年譜長編初稿 (Draft chronological biography of Liang Ch'i-ch'ao). Taipei: Shih-chieh shu-chü, 1962. 3 volumes.

Ting Yu 丁又. "1905-nien Kuang-tung fan-Mei yun-tung" 一九零五年廣東反美運動 (The anti–American movement of 1905 in Kwangtung), *Chin-tai shih tzu-liao* 近代史資料 (Materials on modern historx), 5 (October 1958), 8–52.

Topley, Marjorie. "The Role of Savings and Wealth among Hong Kong Chinese." In I. C. Jarvie, ed., *Hong Kong: A Society in Transition*. London: Routledge and Kegan Paul, 1969. Pp. 167–227.

Tsai Chu-tung. "The Chinese Nationality Law, 1909," *The American Journal of International Law*, 4 (1910), 404–411.

Tsai Shih-shan. "Reaction to Exclusion: Ch'ing Attitudes toward Overseas Chinese in the United States, 1848–1906." Ph.D. dissertation, University of Oregon, 1970.

Tseng, H. P. "China Prior to 1949." In John A. Lent, ed., *The Asian Newspapers' Reluctant Revolution*. Ames, Iowa: Iowa State University Press, 1971. Pp. 31–42.

Tsūsho sangyō-hen 通商產業編 (Commerce and industry). Vol. II in *Nichi-Bei bunka kōshō-shi* 日米文化交渉史 (History of cultural relations between Japan and the United States). Tokyo, 1954.

Tung-ya yin-wu chü 東雅印務局 (Tung-ya press). *Kuang-kao ta-kuan* 廣告大觀 (Great advertisements). Canton, n.d.

Turner, Louis. *Politics and the Multinational Company*. London: Hamilton, 1959.

United Nations Conference on Trade and Development. *Marketing and Distribution of Tobacco*. United Nations: United Nations, 1978.

United States Department of Agricultural Economics. *First Annual Report on Tobacco Statistics*, Statistical Bulletin no. 58. Washington, D.C.: Government Printing Office, 1937.

United States Department of Commerce, Bureau of Foreign and Domestic Commerce. *Tobacco Markets and Conditions Abroad*. Washington, D.C.: Government Printing Office, 1925–1931. Nos. 1–312.

United States Department of Commerce, Bureau of Foreign and Domestic Trade. *Tobacco Trade of the World*, Special Report no. 68. Washington, D.C.: Government Printing Office, 1915.

United States Department of Commerce, Bureau of the Census. *Stocks of Tobacco Leaf*, 1919–1927. Bulletins nos. 139, 146, 149, 151, 157, 159, 163, 165. Washington, D.C.: Government Printing Office, 1920–1928.

United States Department of State. *Records Relating to Internal Affairs of China, 1910–1929*, microfilm, 893.00/351/2–10700, National Archives, Washington, D.C.

Varg, Paul A. *The Making of a Myth: The United States and China, 1897–1912*. East Lansing: Michigan State University Press, 1968.

Vincent, John Carter. *The Extraterritorial System in China: Final Phase*. Cambridge, Mass.: Harvard University, East Asian Research Center, 1970.

Vogel, Ezra F. *Canton under Communism: Programs and Politics in a Provincial Capital, 1949–1968*. Cambridge, Mass.: Harvard University Press, 1969.

Wakeman, Frederic, Jr. *Strangers at the Gate: Social Disorder in South China, 1839–1861*. Berkeley: University of California Press, 1966.

Wales, Nym (Helen Foster Snow). *The Chinese Labor Movement*. New York: The John Day Company, 1945.

Waley, Arthur. *The Life and Times of Po Chü-i, 772–846 A.D.* London: Allen and Unwin, 1949.

Wang Ching-yü 汪敬虞, comp. *Chung-kuo chin-tai kung-yeh shih tzu-liao, ti-erh-chi* 中國近代工業史資料 第二集 (Historical materials on modern Chinese industry, second collection). Peking: K'o-hsueh ch'u-pan she, 1957. 2 volumes.

Wang Hsi 汪熙. "Ts'ung Ying-Mei yen kung-ssu k'an ti-kuo chu-i ti ching-chi ch'in-lueh 從英美煙公司看帝國主義的經濟侵略 (The British–American Tobacco Company as a case study in imperialist economic exploitation), *Li-shih yen-chiu* 歷史研究 (Historical research), 4 (August 1976), 77–95.

Wang Li 王力 (Wang Liao-i 王了一). *Lung-ch'ung ping tiao chai so yü* 龍蟲並雕齋瑣語 (Essays from the Lung-ch'ung ping tiao studio). Shanghai: Kuan-ch'a she, 1949.

Wang Sing-wu. "The Attitude of the Ch'ing Court toward Chinese Immigration," *Chinese Culture*, 9.4 (December 1968), 62–76.

Wang, Y. C. *Chinese Intellectuals and the West, 1872–1949.* Chapel Hill: University of North Carolina Press, 1966.

———"Free Enterprise in China: The Case of a Cigarette Concern, 1905–1953," *Pacific Historical Review*, 29.4 (November 1960), 395–414.

Wang, Y. P. *The Rise of the Native Press in China.* New York, 1924.

The Weekly Review. See *Millard's Review of the Far East.*

The Weekly Review of the Far East. See *Millard's Review of the Far East.*

Who's Who in China. Shanghai: The China Weekly Review, 1926.

Who's Who in China. Shanghai: Millard's Review, 1920. 2 volumes.

Wiebe, Robert. *Businessmen and Reform: A Study of the Progressive Movement.* Cambridge, Mass.: Harvard University Press, 1962.

Wiens, Thomas. "The British American Tobacco Company and The Nanyang Brothers Tobacco Company: A Case Study of the Effects of Foreign Investment on the Development of Domestic Industry, China, 1900–1930." Seminar paper, Harvard University, 1966.

Wilbur, C. Martin. *Sun Yat-sen: Frustrated Patriot.* New York: Columbia University Press, 1976.

———and Julie Lien-ying How, eds. *Documents on Communism, Nationalism, and Soviet Advisers in China, 1918–1927.* New York: Columbia University Press, 1956.

Wilkins, Mira. "An American Enterprise Abroad: American Radiator Company in Europe, 1895–1914," *Business History Review*, 43.3 (Autumn 1969), 326–346.

———*The Emergence of Multinational Enterprise: American Business Abroad from the Colonial Era to 1914.* Cambridge, Mass.: Harvard University Press, 1970.

———"The Impact of American Multinational Enterprise on American–Chinese Economic Relations, 1786–1949." Paper presented at the Conference on the History of American-East Asian Economic Relations, Mt. Kisco, New York, June 1976.

———*The Maturing of Multinational Enterprise: American Business Abroad from 1914 to 1970.* Cambridge, Mass.: Harvard University Press, 1974.

Williams, Edward T. *Recent Legislation Relating to Commercial, Railway, and Mining Enterprises.* Shanghai, 1904.

Willmott, W. E., ed. *Economic Organization in Chinese Society.* Stanford: Stanford University Press, 1972.

Willoughby, Westel W. *Foreign Rights and Interests in China.* Baltimore: Johns Hopkins Press, 1920.

Wilson, Charles. *The History of Unilever: A Study in Economic Growth and Social Change*. London: Cassell, 1954. 2 volumes.

Wilson, David A. "Principles and Profits: Standard Oil Responds to Chinese Nationalism, 1925–1927." Unpublished paper, 1976.

Winkler, John K. *Tobacco Tycoon: The Story of James Buchanan Duke*. New York: Random House, 1942.

Wright, Arnold, ed. *Twentieth Century Impressions of Hong Kong, Shanghai, and Other Treaty Ports of China: Their History, People, Commerce, Industries and Resources*. London: Lloyd's Greater Britain Publishing Company, 1908. 3 volumes.

Wright, Mary C. "Introduction: The Rising Tide of Change." In Wright, ed., *China in Revolution*. Pp. 1–63.

————*The Last Stand of Chinese Conservatism: The T'ung-chih Restoration, 1862–1874*. Stanford: Stanford University Press, 1962.

————ed. *China in Revolution: The First Phase, 1900–1913*. New Haven: Yale University Press, 1968.

Wright, Stanley F. *China's Struggle for Tariff Autonomy, 1843–1938*. Shanghai: Kelly and Walsh, Ltd., 1938.

Wu Ch'eng-lo 吳承洛. *Chin-shih Chung-kuo shih-yeh t'ung-chih* 今世中國實業通志 (An encyclopedia of modern Chinese industries). Shanghai: Shang-wu yin-shu kuan, 1933. 2 volumes.

Wu Ch'eng-ming 吳承明. *Ti-kuo chu-i tsai chiu Chung-kuo ti t'ou-tzu* 帝國主義在舊中國的投資 (Imperialist investment in old China). Peking: Jen-min ch'u-pan she, 1956.

Wu Pao-san 巫寶三. "Chung-kuo kuo-min so-te i-chiu-san-san hsiu-cheng" 中國國民所得，一九三三修正 (Revisions of *China's National Income in 1933*), *She-hui k'o-hsueh tsa-chih* 社會科學雜誌 (Journal of social science), 9.2 (December 1947), 92–153.

Wu Tien-wei. "Chiang Kai-shek's March Twentieth Coup d'Etat of 1926," *Journal of Asian Studies*, 27.3 (May 1968), 585–602.

Yang, C. K. *A Chinese Village in Early Communist Transition*. Cambridge: Technology Press, M.I.T., 1959.

Yang Ta-chin 楊大金. *Hsien-tai Chung-kuo shih-yeh chih* 現代中國實業誌 (Record of industry in contemporary China). Shanghai: Shang-wu yin-shu kuan, 1940. 2 volumes.

Yen Chung-p'ing 嚴中平. *Chung-kuo mien-fang-chih shih-kao* 中國棉紡織史稿 (A draft history of cotton spinning and weaving in China). Peking: K'o-hsueh ch'u-pan she, 1955.

————ed. *Chung-kuo chin-tai ching-chi shih t'ung-chi tzu-liao hsuan-chi* 中國近代經濟史統計資料選輯 (Selected statistical materials on the economic history of modern China). Peking: K'o-hsueh ch'u-pan she, 1955.

Young, Marilyn Blatt. *The Rhetoric of Empire: American China Policy, 1895–1901*. Cambridge, Mass.: Harvard University Press, 1968.

Yü Ta-fu 郁達夫. "Ch'ün-feng ch'en-tsui ti wan-shang" 春風沉醉的晚上 (An intoxicating spring evening). In T'ung Wen 童汶, ed. *Chung-kuo hsien-tai tuan-p'ien hsiao-shuo hsuan* 中國現代短篇小說選 (Selected Chinese modern short stories). Hong Kong: Shang-hai shu-chü, 1960.

Glossary

ai-kuo, ai-kuo, t'u-huo, t'u-huo
愛國，愛國，土貨，土貨
ai ta-ch'ün 愛大羣

ch'an-mien 纏綿
Chang Chien 張謇
Chang Chih-tung 張之洞
Chang Hsun 張勳
Chang Ming-ch'i 張鳴岐
Chang Ping-lin 章炳麟
Chang Tso-chen 張佐臣
chen 鎮
Ch'en Ch'i-chün 陳其均
Ch'en Chin-t'ao 陳錦濤
Ch'en Han-sheng 陳翰笙
Chen-kuang kung-ssu 眞光公司
Ch'en Kung-po 陳公博
Ch'en Lien-po 陳廉伯
Ch'en-pao 晨報
Ch'en Ping-ch'ien 陳炳謙
Ch'en Shun-ch'ing 陳順卿
Ch'en Ts'ai-pao 陳才寶
Ch'en Tse-shan 陳則山
Cheng Po-chao 鄭伯昭
ch'eng-chen 城鎮
cheng-ch'i 爭氣
ch'eng-tu 誠篤
ch'i-ya 欺壓
Chiang Hsia 江蝦
Chiang K'ung-yin 江孔殷
Chiang Shu-ying 江叔頴
chiao-ch'ing 交情
Chien Chao-nan 簡照南
Chien Chien-ch'üan 簡鑑川

Chien Ch'in-shih 簡琴石
Chien Ching-shan 簡靜珊
ch'ien-chuang 錢莊
Chien Han-ch'ing 簡漢青
Chien Jih-hua 簡日華
Chien Jih-lin 簡日林
Chien K'ung-chao 簡孔昭
Chien Ming-shih 簡銘石
Chien Ping-jen 簡秉仁
chien-shang 奸商
Chien Shih-ch'ing 簡實卿
chien-tu ti-wei 監督地位
Chien Yin-ch'u 簡寅初
Chien Ying-fu 簡英甫
Chien Yü-chieh 簡玉階
ch'ih kuei 吃虧
chin 斤
chin-shih 進士
Ch'in Sung-k'uan 秦松寬
Ch'ing (I-k'uang) (See Prince Ch'ing)
Chou Tzu-chi 周自齊
Chou Wei-fan 周維藩
Ch'ü Ch'iu-pai 瞿秋白
Chu Pao-san 朱葆三
ch'üan 權
ch'uan-she 傳舍
Ch'üan yung kuo-huo hui 全用國貨會
chün-tzu 君子
Chung-hua chiu-tien 中華酒店
Chung-hua shu chü 中華書局
chung-jen 衆人
Chung K'o-ch'eng 鍾可成
Chung-kuo ssu-min chih-yen ch'ang
中國四民紙煙廠

Chung-kuo yen-ts'ao kung-ssu
中國煙草公司
Chung-Mei yen-yeh kung-ssu
中美煙葉公司

Erh-shih-ssu hsiao   二十四孝

Feng-shen yen-i   封神演義

Han-yang t'ieh ch'ang   漢陽鐵廠
hang   行
hen-ssu   恨死
Ho Hsun-yeh   何勛業
hong (*see* hang)
hou-she   喉舌
Hsi-liang   錫良
Hsi-yu chi   西遊記
hsiang   鄉
hsiang-jen   鄉人
Hsiang Ching-yü   向警予
Hsiang-tao chou-pao   嚮導周報
hsiao ch'eng-shih   小城市
Hsieh I-ch'u   謝益初
hsien   縣
Hsien-shih kung-ssu   先施公司
Hsin-ch'ao   新潮
Hsing-ta kung-ssu   興大公司
Hsing-yeh yen kung-ssu   興業煙公司
Hsu-ch'ang yen-ts'ao kung-ssu
許昌菸草公司
Hsu Hsueh-yuan   徐學源
Hsu Lo-ting   徐樂亭
Hsu Pai-hao   許白昊
Hsu Shih-ch'ang   徐世昌
hu-chang   虎長
Hua shang yen kung-ssu   華商煙公司
huang-chin shih-tai   黃金時代
Huang Ch'u-chiu   黃楚九
Hung-an ti-ch'an ku-fen yu-hsien
kung-ssu   宏安地產股份有限公司
hung-fang   烘房
Hung-hsi pao   紅錫包
Hung-lou meng   紅樓夢

I-sheng kung-ssu   怡生公司

jen-hsin   人心
Jen Pai-yen   任伯言
jen-shih   人事
jen-ts'ai   人才

Kan   簡
K'ang Pai-ch'ing   康白情
K'ang Yu-wei   康有爲
kang-ch'iang kuo-kan   剛強果敢
K'ang-Ying chi-chin t'uan   抗英急進團
Kawai   川井
k'o-chang   科長
k'o-ya   苛壓
Ku Hsing-i   顧馨一
Kuan Hai-feng   管海峰
kuan-shang ho-pan   官商合辦
kuan-tu shang-pan   官督商辦
Kuang-chi i-yuan   廣濟醫院
Kuang Kung-yao   鄺公耀
k'un   捆
K'ung-ch'iao tien-ying kung-ssu
孔雀電影公司
Kung Ho-ch'ien   龔和軒
K'ung Hsiang-hsi   孔祥熙
kung-kung fu-wu   公共服務
K'ung Shan   空山
kung-ssu ho-ying   公私合營
kung-tsei   工賊
Kung-t'uan lien-ho hui   工團聯合會
kuo-huo   國貨
kuo-huo yun-tung   國貨運動
Kuo-min jih-pao   國民日報
Kuo-min tang   國民黨
Kuo-wu yuan   國務院
kuo-ying   國營

Lao Ching-hsiu   勞敬修
lao hua-t'ou   老滑頭
lao ko-tzu   老哥子
Lao-tung chou-k'an   勞動週刊
li   利
Li Ch'i-han   李啟漢
li-jun tsai-t'ou-tzu lü   利潤再投資率
Li Wen-chung   李文仲
Li Yuan   李援
Li Yuan-hung   黎元洪
Liang Ch'i-ch'ao   梁啟超
liang-ho kung-ssu   兩合公司
Liang Shih-i   梁士詒
Liao Chung-k'ai   廖仲愷
lieh-huo   劣貨
lieh-shen   劣紳
likin (li-chin)   釐金
Liu Chin-sheng   劉金聲
liu-mang   流氓
Lo Chia-lun   羅家倫

lou-chih　漏卮
Lun-ch'uan chao-shang chü　輪船招商局

Ma Yü-ch'ing　馬玉清
mai-pan tzu-pen　買辦資本
mai-pan tzu-pen chia　買辦資本家
Manshū　滿洲
Mao-i kung-ssu　貿易公司
Matsumoto Shonanshi　松本照南子
Meng Shou-ch'un　孟壽椿
Min-kuo jih-pao　民國日報
min-tsu tzu-pen　民族資本
min-tsu tzu-pen ch'i-yeh　民族資本企業
min-tsu tzu-pen chia　民族資本家
mou　畝
Mu Hsiang-yueh　穆湘玥
Murai Kichibei　村井吉兵衛
Murai kyodai shōkai　村井兄弟商會

na-kuan chiu-pan　拿官究辦
Naigai wata kabushiki kaisha
　內外綿株式會社
Nan-pei hang　南北行
Nan-yang hsiung-ti yen-ts'ao kung-ssu
　南洋兄弟煙草公司
Nan-yang hsiung-ti yin-hang
　南洋兄弟銀行
Nan-yang yen-ts'ao chih kung chü-lo-pu
　南洋煙草職工俱樂部
Nan-yang yen-ts'ao ch'ih kung t'ung-
　chih hui　南洋煙草職工同志會
Nan-yang yen-ts'ao kung-ssu chih-tsao-
　ch'ang kung-yu chih-kung t'ung-chih
　hui　南洋煙草公司製造廠工友職工同
　志會
Nei-wu pu　內務部
Nikka bōseki kabushiki kaisha
　日華紡織株式會社
Niu Ch'uan-shan　鈕傳善
Nung shang pu　農商部

Ou Pin　歐彬

pa-chieh　巴結
pai-she　白蛇
pai-tzu t'u　百子圖
p'ai-wai hsin　排外心
P'an Hsing-nung　潘杏農
P'an Hsu-lun　潘序倫
P'an Ta-wei　潘達微
pao-pan　包辦

Peita　北大
Pei-yang yen-ts'ao ch'ang　北洋煙草廠
Prince Ch'ing I-k'uang (Ch'ing ch'in-
　wang I-k'uang)　慶親王奕劻
P'u-i　溥儀
P'u-tung fang-chih kung-hui
　浦東紡織工會
P'u-tung kung-yeh chin-teh hui
　浦東工業進德會

San-ho yen kung-ssu　三和煙公司
San-hsing yen-ts'ao ch'ang　三星煙草廠
San-kuo chih yen-i　三國誌演義
San min chu-i　三民主義
San-pai-liu-shih hang　三百六十行
sha-jen　砂仁
shan-t'ang　善堂
Shan-tung ta-shou　山東大手
shang-chan　商戰
Shang-hai tsung kung-hui　上海總工會
Shang-hai tsung shang-hui　上海總商會
Shang-hai yen-ts'ao kung-hui
　上海煙草工會
shang-pan　商辦
Shang-wu yin-shu kuan　商務印書館
Shao Ping-sheng　邵丙生
She-hui chü　社會局
she-t'uan　社團
Shen-hsin fang-chih wu-hsien kung-ssu
　申新紡織無限公司
Shen K'un-shan　沈昆山
Shen-pao　申報
Sheng Hsuan-huai　盛宣懷
Sheng-t'ai　盛泰
shou-hsu fei　手續費
Shui-hu chuan　水滸傳
ssu-chang　司長
Ssu-shih yun-tung　四史運動
sung-jen li-wu　送人禮物
Sung Tzu-liang　宋子良
Sung Tzu-wen　宋子文
Suzuki kaisha　鈴木會社

Ta ai-kuo p'ai　大愛國牌
ta cheng-lun　大爭論
ta-chü　大局
Ta-hsin hsien-shih kung-ssu
　大新先施公司
Ta-lung chi-ch'i ch'ang　大隆機器廠
Ta-ying　大英
tai chüan ch'ang　代卷廠

tan 擔
T'ang Chi-yao 唐繼堯
T'ang Shao-i 唐紹儀
Tatsu Maru 辰丸
ti-kuo chu-i 帝國主義
ti-kuo chu-i ti ching-chi ch'in-lueh
   帝國主義的經濟侵略
ti-p'i 地痞
T'ien Chün-ch'uan 田俊川
t'ien-hsia 天下
t'o 妥
Tōa 東亞
tou-ping 豆餅
Tsai-tse 載澤
Ts'ao Ts'ao 曹操
Ts'ui Tsun-san 崔尊三
tsung 總
t'u-huo 土貨
t'u-yeh 土業
Tu Yueh-sheng 杜月笙
Tuan Ch'i-jui 段祺瑞
Tuan Hsi-p'eng 段錫朋
t'ung-hsiang 同鄉
t'ung-shui 統稅
tz'u-chang 次長
tzu-pen ying-yeh 資本營業
tzu-yu hsing-tung 自由行動

Wai-wu pu 外務部
Wang Cheng-t'ing 王正廷
Wang Ching-wei 汪精衛
Wang I-hsi 汪毅熙
Wang Li 王力
Wang Shih-jen 王世仁
Wang Shu-kung 王樹功
Wei Fan-shih 魏蕃室
wei-p'o li-yu 威迫利誘
Wen Ying-hsing 溫應星
Wu 吳
Wu-chou ku-pen tsao yao ch'ang
   五洲固本皂藥廠
Wu-Hua chih-yen ch'ang 物華紙煙廠
Wu K'o-chai 吳克齋
wu-lai 無賴
wu-shih 武士
Wu T'ing-sheng 鄔挺生

ya-p'o 壓迫
Yang Chi-wu 楊輯五
Yang Chih-hua 楊之華
Yang Hsiao-ch'uan 楊小川
Yang Kuei-fei 楊貴妃
Yang Teh-fu 楊德富
Yang Yin 楊殷
Yeh Ch'u-ts'ang 葉楚傖
yen 円
yen-k'ung pi-ta 眼孔必大
Yen-ts'ao kung-hui 煙草工會
Ying-Mei yen-ch'ang kung-hui
   英美煙廠工會
Ying-Mei yen-ts'ao kung-ssu
   英美煙草公司
yü-fu 迂腐
Yü Hsia-ch'ing 虞洽卿
yü-pang 餘磅
Yu Shao-tseng 尤少增
Yü-sheng-ho hao 玉盛合號
yuan 元
Yuan Shih-k'ai 袁世凱
yuan-tsu cheng-chih fei 援助政治費
Yueh Fei 岳飛
Yung-an t'ang 永安堂
Yung-fa yin-wu kung-ssu 永發印務公司
Yung-li chih chien kung-ssu
   永利製鹼公司
Yung-t'ai-ho kung-ssu 永泰和公司
Yung-t'ai-ho yen-ts'ao ku-fen yu-hsien
   kung-ssu 永泰和煙草股份有限公司
yung-yuan tsung-li 永遠總理

Index

Achison, Lord, 149

Advertising: dissemination by BAT, 19-22, 30, 31, 34, 73, 103-104, 134, 198-199; BAT's use of Chinese imagery in, 35-38, 218, 219; of BAT in response to nationalism, 62-65, 110-113, 158, 179, 181-182; dissemination by Nanyang, 70-74, 103-104, 158-160, 169; of Nanyang in response to nationalism, 66-70, 76-77, 88, 93, 115, 150, 178, 187, 218, 219; by other Chinese companies, 51, 114; illustrations of, 20-21, 37, 67, 132-133, 136, 180. *See also* Boycotts; Motion pictures; Philanthropy

Advertising agencies, 134-135

Agents: in BAT, 27-32, 54, 70, 104-108, 111-113, 131; in Nanyang, 72, 76, 214; competition for, 72, 74. *See also* Compradors; Tobacco; *name of individual*

Allen, George G., 15, 146

Alliance of Workers, Merchants, Intellectuals, and Peasants of Kwangtung Province, 186

American Cigarette Company, 14

American government, 45, 49, 127, 194, 205

American Tobacco Company, 10-11, 13, 24, 25, 41, 206

American Trading Company, 46

Americans: in BAT management, 10-16, 39, 124-126, 138, 163-166; in BAT marketing, 16-17, 22, 24-30 passim, 38-39, 148-149, 165, 181-182, 193, 212-213; in BAT tobacco purchasing, 22,

26. *See also* Representatives; *name of individual*

Anfu clique, 119

Anglo-Chinese College, 28

Anhwei, 34, 182; tobacco-growing in, 75, 85, 130, 141, 160, 161, 199, 204

Anking, 19

Anshan, 4

Anti-American boycott, 45-52, 58, 114

Anti-British boycott, 177-182

Anti-foreign attitudes, 58-59, 60-61, 62-65, 88-92, 99, 112. *See also* Racism

Anti-Japanese boycott, 58-60, 62-65, 68-69, 103, 196

Argentina, 2

Arnold, Julean, 22, 48, 186

Asahi, 58

Asiatic Petroleum Company, 180, 205

Association for the Protection of Chinese Tobacco Interests, 44

Association of Overseas Chinese in Singapore, 118, 119

Australia, 15

Bangkok, 55-56, 63, 90

Bank of Communications, 113

Banks, 50, 152, 179, 191, 207. *See also name of individual bank*

Bassett, Arthur, 79, 86

Batavia (Djakarta), 198

Bergère, Marie-Claire, 123, 210

Bhamo, 19

Bonsack, James, 13, 163

Borneo, 58

HARVARD STUDIES IN BUSINESS HISTORY

*Out of print